P9-AOW-347

HALLUCINOGENS

HALLUCINOGENS

A Forensic Drug Handbook

Edited by
Richard R. Laing

with contributions by
Barry L. Beyerstein
Terry A. Dal Cason
Edward S. Franzosa
John Hugel
Mark Kalchik
Richard Laing
David Lankin
John Meyers
Alexander T. Shulgin

Series Editor
Jay A. Siegel

Amsterdam Boston London New York Oxford Paris
San Diego San Francisco Singapore Sydney Tokyo

This book is printed on acid-free paper.

ACADEMIC PRESS
An imprint of Elsevier Science
84 Theobald's Road, London WC1X 8RR, UK
http://www.academicpress.com

ACADEMIC PRESS
An imprint of Elsevier Science
525 B Street, Suite 1900, San Diego, California 92101-4495, USA
http://www.academicpress.com

ISBN 0-12-433951-4

Library of Congress Control Number 20021144921
A catalogue record for this book is available from the British Library

Typeset by Kenneth Burnley, Wirral, Cheshire, England
Printed in Great Britain
03 04 05 06 07 08 MP 9 8 7 6 5 4 3 2 1

CONTENTS

PREFACE

"Tune in, turn on and drop out" was the mantra of a generation embodying the ideals of the psychedelic subculture of the turbulent 1960s, to the chagrin of Western civilization. These words caused lines to be drawn between those who sought "enlightenment" through psychedelic drugs and those who believed that the cost for this "enlightenment" was too high in health and safety terms. Nevertheless, throughout history humankind in numerous civilizations has used psychedelic drugs for religious and spiritual purposes, and the question "Are these purposes any different from those professed in 1960s?" remained unanswered. The argument still rages on with a new generation. Lysergic acid diethylamide (LSD) was the focus of the 1960s while today it is with the "Entactogenic" or "Empathogenic" 3,4-methylenedioxy-N-methylamphetamine (MDMA) family of drugs. There is no mantra now, but rather a social event whose significance is on the same scale as those immortal words of the 1960s. Today's setting is that of a warehouse-style all-night dance party called a "Rave," where drug use is rampant and socially accepted within this subculture. Again, the self-professed justification for using these drugs socially is to the chagrin of Western civilization. Are the reasons for this subculture's use of MDMA and psychedelic drugs today any different from the 1960s? While this book cannot answer these important questions it does provide information behind the arguments. It is a compilation of writings from scientists and forensic chemists so that the reader will gain valuable insight in to the history, forms, effects, manufacture and the analysis of LSD, indolealkylamines, phenylalkylamines and other hallucinogens.

R. Laing

ABOUT THE AUTHORS

RICHARD R. LAING

Richard Laing is the manager of the Drug Analysis Service Laboratory in Bumaby, BC, serving Western Canada. For the past 12 years he has been running the clandestine laboratory program line where he has established himself as an expert in assisting the seizure of more than 100 clandestine laboratories. He has participated in the seizure of laboratories manufacturing Amphetamine, Methamphetamine, LSD and the designer LSsecB, DMT, 2C-B, Mescaline, 5-Methoxy-N,N-Diisopropyltryptamine, MDA, MDMA and cocaine processing, to name a few. Richard received his BSc in Chemistry in 1986 from Concordia University in Montreal, and his MSc in Food Chemistry in 1991 from Laval University in Quebec City while working full time with Agriculture Canada in St Hyacinthe, QC. Richard's interests in research revolved around the application of novel Mass Spectrometric and NMR analyses in food research. In 1990 Richard joined Health Canada to start working in the area of clandestine laboratories.

Richard is a Past President of the Association of Clandestine Laboratory Investigating Chemists and is a contributing Regional Editor to this association's journal. He has written many publications, most of which deal with illicit drug manufacturing, synthetic techniques and analytical techniques. Richard was selected to represent Health Canada on the Scientific Working Group Drugs (SWGDrug) and has contributed to the establishment of recommended minimum standards in the analysis of seized forensic drugs. He also teaches at the British Columbia Institute of Technology in the program of Forensic Science.

BARRY BEYERSTEIN

Barry Beyerstein is Associate Professor of Psychology and a member of the Brain Behavior Laboratory at Simon Fraser University. A native of Edmonton, Alberta, he received his bachelor's degree from Simon Fraser University and a PhD in Experimental and Biological Psychology from the University of California at

Berkeley. Dr Beyerstein's research has involved a number of areas related to his primary scholarly interests: brain mechanisms of perception and consciousness and the effects of drugs on the brain and mind. He is an elected member of the Council for Scientific Medicine and a founding board member of the group supporting evidence-based medicine, Canadians for Rational Health Policy. He is associate editor of the journal, *Scientific Review of Alternative Medicine* and on the editorial board of the journal, *Scientific Review of Mental Health Practice.*

Dr Beyerstein's publications in psychopharmacology have included areas such as drug effects on mental processes, mechanisms of addiction, environmental effects on drug use, and social consequences of drug use. He has also been involved in issues related to how scientific data should inform drug policy and various aspects of legal approaches to drug regulation. Dr Beyerstein has been a member of the Advisory Board of the Drug Policy Foundation (Washington, DC) and a founding board member of the Canadian Foundation for Drug Policy (Ottawa, Ontario). He is a former contributing editor of the *International Journal of Drug Policy* (Manchester, England). Dr Beyerstein has testified as an expert witness in numerous civil and criminal cases and was invited to address the House of Commons Standing Committee on Health during their discussions leading up to passage of the Controlled Drugs and Substances Act, the bill which replaced the Narcotics Control Act.

Dr Beyerstein's teaching interests include courses on brain research, drugs, sensory psychophysiology and consciousness, and the history and philosophy of psychological research. His awards include a Woodrow Wilson Fellowship, the gold medal of the BC Psychological Association, and the Donald K. Sampson Award of the BC College of Psychologists. He has also held a visiting professorship at Jilin University in the People's Republic of China where he had the opportunity to interact with various practitioners of Traditional Chinese Medicine.

TERRY A. DAL CASON

Terry A. Dal Cason is a senior forensic chemist with the US Department of Justice, the Drug Enforcement Administration (DEA). He graduated from Rockford College in 1968 and received his MSc degree from Roosevelt University in 1978. From mid-1968 through early 1970, he served in the toxicology division of the First US Army Medical Laboratory at Fort Meade, Maryland. He began his career with the Justice Department as a forensic chemist with the Bureau of Narcotics and Dangerous Drugs (BNDD) in June of 1971 and retained this position when the BNDD became the DEA in 1973. Since that time he has published 20 articles in refereed journals and served as a guest reviewer for three journals. Mr Dal Cason is Past President of the Clandestine Laboratory

Investigating Chemists Association (CLIC) and a fellow of the American Academy of Forensic Sciences (AAFS).

EDWARD SYKES FRANZOSA

Edward Sykes Franzosa received his BS in Chemistry from Rensselaer Polytechnic Institute, Troy, NY, in 1967; and his PhD in Physical Chemistry (Molecular Spectroscopy) from the State University of New York at Binghamton, NY, in 1975. He served in the US Navy in Vietnam from 1967 to 1969; and he was a police officer in New York for five years. He joined the Drug Enforcement Administration in 1975 as a Forensic Chemist in the Special Testing & Research Laboratory, now at Dulles, VA. He is the Senior Forensic Chemist in DEA's Source Determination Program and author of *The Logo Index for Tablets and Capsules* (currently in its 6th edition).

JOHN HUGEL

John Hugel has been analyzing suspected illicit drug exhibits and investigating clandestine laboratories for over 27 years. He graduated in 1974 with a BSc with a specialization in chemistry from the University of Toronto. He began working with Health Canada in June of that year. After six months analyzing pharmaceuticals to determine compliance with legislation, he transferred to the Drug Analysis Service where he is employed to this day. Over the course of his career with the Drug Analysis Service, he has analyzed over 10,000 suspected illicit drug exhibits and provided assistance in the investigation of about 100 clandestine laboratories. Because of his experience, John is regarded as one of the most knowledgeable persons in the investigation of clandestine laboratories in Canada. His willingness to share his knowledge has led him to spend a significant portion of his time as a guest lecturer at the Canadian Police College in Ottawa, Canada.

John has published several articles in the journal of the Canadian Society of Forensic Sciences, *Microgram*, and *Journal of the Clandestine Laboratory Investigating Chemists Association*. He has served on the executive of the Clandestine Laboratory Investigating Chemists Association in several capacities including the position of President in 2002. John regularly presents at the annual meeting of this chemists' group. He is a member of the team which has produced the Drug Yield Calculator – a computer program to calculate illicit drug yields given an amount of precursor and a known reaction. John is a member of the Canadian Society for Chemistry and the Canadian Society of Forensic Sciences, and is a Chartered Chemist in the province of Ontario.

MARK F. KALCHIK

Mark F. Kalchik received his BS degree in Chemistry from the University of California at Davis in 1971. After graduation he worked for two years as an agricultural chemist, primarily testing plant and soil samples for nutrient requirements. Mark worked for approximately a year as a laboratory technician for the State of California Department of Justice at the Redding, California Laboratory. He worked at the state's San Luis Obispo Laboratory for about five years, becoming a Criminalist while concentrating on the drug analysis field. Since 1979 he has worked at the state's Fresno Laboratory as a Senior Criminalist. His current responsibilities include clandestine laboratory investigations, drug analysis of plant material, analysis of larger drug cases and working on impurity profiling in methamphetamine cases. He has taught drug analysis and clandestine laboratory investigation to newer criminalists and has been involved with legal requirements of drug control in California.

DAVID C. LANKIN

David C. Lankin earned his PhD at the University of Cincinnati in 1972 (Hans Zimmer). Following a two-year post-doctoral at the University of New Orleans (Gary W. Griffin/Photochemistry), he joined Borg-Warner Chemicals. During that time he was involved in research in organophosphorus chemistry and developing NMR capability for the Borg-Warner Research Center in Des Plaines, Illinois. In 1985 he joined Varian Associates as an NMR Applications Chemist and was responsible for the operation of the Chicago Applications Laboratory. In 1988 he joined the Physical Methodology Department of G.D. Searle (now Pharmacia Corporation) where he is currently Principal Research Scientist and Supervisor of the NMR Laboratory. He has co-authored more than 50 papers in the field of organic chemistry, with recent emphasis on the structural applications of NMR.

JOHN A. MEYERS

John A. Meyers received a Bachelor of Science degree in Chemistry from The American University (Washington, DC) in 1969. He joined the Bureau of Narcotics and Dangerous Drugs (BNDD) Chicago Laboratory in 1970. In 1973, BNDD was merged and became a part of the Drug Enforcement Administration (DBA). In 1989, John became a Senior Forensic Chemist for DEA with a specialty area of NMR. He has worked with NMR since 1976. He has co-authored four papers relating to Forensic NMR use. He has presented several papers related to Forensic Drug Analysis in general and Forensic NMR analysis. He is involved with the evaluation and selection of new instruments for the laboratory system.

ALEXANDER T. SHULGIN

Alexander T. Shulgin was born in Berkeley, California in 1925. His undergraduate education started with the Class of 1946, at Harvard University in Cambridge, MA, but was aborted by his entry into the US Navy during World War II. His military service was largely in the North Atlantic on the US *Pope*, DE-134 and in the Pacific area on the US *Vigilance*, AM-324. He received a BA in Chemistry from the University of California at Berkeley in 1949 and a PhD in Biochemistry in 1954. Post-doctorate studies were at the UC Medical School in San Francisco (1967–68) in the departments of Pharmacology and Psychiatry.

Alexander Shulgin's professional connections were with BioRad Laboratories as Research Director from 1954 to 1955 and at The Dow Chemical Company as a Research Chemist from 1955 to 1961, and as a Senior Research Chemist from 1961 to 1966. From 1968 to the present, he has been an independent Scientific Consultant and has functioned out of his own private laboratory.

Over this period of time, Alexander's research has been devoted to the design, synthesis and evaluation of the action of drugs that may serve as research tools for the study of the human mental process. This has resulted in the awarding of some 20 patents, the publication of over 200 papers in peer-reviewed scientific journals, and the writing of four books.

To Chelsea, Brittany and Morgan

CHAPTER 1

HISTORY OF THE PSYCHEDELIC EXPERIENCE

Barry L. Beyerstein
Mark Kalchik

1.0 ORIGIN

> The passionate desire which leads man to flee the monotony of everyday life has made him instinctively discover strange substances. He has done so, even where nature has been most niggardly in producing them and where the products seem very far from possessing the properties which would enable him to satisfy this desire.
> (Louis Lewin, 1931)

1.0.1 DISCOVERY AND ENCULTURATION OF HALLUCINOGENS

The longing to experience altered states of consciousness has remained so constant across the expanse of history and geography that the psychopharmacologist Ronald Siegel (1989) has likened it to the physiological drives for food and sex. Indeed, ethnologists have even found instances of non-human animals that voluntarily ingest plants in the wild that are known to contain psychoactive chemicals (Siegel, 1989). Archeological excavations have revealed remnants of plants, paraphernalia, and artwork, attesting to the fact that cultivation and ritualized use of mind-altering substances reaches far back into prehistory (Devereux, 1997; Furst, 1976, pp. x–xi, 4; LaBarre, 1972; Emboden, 1979, p. xiv).

The discovery of the hallucinogenic properties of most well-known psychoactive plants occurred before the advent of written record-keeping. It stands to reason, however, that trial-and-error experimentation in the process of searching for new sources of food must have played a large role. Food being perennially scarce, ancient hunter-gatherers were inclined to sample a wide variety of vegetation for its possible nutritive value. Occasionally, such experimentation yielded far more than sustenance for the body.

Upon ingesting certain fruits, seeds, leaves, bark, or roots, there welled up a stunning kaleidoscope of visions, emotions, and insights of seemingly cosmic significance. Time seemed to become elastic, stretching and collapsing in astonishing ways. The sense of self might seem estranged from the body or to disintegrate altogether. At times, it felt as though some all-powerful entity was usurping the seat of the will or that profound truths were being imparted from some universal fount of knowledge. Though all this would pass within hours, and everyday reality resume once more, the interlude often left the experiencer feeling fundamentally changed (Huxley, 1956; Grinspoon and Bakalar, 1997, Ch. 4).

Although accidental ingestion of mind-altering plants almost certainly gave rise to their future ritualized use, it is also well to note that some hallucinogenic plants dear to indigenous cultures require extensive preparation before they are usable (often to render dangerous toxins inert) or must be mixed with other substances before they become psychoactive. This suggests that once an association between plants and ecstatic states had become established in a culture, considerable effort must have been expended thereafter on intentional experimentation to expand the repertoire.

As migration occurred, roving bands would have encountered new vegetation to be sampled. Some would appear quite novel, to be approached cautiously. Others would resemble familiar plants from the band's previous environs. This familiarity can be deceiving, however, in that the plant kingdom contains many innocuous and highly toxic species that look disarmingly alike. It is a tribute to the value indigenous peoples have placed on experiencing altered states of consciousness that they have willingly and repeatedly accepted such risks.

When a newly sampled plant product proved not deadly but to have dramatic mind-altering properties, it would quickly be woven into the culture for magical and ceremonial purposes, usually under strict shamanic (i.e., priestly) control. Such oversight is crucial because some of these agents lack a reasonable margin of safety. For instance, cytisene, the active ingredient in mescal beans (not to be confused with mescaline) is poisonous at levels only slightly above its psychoactive dose (Schultes and Hofmann, 1973, p. 99). They were used, nonetheless, for inducing visions in divination and initiation rites by various North American aboriginals. Mescal beans were used, for example, in the impressive ceremony known variously as the "red bean dance," the "Wichita dance," or the "Deer Dance." Because of their inherent danger, though, the beans were gradually abandoned in favor of peyote (see below) by most tribes.

Observation of the eating habits of animals likely also played a part in the human domestication of psychoactive plants. Within the era of recorded history, we have the (possibly apocryphal) example of the discovery of the stimulant properties of coffee. According to legend, an Abyssinian goat herder observed heightened activity in his herd after they had ingested the red beans of the coffee bush that was native to the area. The herder learned to make an energizing beverage from an infusion of the beans. Over time, the process was refined to include fermentation and roasting of the ripened beans. Rather quickly, the drink became favored by devout Muslims who were required to maintain long prayer vigils. Unlike alcohol, which is prohibited by the teachings of Islam, coffee use flourished in Muslim cultures and eventually spread to Europe. Its popularity there led the European colonial powers to establish coffee cultivation in their conquered territories in the Far East and Central and South America. Today, caffeine-containing beverages (including coffee, tea, cocoa, and cola drinks) are the most widely consumed psychoactive drugs in the world (used daily by up to 80% of the world's population) (Julien, 2001, p. 220). Coffee production is estimated to be over six billion kilograms per year; it is grown in more than 50 countries, involving more than 25 million workers. It is second only to the oil industry in the value of its annual sales.

1.0.2 THE ALLURE AND MEANING OF HALLUCINOGENS

Our early ancestors lived in an animistic world where natural phenomena, such as lightning, growth of vegetation, or the setting of the sun, were presumed to behave in their characteristic ways because hidden spirits within had willed it so (Frazier, 1890/1993). Living as they did in what the late Carl Sagan (1995) called "a demon-haunted world," it is understandable that our forebears would assume that when ingesting a new plant unleashed such dramatic fits of ecstasy, some sacred inhabitant of the plant must be responsible. Veneration of plant deities who revealed themselves in such spectacular ways was entirely predictable. We see this reflected in the names given to many of the substances to be discussed below, e.g. coca, "the divine plant" of the Incas (Mortimer, 1901/1974); peyotl, "the divine messenger" of the Huichols (Furst, 1972b); and *teonanacatl*, "God's flesh" according to the ancient Aztecs (Furst, 1972a).

As nomadic hunter-gatherers gradually settled into villages, and various crops and livestock were domesticated, these bands retained a keen interest in the surrounding flora and fauna. Knowledge of the special powers of certain plants was highly prized because of their ability to produce ecstatic visions, complete with glimpses of paradise and visitations of gods and ghostly ancestors. And, of course, an occasional peek through the gates of hell never hurt to keep the more unruly factions in line. Other plants were found to possess the equally miraculous ability to assuage pain, ease anxiety, reduce hunger, enhance alertness and endurance, affect fertility, or promote the healing of wounds. On the darker side, some natural products yielded powerful poisons, useful for dealing with enemies, prey, or pests – both animal and human (Martinetz and Lohs, 1987). And not least, societies have always cherished substances that could ease social interactions or merely provide a brief respite from an often harsh and threatening reality.

As bodies of such knowledge accumulated, they conferred great power on the holder and the secret lore was passed down in preliterate societies from generation to generation of "medicine men" or, not atypically, "medicine women." The power to define consensual reality by altering consciousness was held in such esteem that the offices of shaman (keeper of this knowledge) and political leader were often combined in the same individual (LaBarre, 1972; Furst, 1976; Furst, 1972b, pp. 140–141; Norman, 1977; Grob *et al.*, 1996). The word "shaman" is derived from the Siberian Tungus tribe's word "saman," meaning "medicine man" or woman. Shamanistic religions have typically been more open to sharing power with women than other more patriarchal religions. The shaman also performs, in a religious context, medical services and the sort of counseling we would recognize as psychotherapy.

In all cultures, religious teachings have traditionally supplied the answers to great metaphysical questions such as the origins of the universe, the descent of one's ancestors, the meaning of life, the foundations of morality, what happens after death, etc. The power of psychoactive plants to suggest hidden planes of existence, to conjure up visions of deceased elders, and to produce transcendent awe or abject terror ensured that their effects would be explained in supernatural terms and that their regulation would fall into the hands of the priestly class. The ability of drug-induced reveries to provide divine guidance, prophesy the future, or reveal what lies beneath mundane appearances is a recurrent theme in legends and folklore from around the world. Anthropologists have traced many religious beliefs that survive to this day to probable origins in the drug-taking rituals of ancient peoples (Furst, 1972a, 1976; Smith, 1964). It has been suggested that sharing of transcendent experiences (which can be induced by behavioral as well as chemical means – see, e.g., Beyerstein, 1996a, b, c) played an important role in forging alliances that were essential to the survival of early groups of hunter-gatherers (Hayden, 1987).

Following the European conquest of the Americas, colonial officials expended considerable effort attempting to stamp out the ecstatic rituals of the native populations. Anthropologists have noted that when missionaries discovered that the Aztec word for sacred mushrooms meant "flesh of the gods," this similarity to the Christian communion service made the indigenous ceremonies particularly offensive to the newcomers. Despite the horrific cruelties inflicted by the missionary clergy upon aboriginals who attempted to go underground with the ecstatic intoxicants of their native religions, the colonial powers never succeeded in eradicating these practices (Furst, 1976, p. 19). In many instances, the subjugated people merely grafted their pre-European rites onto the Christian liturgy so that they could claim adherence to the new religion while continuing to seek guidance with the shamanistic inebriants of their forefathers.

In modern industrial societies, some of the most vociferous objections to the use of mind-altering substances continue to come from religious quarters. For if consciousness is considered essentially equivalent to the soul, all but doctrinarily sanctioned means of altering mental states must be the work of the devil. Despite longstanding attempts at suppression, however, some present-day aboriginal people in North America have managed to retain the hallucinogenic practices of their pre-European religions to the present day. The Native American Church has annexed the ritual use of peyote onto a mix of Christian doctrine and other spiritual beliefs. Many observers agree that the experiences shared in these ceremonies have helped these communities cope better than some with the social strains of their beleaguered way of life (Clark, 1970; Barron *et al.*, 1964). Similar hallucinogen-based religions have grown up among non-aboriginals in more urbanized regions as well (Lyttle, 1988).

While the shared experiences induced by these sacred plants, and the rituals that emerged to regulate their usage, have undoubtedly served socially cohesive purposes throughout history, the potential of such powerful mind-altering substances to trigger perceptions and feelings that might be inimical to the established social order must also have been apparent to early experimenters. Not surprisingly, every culture has evolved strict folkways, mores, and customs (backed up by stern penalties for violators) aimed at regulating the use of psychoactive substances (Zinberg, 1984). The use of the criminal law to try to regulate such practices is, however, a surprisingly recent development (Szasz, 1975; Trebach, 1982). With the possible exception of sex (also a powerful source of both cohesion and strife), no other aspect of human behavior has had as much attention devoted to rules governing when, how much, and with whom indulgence should occur, and as rigid codes as to what is required, permitted, and forbidden at such times.

Ethnologists and botanists have surveyed the world's cultures with respect to their discovery and usage of naturally occurring plant products for religious, social, and medicinal purposes (Lewin, 1931; Efron *et al.*, 1967; Schultes, 1969, 1972; Lewis and Elvin-Lewis, 1977; deSmet and Rivier, 1989; Davis, 1996). Evidence of such usage by native populations has been found throughout the Oceania region and at least five of the seven continents – the two exceptions being Antarctica, which had no indigenous peoples, and Australia, whose aboriginal culture stands nearly alone in not having established ritual use of psychoactive plants.

It is interesting to note that, of all the regions studied in this respect, Central and South America stand out as having produced by far the most varied and intricate patterns of usage. LaBarre (1972) argues that this stems not from greater abundance of suitable plants in this region but rather from important cultural functions served by this form of ritualized behavior. Of greatest significance was the survival in the New World of the Mesolithic shamanistic religions of their Eurasian ancestors. These practices had been essential to the social organization of their cultures when they had been bands of big game hunters, and as migration brought these groups to the Americas, they retained what has been called the "cultural programming" of their people for personal ecstatic experience (Furst, 1976, p. 2; Hayden, 1987).

Shamanistic religions embrace the priestly use of ecstatic trances as a conduit between the natural and spiritual worlds. This sort of divine inebriation had largely died out among Europeans by the time they began to explore the New World. Though many of their own religious practices are probably traceable to similar origins, the actual use of the inebriants themselves had long since been replaced, at least in Christian rituals, by symbolic remnants that provoke milder alterations of consciousness through the suggestive powers of liturgy, music, exhortation and ceremonial setting.

1.1 IDENTIFYING NATURAL SOURCES OF DRUGS AND THEIR ACTIVE CONSTITUENTS

Throughout history, people have scoured their environments for plants (and occasionally animal tissues as well) that have medicinal value or alter consciousness in ways they find pleasurable or beneficial (Wall, 1927; Lewin, 1931; Efron *et al.*, 1967; Schultes and Hofmann, 1972; Emboden, 1979; Talalay and Talalay, 2001). As the explorers, the conquerors, and the missionaries extended European governance to the far reaches of the world, it became apparent that indigenous peoples often possessed sophisticated knowledge of economically and medicinally valuable plants in their vicinity. To understand the hold of the native religions the missionaries were hoping to supplant, it was necessary to understand the centrality of the mind-altering plants to indigenous cosmologies and their related spiritual beliefs. Many informative manuscripts sent home by early missionaries describe these practices, with a mixture of fascination and revulsion; they still make interesting reading today – see, e.g. the chronicles of Fr Bernardino de Sahagun (Furst, 1972b).

As the missionaries also tended the sick, they soon became aware of the medicinal value of folk remedies such as *chinchona* (quinine) in Andean culture and *Strychnos toxifera* (curare) in the Amazon basin or *Rawolffia serpentina* (reserpine) on the Indian subcontinent. As this information became apparent to colonial administrators, scientific expeditions were soon dispatched from the European capitals in hopes of exploiting yet another source of riches. Interestingly, many sought-after spices that spurred early European exploration and trade have, if taken in larger doses, psychedelic properties, e.g. mace and nutmeg, discussed below (Schultes and Hofmann, 1973, pp. 65–70).

Out of these early expeditions there emerged a new scientific discipline, a melding of anthropology, biology, medicine, and psychology. Today, different subspecialties of this diverse field are known by various names, e.g. ethnobiology, ethnobotany, ethnomycology, ethnopharmacology, pharmacognosy, and phytochemistry (Lewin, 1931; Lewis and Elvin-Lewis, 1977; Houghton and Bisset, 1985; Tyler, 1993; Talalay and Talalay, 2001). After learning from native informants what plants were likely to be useful for what purposes, these ethnobotanical pioneers returned specimens for cultivation back home and engaged specialists in the newly bourgeoning specialty of analytic chemistry to extract the active ingredients from the crude preparations of the plant or animal products. This, in turn, prompted the development of yet another specialty, psychopharmacology, for whenever psychoactive drugs are involved, someone must assess, under controlled conditions, just what the effects of the active ingredients in these biological products are (Witters *et al.*, 1992; Leavitt, 1995; McKim, 2000; Julien, 2001). It soon became apparent that the psychological, social, and physical conditions surrounding any instance of psychoactive drug use, exert an enormous influence on the particular subjective and behavioral effects the drug user will experience (Zinberg, 1984).

1.1.1 NAMES AND CLASSIFICATION SCHEMES

In the remainder of this chapter the word "drug" or "substance" will be used to refer to naturally occurring plant or animal products with psychoactive properties, whereas "active ingredient" or "chemical" will be used to designate their purified neuroactive components.

Among psychopharmacologists, debate still occurs as to the proper class name for this diverse group of mind-altering substances. Because of their heterogeneity, their wide-ranging ability to affect cognition, memory, motivation, mood, perception, and motor skills, not to mention their ability to alter one's view of reality, many different names have been suggested over the years. The term "hallucinogen" is widely used in the psychopharmacology literature because of the undisputed ability of this class of drugs to distort sensory experience (although debate still rages in the field as to whether these perceptual contortions constitute true hallucinations or not). The term '"psychotomimetic" is still seen occasionally because the cognitive disruptions caused by this class of drugs in some ways resemble the subjective experiences of psychoses. This term largely disappeared, however, after researchers pointed out considerable differences between drug-induced states and those of, say, schizophrenia (Hollister, 1962). Commentators who wish to emphasize the drugs' disruptive effects on normal cognition have favored the expression "psychodysleptic." Those more interested in the magico-religious motivations for using mind-altering drugs have preferred the term suggested in 1924 by Louis Lewin (the "father of ethnopharmacology"): "phantastica." Along the same lines, another term, "entheogen," has emerged more recently in the literature (Ott, 1993), emphasizing once again the experiential/religious/cultural functions of such practices. Researchers who have been more interested in the use of these agents as a means of self-exploration or gaining experiences that could be a spur to artistic creativity have tended to prefer the class name suggested by Osmond in 1957: "psychedelic" (from Greek roots meaning "mind expanding"). In this chapter, we shall use the terms "hallucinogen" or "psychedelic" more or less interchangeably, depending on the aspects of a given drug's effects we wish to emphasize at the moment.

Like the profusion of names, there has been a surfeit of classification schemes proposed as well. None is inherently preferable – each is useful at different times and reflects primarily the background of the writer and the particular expository task he or she has undertaken. In addition to classifying naturally occurring hallucinogens according to their ethnic or geographic correlates (plus the few that are of animal origin), the following organizing systems have been widely used in the literature.

Chemical classification

From a biochemical standpoint, Schultes (1972) divided hallucinogenic plants according the active ingredients they contain. He first sorted them into groups he called "nitrogenous" and "non-nitrogenous." The former (mostly alkaloids or related compounds), he subdivided into the following chemical groups: (1) β-carbolines, (2) ergolines, (3) indoles, (4) isoquinolines, (5) isoxazoles, (6) β-phenylethylamines, (7) quinolizidines, (8) tropanes, and (9) tryptamines (Schultes, 1972, p. 6). On the non-nitrogenous side, Schultes includes the dibenzopyrans and phenylpropenes.

Most, though not all, of the psychoactive plants we shall be considering derive their mind-altering properties from organic nitrogen-containing constituents called alkaloids. Inasmuch as most alkaloid-containing plants are highly toxic, more than a few of our more adventuresome ancient relatives must have perished in this process of experimentation. Indeed, the fact that bitter tastes are almost universally aversive to mammals is seen as a naturally selected adaptation to avoid poisoning by alkaloid-containing plants. The fact that some plants produce alkaloids is somewhat of a puzzle in itself (Robinson, 1974). The synthetic pathways involved are metabolically expensive, drawing considerably on the plants' resources, but production of alkaloids seems to be unrelated to most other vital aspects of plant physiology, such as photosynthesis for instance (though some other possible functions in the plants' internal economy have been suggested). The best, though not universally accepted, guess is that alkaloid-containing plants evolved the ability to produce these chemicals as protective measures to fend off predators, parasites, or competitors.

Botanical classification

Classifying psychoactive substances by botanical origin, Schultes (1972) employs the following categories (in alphabetical order with typical examples following in parentheses):

- Agaricaceae – the agaric family (e.g., the psilocybin-containing mushrooms)
- Cactaceae – the cactus family (e.g., peyotl cactus)
- Convolvulacae – the morning glory family (e.g., ololiuqui)
- Labiatae – the mint family (e.g., salvia, pennyroyal)
- Leguminosae – the pea family (e.g., cohoba snuff, mescal beans)
- Lythraceae (e.g., sinicuichi)
- Malpighiaceae (e.g., ayahuasca)
- Myristicaceae (e.g., virola, nutmeg)
- Solanaceae (e.g., Datura, tobacco)

By neurotransmitters affected

With advances in modern neuroscience, it has been possible to elucidate the sites and modes of action of psychedelic drugs at the cellular level in the nervous system. This has led to a classification of psychedelic drugs according to which neurotransmitters they primarily interact with to affect consciousness and behavior (Houghton and Bisset, 1985). Thus we have, for example, hallucinogens such as scopolamine that affect the brain's acetylcholine system; mescaline which acts on the catecholamine systems (i.e., norepinephrine and dopamine) and LSD and certain "magic mushrooms" that are serotonin affecting psychedelics (Julien, 2001, p. 332).

By conscious and behavioral effects

Finally, psychoactive drugs of plant origin can be classified according to their psychological effects. Thus, Houghton and Bisset (1985) divide the domain into the following categories: psychostimulants, psychodepressants (anxiolytics and sedatives), drugs that affect consciousness, and drugs with possible psychotherapeutic value. Leavitt (1995), for example, chose to organize his psychopharmacology text under behavioral headings (rather than the more usual classification by substance type), placing in the same chapter diverse drugs that can be meaningfully bundled together because they are all agents that affect memory, creativity, sexual behavior, aggression and violence, sleep and dreams, etc.

1.2 WHAT WE HAVE NOT INCLUDED HERE

1.2.1 ALCOHOL

Although ethyl alcohol has a longer documented history of human use than any other psychoactive drug and has arguably had more impact on human history as well (Lucia, 1963; Chaffetz, 1965; Kobler, 1974), it is a sedative and anxiolytic (i.e., central nervous system depressant), not a hallucinogen, and thus beyond the scope of the present chapter. In addition, alcohol consumption has tended to be somewhat less restricted to ceremonial use, at times drunk by children and adults alike in preference to the almost certainly contaminated sources of drinking water.

1.2.2 TOBACCO

Likewise, although the cognitive effects of nicotine are greater than generally believed (Ashton and Stepney, 1982) and more potent strains of tobacco (e.g., that called *picétl* by the Aztecs of Mexico) have been used by indigenous cultures for their psychedelic properties (Furst, 1976, Ch. 2; Wilbert, 1972; Emboden, 1979, pp. 35–43), its current reasons for and patterns of usage in the industrialized world place it outside our immediate areas of concern.

1.2.3 NATURAL PSYCHOSTIMULANTS

There are also a number of naturally occurring stimulants, e.g., coca, and strychnine (Houghton and Bisset, 1985; Johanson, 1988). Their use tends to be more for their effects as euphoriants or mood affecters (Nesse and Berridge, 1997) which also places them outside the realm of the present chapter as well.

1.2.4 CANNABINOIDS

Although the products of *Cannabis sativa* (e.g., marijuana and hashish) can definitely be hallucinogenic if taken in sufficient doses, most users stop short of this, content to experience the euphoria, entertaining mental associations, and sensory enhancements obtained at the doses used socially in most cultures today. Unusual because it is not an alkaloid-based psychoactive drug, cannabis contains several cannabinoids, notably Δ-9-tetrahydrocannabinol, that have been shown to account for its effects on consciousness and behavior. Recently, it has been discovered that it exerts these effects by interacting with a hitherto unsuspected neurotransmitter named anandamide. Three excellent reviews of cannabis' history, pharmacology, and psychological and medical effects have been published recently (Zimmer and Morgan, 1997; Joy *et al.,* 1999; Swiss Federal Commission, 1999).

1.2.5 OPIATES

And finally, though they have a history of human use and cultivation that spans more than 6000 years – attested to, for example, by emblems on ancient Sumerian tablets (Witters *et al.*, 1992, Ch. 8) – and they certainly have extensive effects on consciousness (Snyder, 1989), the opiates are not hallucinogenic as such and will thus not be considered further in this chapter.

We now turn to a discussion of the effects of a variety of naturally occurring drugs from various regions of the world.

1.3 PSYCHEDELIC PLANTS

When Europeans arrived and spread through the New World toward the end of the fifteenth century, they observed that native populations used a wide variety of consciousness-transforming plants for magico-religious purposes. Depending on the region and its growing conditions, colonizers found different varieties of sacred mushrooms, morning glory seeds, tobacco and other snuffs, datura, hallucinogenic cacti, etc., woven into the supernatural worldview of the inhabitants.

1.3.1 LSD

Though not, strictly speaking, a plant, no discussion of mind-altering drugs would be complete without the mention of LSD (lysergic acid diethylamide-25). LSD does not occur in nature, though it has naturally occurring relatives. More than any other substance, it has come to epitomize the class of psychedelic agents because it was the first psychedelic to have a major impact on modern Western societies (Lee and Schlain, 1985).

The grain parasite, ergot (*Claviceps purpurea*) was known to ancient Europeans and probably figured in various religious rites (Wasson *et al.*, 1978/1999). By 1920, ergot had been found to contain the compounds lysergic acid and lysergic acid amide, or ergotomine, which were subsequently synthesized by the chemist Arthur Stoll. This fungal parasite appears throughout the history of European witchcraft and was also known to midwives of the time as a useful aid to stimulate uterine contractions. It was also likely responsible for several episodes of unintentional mass poisonings in the Middle Ages and on into modern times, i.e., whenever the local grain supply became infested (Matossian, 1989). The active ingredient, ergotomine, survives milling and baking of the flour into bread and can have milder psychedelic properties akin to those of LSD. When the infestation struck, whole towns were thought to have suddenly succumbed to demonic possession. It has been suggested that the events that led up to the infamous Salem witch trials in Massachusetts, in 1692, may have also been an attack of ergotism due to crop infestation (Matossian, 1989), though some have raised doubts. Ergotamine by-products are still among the most effective drugs for treatment of migraine. In the late 1930s, Dr Albert Hofmann of the Swiss pharmaceutical company Sandoz was synthesizing and testing a series of ergotamine derivatives in hopes of improving their effectiveness as perinatal drugs, especially for mothers who experienced excessive bleeding during delivery. The twenty-fifth alteration of the molecule (LSD-25) appeared unpromising in animal tests (which, of course, were unable to reveal its psychedelic properties) so it was set aside. In a series of serendipitous moves, Hofmann decided to return to the molecule in 1943. At that time, he accidentally ingested a minute amount and the world's first LSD "trip" was under way (Hofmann, 1983). Though occasionally terrified by thoughts that he was going crazy and by the bizarre sights, sounds and emotions he was experiencing, Hofmann was nonetheless intrigued to the point that he tried the substance again the next day, and many times thereafter, to investigate its astonishing properties. Thus Hofmann helped found a new area of psychopharmacology and went on to become one of the leading authorities on the chemistry and psychology, as well as the philosophical and social impacts, of psychedelics. As he tried them systematically himself, Hoffman provided highly literate as well as scientifically important descriptions. LSD has been found to be extremely potent, with mild psychedelic effects being felt with doses as small as 25 micrograms. The more typical dose, with dramatically stronger effects, is about ten times greater. Other detailed descriptions of the effects of LSD and related hallucinogens can be found in Grinspoon and Bakalar (1997, Ch. 4) and Snyder (1986, Ch. 7).

Unable to find a medical use for LSD-25, Sandoz made it available for researchers and psychotherapists (Grinspoon and Bakalar, 1997; Lee and Schlain, 1985). By the early 1960s, LSD had found some limited uses in the treatment of alcoholics, in hospices, and as a way for mental health professionals to experience some of the warpages of mind their patients were experiencing spontaneously. At the same time, it became a recreational drug among members of the artistic and social élite of America who used it with little fanfare and few adverse effects (Huxley, 1956; Lee and Schlain, 1985). Among those who held frequent LSD parties in their mansions for the benefit of this aristocracy were the publishing magnate Henry Luce and his wife, Claire Booth Luce, who, among other claims to fame, placed Barry Goldwater's name in nomination at the Republican National Convention that selected him as their 1960 presidential candidate.

Also at this time, LSD figured in the infamous "MK Ultra" mind-control experiments conducted by the CIA (Marks, 1979; Lee and Schlain, 1985). The intelligence communities on all sides in the Cold War expended much money and effort in ethnopharmacological research in (fortunately vain) hopes of developing drugs that would effect political conversions or at least make adversaries malleable and vulnerable to interrogation. The public of the day remained unaware of these unethical experiments conducted by the intelligence agencies and it was not until Timothy Leary was fired by Harvard University for involving his students in his LSD experiments that the drug became a *cause célèbre.* It remained legal until its use among disaffected youth in the 1960s "Counterculture" spawned a moral panic that led to its prohibition. Although short-term adverse reactions sometimes occurred, especially among novice and apprehensive users, and more serious effects were seen in some users who were already susceptible to mental disorders – and many people have found the experience sufficiently disquieting to deter further use – the incidence of long-term adverse effects has been found to be quite low (Strassman, 1984; Halpern and Pope, 1999).

The neuropharmacology of LSD is still far from completely understood, despite important advances in the field (Aghajanian and Marek, 1999; Vollenweider *et al.*, 1999). All agree that the shared indole structure of LSD (and the dimethyl tryptamine [DMT] found in several other hallucinogens to be discussed shortly) and the neurotransmitter serotonin is somehow central to its effects at the cellular level. However, simple serotonin mimicry or antagonism cannot account for LSD's effects. Complex interactions of these serotonin effects with the brain's dopamine, and possibly GABA and histamine systems, has been suggested to account for LSD's diverse properties. Differential effects on one or another of the several subtypes of serotonin receptor in the brain have also been proposed. Julien (2001, Ch. 12) offers a concise summary of the current status of this scientific debate.

1.3.2 DATURA

Datura would not have been totally new to the early European colonialists entering the Americas, for it had played a role in their own cultures' medieval practices of medicine and witchcraft (Satina and Avery, 1959). Incorporated into the nature-based and matriarchal spirituality of the ancient Druid and Wiccan religions, datura and its botanical cousin, belladonna, were employed for their visionary properties. Notions such as the witch's cauldron and flying on broomsticks are probably traceable to this era. Wicca (from which we get the modern English word "witch") revered female fertility goddesses and employed women priests. The latter would boil various herbal concoctions to make salves that were applied with broom-like appliances to the genital areas.

There the active alkaloids were readily absorbed through the mucous membranes. The dreamy state of delirium, accompanied by vestibular effects and extreme muscle relaxation, gave these women the subjective impression of flying, along with other ghostly visions.

A member of the Solanaceae family, which includes potatoes, tomatoes, and chile peppers, as well as mandrake and deadly nightshade (*Atropa belladonna*), the genus *Datura* contains about twenty species. Mentioned in ancient texts from India, China, and the Middle East, *D. metel* has a long history as a hallucinogen in the eastern hemisphere (Schultes and Hofmann, 1973, p. 166).

On the eastern seaboard of North America, European settlers encountered *D. stramonium*, which has come down to us as "Jimson Weed" – a corruption of "Jamestown" (Virginia), an early settlement near which a troop of English redcoats went on a frolicking rampage after they (allegedly) unknowingly ingested what appeared to be edible flavoring herbs (Furst, 1973, p. 142). Among the Algonquins and other tribes in what are now the northeastern border regions of Canada and the US, concoctions of the datura variety known as "thorn apple" were widely used in adolescent rites of passage that lasted for two to three weeks. Kept in a state described by European settlers as agitated, raving madness, native boys threw off the persona of childhood and replaced it with the status of manhood. The common theme of death and rebirth was a feature of many such rites around the world in which the young man supposedly forgot all memories of boyhood.

For curative purposes, native Americans used *Datura* both as a means by which the shaman could solicit divine help in diagnosing illnesses and as a medicine for the patient (applied both externally and internally). Concoctions of the plant were used to alleviate pain and, in higher doses, to render the patient unconscious in order to perform surgery, set broken bones, lance infected wounds, etc. (Furst, 1973, p. 139). It remains a constituent of some asthma medications today.

In the American southwest and Mexico and throughout Central and South America, the hallucinogenic use of *D. inoxia* (*meteloides*) was encouraged, under priestly auspices. It was used as a means of seeking supernatural advice or divine intercession, contacting the ancestors, and in adolescent initiation rites. Datura was claimed by Davis (1988) to be a component (along with the pufferfish paralytic, tetrodotoxin) of the infamous Haitian "zombie poison." It was said that this concoction could leave obstreperous target individuals in such a low state of metabolism as to be mistaken for death. Upon revival, the victims were often left brain damaged and addled to the point they might resemble the hulking louts of the Hollywood movie stereotype. This intriguing interpretation has been challenged by some researchers, however (Lawless, 1989; Garlaschelli, 2001). Regardless, the Hatian zombie saga told by Davis points out once again that it is cultural rather than legal sanctions that are most important in regulating the use of such substances, a theme we shall return to at the end of this chapter.

The medicinal use of datura in meso-America was similar to that of the more northerly tribes, but here the drug found additional use as a stupefacient prior to human sacrifice, and in the case of the Mapuches of Chile, the drug is used as a correctional tonic for unruly children – it is thought that in the ensuing visions the ancestors warn wayward children to mend their ways (Schultes and Hofmann, 1973, p. 171).

In South America there is a variant of the *Datura* family known as "tree Datura." Unlike the rest of the genus which grows as small bushes, the Brugmansia variant reaches the stature of a modest-sized tree with trumpet-shaped, brightly colored flowers. It too is harvested by native people for religious purposes.

The effects of *Datura*, like all psychoactive substances, are dose dependent. These include, as dosage increases, feelings of lassitude, mild euphoria, muscle relaxation, delirium, and dreamless unconsciousness. It does not enhance perceptual novelty and interest like cannabis does, but at higher doses it can produce hallucinations and profound amnesia. Intoxication can be accompanied by extreme agitation and disorientation. At very high doses, it is prone to cause toxic psychotic-like reactions and even coma and death. Tribes that use *Datura* for ceremonial purposes are well aware of its possible adverse effects and regulate doses and settings of use carefully. Some that find it culturally useful nonetheless still assume that its indwelling spirit can have a malevolent nature at times.

In Asia, Datura has been used more for medicinal than religious purposes, but both have occurred in the past. At certain times datura created social problems in India because bandits called the Thuggee (from which we get the word "thug") used it to incapacitate victims before robbing them.

Chemical analysis has revealed that, in varying concentrations and relative proportions, the twenty or so species of *Datura* contain the tropane alkaloids hyocyamine, norhyocyamine, and scopolamine (Furst, 1976, p. 140). These alkaloids are potent inhibitors of the neurotransmitter acetylcholine and thus depress the activity of the parasympathetic branch of the autonomic nervous system, producing many symptoms throughout the body. In the central nervous system, this anticholinergic effect also accounts for its profound disruptions of memory. It enjoyed a brief vogue during the heyday of CIA drug experimentation as a so-called "truth serum," but it was found to be no better than any other depressant, such as alcohol, for loosening lips. In modern medicine, scopolamine is prescribed primarily to treat motion sickness, asthma, and certain muscle tremors.

1.4 "MAGIC" MUSHROOMS

1.4.1 FLY AGARIC

The use of fly agaric (*Amanita muscaria*) has a long history in northern Europe, Asia and North America – so much so that its shape has become the sort of generic mushroom depicted commercially, in cartoons, etc. It has a red cap speckled with white flakes. Some sources claim that it acquired its common name from the practice of mixing it with milk and leaving it about as a means of poisoning houseflies; others say it is because it causes the mind to "buzz around like a fly." Most members of the *Amanita* family are extremely poisonous and the differences in appearance between the desired and toxic ones are quite subtle – again pointing to the motivation and bravery of the early people who made these distinctions by trial and error. Fly agaric mushrooms grow in a symbiotic relationship with certain northern tree species, a fact reflected in artwork and folklore surrounding their religious use (see Plate 1.1).

Effects of *Amanita* include euphoria, inability to concentrate (a vivid flight of random associations), and hallucinations, lasting two to three hours. Ceremonially, *Amanita* has long been used by the shamanic cultures of northern Eurasia, a practice thought to have been brought to the Americas with early migrants across the Bering land bridge.

The mushroom is found along the Pacific coast of Canada and various other parts of the North American continent. It was used ceremonially by, among others, the west coast aboriginal cultures as well as by the Ojibway of central Canada and the US. According to Wasson (1968, 1972b), *Amanita* may also have been the "god plant," soma, of the ancient Aryans of the Indian subcontinent. Soma is described poetically in the ancient *Rig-Vedas,* the Sanskrit hymns that form the foundations of Hinduism. It has been argued that *Amanita* was carried to India from Siberia or the Near East in prehistoric times, because it does not grow south of the Himalayas. After its ceremonial use had been established there, supplies seem to have been lost for some reason, leaving only the cryptic lyrical references in the Rig-Vedas, the cornerstone of the holy teachings of Hinduism. On a more speculative note, the scholar of ancient languages, John Allegro (1970), has linked *Amanita*-using drug cults of the Middle East to the early origins of Christianity. Although Allegro is a respected Old Testament authority (a member of the original team selected to translate the Dead Sea Scrolls), his thesis in this specific instance has been strongly disputed.

Originally thought (mistakenly) to exert its psychotropic properties through interaction with the brain's acetylcholine system, *Amanita*'s active ingredients, the isoxazole derivatives ibotenic acid and muscimol, are now known to affect brain function by stimulating receptors for the neurotransmitter gamma-amino-butyric acid (GABA). GABA is the brain's major inhibitory neurotransmitter, found in numerous areas of the central nervous system. Festi and Bianchi (1985) have suggested a possible route by which stimulation of the inhibitory actions of GABA could alter neural activity in other neurotransmitter tracts that are known to be influenced by the serotonin-affecting hallucinogens.

Unique among the hallucinogens, the active chemicals in *Amanita* are passed through the body unchanged in the urine. Among the Koryak tribe of Siberia, for instance, ceremonial ingestion of the drug by the shaman was followed by collection of his urine, which was then drunk by himself or passed on to the person of next highest status to drink. This would be repeated many times, allowing more of the tribe to experience the sacred visions central to their religion. It has been suggested that in this way the shaman provided the added service of using his own body chemistry to make the initial stages of these ecstatic reveries less disagreeable for those who acquired the active chemicals second-hand. Chemical modification in the body of the primary user renders less toxic some of the ingredients responsible for the drug's more unpleasant somatic effects.

1.4.2 PSILOCYBE

In addition to the *Amanita*-type of sacred mushroom, described above, there are several other genera of psychedelic mushrooms that have a long history of use in religious settings (Ott and Bigwood, 1978). As we have seen, *Amanita* can be found as far north as Siberia, whereas the lusher climates of Mexico and Central America are the primary habitat for the genera *Panaeolus, Strophoria, Conocybe* and *Psilocybe.* The name *Psilocybe* is derived from the Greek for "bald head," a rather less poetic appellation than the Nahuatl word from ancient Mexico, *teonanacatl,* meaning "flesh of the gods." *Psilocybe mexicana,* which can be consumed either fresh or in dried form, has a longer history of human usage than any other psychedelic botanical. In Central America, religious sculptures of mushrooms with visages of gods etched on their stems date back more than 2500 years. Early descriptions of their ritual use were recorded by the priests who accompanied Cortez on his expeditions to conquer the Aztec nation (see Figure 1.1). Because persecution of these mushroom cults by the Spanish missionaries succeeded in driving them far into the hinterlands, relatively little was known about them by Americans of European descent until mid-way into the twentieth century.

Figure 1.1

A sixteenth-century Aztec drawing depicting the ritual use of mushrooms (Florence: Biblioteca Medicea Laurenziana, ms, with permission of Ministero per i Beni e le Attività Culturali. Any further reproduction through any means is forbidden.)

In the early 1950s, the Wall Street investment banker-turned-ethnomycologist, Gordon Wasson and his physician wife, Valentina, began to investigate stories of ritual use of mind-altering mushrooms by the Indians of Oaxaca, Mexico (Wasson, 1972a). The Wassons succeeded in gaining the trust of the Maztec users of these fungi who eventually permitted the American visitors to try them and to take samples home for analysis. The particular variety was designated *Psilocybe mexicana.* After several unsuccessful attempts to identify the active chemical in the fungus, the Wassons and their French collaborator, Roger Heim, turned to Albert Hofmann, the chemist who had discovered the psychedelic properties of LSD. In remarkably short order, Hofmann succeeded in identifying the psychoactive components which he dubbed psilocin (4-hydroxy-dimethyltryptamine) and psilocybin (4-phosphoryl-dimethyltryptamine). It turned out that psilocin is the actual neuroactive chemical – psilocybin is converted to psilocin by enzymes in the body before it affects brain tissues. Unlike DMT itself (see below), the active chemicals in these mushrooms remain so when taken orally, the usual route of administration.

Hofmann tried the mushrooms' purified chemicals himself, finding their effects very much like those of LSD. The structural similarity among psilocin, the neurotransmitter serotonin, and his previous discovery, LSD, was immediately apparent to Hofmann. On a weight-for-weight basis, psilocin and psilocybin have only one-one-hundredth the potency of LSD and their duration of action is only about half that of the approximately 8–12-hour duration of the LSD trip (Snyder, 1986, p. 190). Not surprisingly, their effects are very much like those of LSD, for the psychoactive molecules in both cases are closely related. It was an encounter with *Psilocybe mexican,* shown in Plate 1.2, while on

Plate 1.1

The fly agaric (Amanita muscaria) *mushroom (Reproduced by permission of C. Ratsch, Hamburg, Germany, from* Plants of the Gods*)*

Plate 1.2

A flush of "psilocybe" mushrooms growing on a straw "log"

Plate 1.3

A 500 AD mural from Teotihuacan, Mexico, depicting hallucinogenic intoxication of Ololiuqui (Reprodued by permission of Peter T. Furst, PT, University of Pennsylvania, Albany, PA, USA, from Plants of the Gods*)*

Plate 1.4

Grafted peyote cactus

Plate 1.5

A diagram from the Huichol Indians depicting the visionary state induced by peyote (Reproduced by permission of C. Ratsch, Hamburg, Germany, from Plants of the Gods*)*

vacation in Mexico that set Timothy Leary onto the path that led from Harvard psychology professor to LSD advocate, counterculture guru, and eventually, incarcerated felon.

1.4.3 OLOLIUQUI

A chemically related, naturally occurring hallucinogen encountered by the early Spanish in Central and South America was contained in the seeds of the morning glory, *Peptidina.* The Spanish royal physician, Francisco Hernández (who, from 1570 to 1575, worked alongside Cortez's expedition in Mexico) described its ritualized use and effects, which are much like those of LSD (Schultes, 1969) (see Plate 1.3).

The reasons for this became apparent when the active chemical was finally isolated in the early 1960s. Yet again, the honor belongs to the ubiquitous Albert Hofmann, who discovered that morning glory seeds contain lysergic acid amide, a close but not identical molecule to lysergic acid diethylamide (LSD-25). Although its effects are similar, its potency is also somewhat less than that of the incredibly powerful LSD molecule.

1.5 PEYOTE

Peyote (*Lophophora williamsii*), shown in Plate 1.4, is a squat, grayish-brown spineless cactus native to Mexico and the southwestern United States (Furst, 1972b; Schultes and Hofmann, 1973). It grows close to the ground, often amid grasses that make it easy to overlook.

Spanish Conquistadors found that the native peoples of Mexico had elaborate rituals for harvesting and ingesting the brown "buttons" that top off this modest cactus. Collection began with a devout pilgrimage to the growing regions. To this day, the harvesting ceremony starts with solemn singing of sacred songs, followed by careful removal of the tops of the cacti (Norman, 1977). Only the crown of the plant was taken, in order to allow repeated harvesting as new caps grew from the base. The collected tops were strung like beads and dried to prevent spoilage. This yielded a number of large, dark brown "buttons" that could be easily transported and stored for long periods. These peyote buttons were then reserved for religious ceremonies (see Plate 1.5) in which they were treated as a bridge to the spirit world. Much of what we know of the indigenous use of peyote is due to observations made at the time of the European conquest by the court physician to the king of Spain, Francisco Hernández, mentioned above.

The subjective effects of ingesting peyote buttons are essentially similar to those of LSD or psilocybin, described earlier. This similarity created explanatory problems for psychopharmacologists when it was discovered that the active chemical in peyote, mescaline, affects the catecholamine neurotransmitters dopamine and norepinephrine, rather than serotonin as in the case of LSD or *Psilocybe mexicana.* Mescaline was identified in 1896 by the German chemist A. Heffter as the major psychoactive component of peyote. The chemical structure of the compound was determined in 1918, by E. Spaeth, also in Germany (Snyder, 1986, Ch. 7).

The common household spices nutmeg and mace contain the compounds myristicin and elemicin, which are structurally similar to mescaline and share its psychedelic effects (Schultes and Hofmann, 1973; Efron *et al.,* 1967, pp. 185–233). They are the products of the tropical plant *Myristica fragrans.* Psychedelic use of these spices requires ingesting large quantities, however. This is likely to produce many undesirable bodily effects such as nausea, vomiting, tremors and general malaise – sufficient to deter most users today.

Several hypotheses have been advanced to resolve the apparent inconsistency that catecholamine-affecting drugs such as the foregoing could have effects so similar to those of serotonin affecter such as LSD or psilocybin. One is that the catecholamine-like psychedelics also stimulate one of the many subtypes of serotonin receptor, the 5-HT_{2A} type. An earlier explanation was that mescaline-like drugs might affect catecholamine-containing cells in the brainstem that might, in turn, modulate the action of serotonin-containing cells whose connections spread through the higher centers of the brain. In this way, the indirect modulation of serotonin circuits might be achieving, secondhand, what LSD accomplishes more directly in the serotonin system itself. At this point, no one can say for sure how either the catecholamine-affecting or the serotonin-affecting psychedelics achieve their effects in the brain, but these hypotheses at least suggest possible ways such different chemicals could produce such similar subjective effects. The balance between serotonin and norepinephrine in the brain has been implicated in the regulation of sleep and dreaming, inviting the speculation that the serotonin- and catecholamine-affecting psychedelics might produce hallucinations essentially by triggering "waking dreams."

The visual effects of peyote consist of brightly colored lights, scintillating geometric designs, and visions of animal and sometimes human figures. Ordinary perception of color and space is impaired but otherwise conscious awareness is preserved. Auditory as well as tactile hallucinations are also common. Sensations of weightlessness and feelings of depersonalization, doubling of the ego, and distortions of time are also typical (Schultes and Hofmann, 1973, pp. 121–123). In their peyote ceremonies, the people often experience the voices of the elders from whom they seek guidance. The extreme upwelling of emotion adds cosmic significance to these experiences.

The striking visual effects of peyote have most likely contributed to the distinctive and very beautiful genre of art produced by the Huichol people of Mexico (Norman, 1977). Therapeutically, the Huichols have also used peyote for various medical complaints (Furst, 1972b).

The concentration of mescaline in peyotl cactus is fairly low and, weight for weight, mescaline has only 1/2000 the potency of LSD (McKim, 2000, p. 330). This necessitates the ingestion of enough of the dried buttons to cause nausea and other unpleasant somatic symptoms along with the desired psychotropic effects. Thus the chemist Alexander Shulgin was prompted to begin experimenting with the mescaline molecule, a chemical relative of both the neurotransmitter dopamine and the synthetic stimulant amphetamine. In the process of systematically modifying the phenylethylamine core of the molecule and testing the proceeds on himself, Shulgin produced the first series of so-called "designer drugs." With the right chemical substitutions on the benzene ring at one end of the molecule, one gets varying degrees of psychedelic effects, rather than the typical "speed" reactions of amphetamine. "Ecstasy" or MDMA is the currently most loved or reviled member of this group, depending on one's point of view. Shulgin and his wife, Ann, describe their daring psychopharmacological adventure in their contribution to the present volume.

1.6 OTHERS

1.6.1 HARMINE-CONTAINING PSYCHEDELICS

The psychedelic harmala alkaloids are found in species as diverse as *Peganum harmala*, the Syrian Rue native to the Middle East, and *Banisteriopsis caapi*, that grows in the tropical rain forests of South America (Schultes, 1972; Schultes and Hofmann, 1973, pp. 101–116; Davis, 1996). *Yagé* or *Ayahuasca* (a.k.a. *Hoasca*) is a drink prepared from the bark of the vine of one of several Banisteriopsis species. It has been found to contain the psychoactive β-carboline alkaloids harmine, harmaline, and tetrahydroharmine (Grob *et al.*, 1996). These alkaloids act as potent inhibitors of monoamine oxidase (MAO) – an enzyme that breaks down excess serotonin, norepinephrine and dopamine in the brain. People of the Amazon basin have long used Banisteriopsis alone in their sacred drinks, but they have also found that its combination with certain plant substances containing dimethyltryptamine (DMT, e.g., in snuffs made from species of *Virola*), makes these DMT-containing preparations active orally, which they would not have been without the MAO-inhibiting properties of *Banisteriopsis* (Strassman, 2001).

Continuous ritualistic use of the harmala plants from prehistoric to modern times has survived the, by now familiar, colonial attempts at eradication. The visions produced in these strongly religious settings serve to strengthen group cohesion and reinforce shared metaphysical belief systems. Jaguars, snakes and other predatory animals tend to populate these vivid hallucinations, as do visions of distant persons, living and dead, and feelings of being in contact with an alternate reality. Users interpret this as a strong contact with the supernatural. It is also used in rites of initiation. Use of *Ayahuasca* is permitted under Brazilian law where it has become the basis of a rapidly growing syncretist religion. It is becoming increasingly popular among spiritually inclined citizens of the northern hemisphere as well (Montogmery, 2001).

1.6.2 DIMETHYLTRIPAMINE-CONTAINING PLANTS

As mentioned above, N,N-dimethyltryptamine (DMT) is structurally related to serotonin and LSD. Its effects are similar to those of the latter, as are its binding patterns in the brain, but its duration of action is much shorter (Strassman, 1996, 2001). DMT is found in ritually-used plants around the world. Best known is the South American species *Virola calophylla* and *Mimosa hostels* (Schultes and Hofmann, 1973). Because DMT is destroyed by stomach acid, its various preparations must be taken as snuffs or by smoking. The powers of the widely used snuffs from this region, Elena ("semen of the sun") and Cohoba are attributed to sun deities. The former is stored in containers whose name translates as "the sun's penis." These sacred snuffs play an important role in the spiritual lives of the indigenous people of South America. They were also used to prepare warriors for battle. Because the effects of DMT can be over in as little as 30 minutes, the psychedelic subculture of the 1960s dubbed it "the businessman's lunch." Breakdown and deactivation of DMT is accomplished by the enzyme monoamine oxidase (MAO).

1.6.3 IBOGA

Tabernathe iboga, a member of the Apocynaceae family, is a shrub native to central western Africa. It can grow to a height of over two meters and has yellow blossoms and edible orange-colored fruit. For centuries, the bark of the root has been scraped off and dried to produce a powder that is eaten (Fernandez, 1972). At low doses, it combats fatigue and hunger, and at high doses it has powerful, long-lasting psychedelic effects. Native users also credit it with an aphrodisiac effect. Traditionally, several tribes in the region worshiped the plant as a source of spiritual knowledge, especially those in the Congo, Gabon and Cameroon. The user is said to travel to "the other side" to meet the ancestors and the great spirit "Bwiti," the creator, who is thought of as a superconscious entity that guides human affairs. Its transformative use in the initiation rites of the Bwiti cult is reflected in the fact that "Iboga" is referred to as "the plant that breaks open the head." Initiates are said to be able to foresee the future and heal the sick as well as commune with the ancestors.

Trust and rapport built up during Iboga ceremonies has been credited with forging peaceful alliances among warring tribes, and the ceremonial use of Iboga has been a factor in the greater than usual resistance to the proselytizing activities of Christian and Muslim missionaries in the area (see Figure 1.2). In recent years, Iboga has been touted in industrialized societies as an agent that reduces craving in persons addicted to various substances, though this claim remains controversial.

Figure 1.2
Eboka (Tabernathne iboga) *is consumed so that ancestors can be met (reprinted from* Psychedelics Encyclopedia *by Peter Stafford, by permission of Ronin Publishing, Berkeley, www.roninpub.com)*

The plant contains a number of alkaloids, the principal active one being ibogaine, first isolated in 1901. Ibogaine has an indole structure that resembles those of serotonin and harmine. Pharmacologically, it is known to inhibit cholinesterase, the enzyme responsible for breaking down the neurotransmitter acetylcholine once it has finished stimulating post-synaptic receptors. In addition to serotonergic effects, ibogaine may also interact with kappa-type opiate receptors. One suggested explanation for ibogaine's psychedelic effects is that it might block N-methyl-d-aspartate receptors in the brain in a manner analogous to phencyclidine (PCP or "angel dust") (Javitt and Zukin, 1991; Julien, 2001, p. 387).

1.6.4 QAT (KHAT, KAT, CHAT, MIRAA, KUS-ES-SALAHIN, ABYSSINIAN TEA, ETC.)

Catha edulus is a flowering evergreen shrub that can grow to full tree size. Today, it is often grown interspersed with coffee plants. Qat originated in Ethiopia but its recreational and sacramental use has spread it through much of Africa, the Arabian Peninsula, and the Middle East. Its cultivation and use was described by Lewin in 1931. The fresh leaves are typically chewed like tobacco or brewed into a tea. Qat was used for religious purposes by the ancient Egyptians who recorded that it was able to make mere mortals experience the sense of power and grandeur usually reserved for the gods. Qat subsequently fell out of favor, but interest revived after the founding of Islam which considered it one of the few permissible drugs. Its stimulant properties made it quite useful during long prayer vigils.

Chewing Qat produces sympathomimetic somatic effects followed by a euphoric "high" similar to, but much milder than that of cocaine or amphetamine (Kalix, 1990). Absorbing the active ingredients in this slower fashion does not produce the intense "rush" common to other psychostimulants and neither does it seem to produce the edginess or suspiciousness typically reported with these other agents. Qat users report elevated mood and energy and an alert, focused state of mind that improves concentration and the incisiveness of one's thinking. Users also tend to become more active and loquacious. As the effects wear off, they experience a gentler version of the depressing "crash" reported by cocaine and amphetamine users. Like other stimulants, it suppresses appetite and can disrupt sleep patterns. As with other stimulants, prolonged, high-dose use has been reported, in some cases, to lead to psychotic-like symptoms.

The stimulant properties of Qat are understandable, given that the principal active ingredient is cathine (d-norisoephedrine), also found in *Ephedra vulgaris,* a mainstay of traditional Chinese medicine. Chemical manipulation of its active constituent, ephedrine, ultimately led to the development of the synthetic stimulant amphetamine (Snyder, 1986, Ch. 5). Similar to the body's own neurohormone epinephrine, ephedrine is related to the therapeutic agents in certain decongestants and anti-asthma drugs available in developed countries. The stimulant properties of these remedies have led to their inclusion on the list of banned substances in international sporting competitions.

Until the advent of rapid transcontinental flight, the use of Qat was restricted to its areas of cultivation, because the main active ingredients lose their potency soon after harvesting the plant. With modern transportation facilities now available, the use of Qat has become a worldwide phenomenon. It is currently illegal in the US to possess Qat, but clandestine labs have been producing a synthetic analogue of cathinone, another of its active ingredients, called "methcathinone." Its street name is "cat."

1.6.5 KAVA

Captain James Cook's expeditions to the South Pacific in the late 1700s found the islanders using the root of a vine, *Piper methysticum*, to brew a drink called kava or kava kava (Efron *et al.*, 1967, pp. 105–184; Foo and Lemon, 1997). Native to the Oceania region, the vine is related to the common black pepper plant, *Piper nigrum*. Originally, the root was chewed in ritual fashion and spat back into a bowl. Salivary enzymes helped break down the fibrous mass and the resulting pulp was brewed into a drink used ceremonially and medicinally. It was traditionally imbibed in group settings. The beverage has an anxiolytic and relaxing effect that promotes social interaction, much like alcohol does in our own culture. It was also used as an aid to dispute resolution and in decision-making forums, and as a sleep tonic. At the time of European contact kava was used strictly for these religious, medicinal, and cultural purposes. It has since become primarily a social drug (see Figure 1.3).

Figure 1.3
Kava-Kava extract is a socially accepted intoxicant in some South Pacific cultures (reprinted from Psychedelics Encyclopedia *by Peter Stafford, by permission of Ronin Publishing, Berkeley, www.roninpub.com)*

There have been many psychoactive compounds identified in kava. Those of primary interest have been the alpha-pyrones known as the kava lactones. They have been shown to bind in various parts of the multi-faceted receptor complex for GABA. Ethyl alcohol, the barbiturates, and the benzodiazepines also exert their sedating effects by potentiating the effects of this inhibitory neurotransmitter. This is consistent with the sedative and anxiolytic effects of kava. It is also capable of producing all the typical side-effects of central nervous system depressants. The kava lactones have also been found to inhibit the B-variant of monoamine oxidase (MAO-B). It is MAO-A that is responsible for the drug's antidepressant effects, so the significance of this action of kava is uncertain at this time.

Kava extracts are widely available in health food stores for the self-medication of anxiety. Given its ability to impair attention and performance along with reducing anxiety, Julien (2001, p. 592) remarked that ". . . it is interesting that barbiturates and benzodiazepines are restricted to prescription use, alcohol has age restrictions, marijuana is illegal, but kava is available without restriction." Treating so-called "natural" products as if they have only beneficial effects and can do no harm is a serious misconception (Beyerstein and Downie, 1998; Talalay and Talalay, 2001; Markus and Grollman, 2002).

1.7 ANIMAL-DERIVED HALLUCINOGENS

As part of the endless co-evolutionary struggle between predators and prey, some species have evolved the ability to produce chemicals that taste bad, sting the mouths of attackers, or cause them aversive effects such as nausea, disorientation or perceptual aberrations. Species so endowed usually signal their unpalatability with bright coloring that advertizes the fact to would-be predators. This, in turn, leads other non-toxic species to mimic the color or markings of those that are genuinely noxious. Occasionally, however, this has led humans to prey on such species, precisely for the medicinal or consciousness-affecting properties of their exudates.

Along with the bats' wings and babies' skulls that conferred little beyond their placebo value, toad skins figure prominently in the lore of medieval witches' brews. In the latter case, these proto-pharmacologists may actually have been on to something (Strauss, 1994). Special glands in the skin of certain toads produce the hallucinogenic chemical bufotenine (5-hydroxy-DMT) which can be collected and dried for human use (McKenna and Towers, 1984; Strassman, 2001). The chemical's name is derived from the designation for the toad genus, *Bufo.*

Like psilocin, which it resembles structurally, bufotenine is a potent serotonin agonist, affecting several of the subtypes of serotonin receptor. It is not active orally unless it is taken in conjunction with something containing a monoamine oxidase-inhibitor; otherwise the bodily enzyme MAO will neutralize it. Bufotenine can be taken by smoking or insufflation ("snorting") where it is absorbed through the lungs and mucous membranes, respectively.

From time to time, scare stories concerning bufotenine have erupted and been spread by an uncritical press. Thrill-seeking youngsters are claimed to be placing themselves at grave risk as they go on toad-licking rampages, hoping to get high. Like earlier stories of LSD users staring into the sun until they went blind, or fiendish drug-pushers who (for some incomprehensible reason) allegedly impregnated children's lick-on tattoo transfers with LSD, these tales are classic examples of "urban legends" (Brunvand, 1993, pp. 109–112). They are typical of the "pious lies" spread during moral panics where well-meaning moralists exaggerate dangers, assuming that the ends justify the means. The media, who should have learned from previous debunkings of the same old horror stories, often recycle them repeatedly.

1.8 SOCIAL AND LEGAL CONSIDERATIONS

Psychoactive drugs affect the mind, which the world's religions have traditionally considered a spiritual, and hence sacred, phenomenon – i.e., something to be approached only in narrowly sanctioned ways without fear of engendering a strong backlash (Beyerstein 1996a, b, c). Mind-altering substances are capable of creating states of consciousness so intense and meaningful to the experiencer that his or her metaphysical beliefs, indeed whole life, can be changed in fundamental ways (whether these "revelations" are objectively true or not). All societies have a stake in regulating the usage of anything with such power for good or ill. We should not forget that, used unwisely or by psychologically unstable individuals, these powerful mind-altering chemicals can be the catalyst for foolish and harmful behaviors.

At the same time, it is demonstrable that most hallucinogen users indulge without causing much, if any, harm to themselves or society and feel that their lives have been enriched by their drug experiences (Zinberg, 1984; Alexander, 1990; Grinspoon and Bakalar, 1997). Their drugs of choice have low addiction potential and most people indulge only occasionally. Understandably, these non-problematic users bristle at the thought that it is anyone else's business, least of all the government's, as to if and how they choose to alter their consciousness (provided they do so responsibly, on their own time). Because these non-harmful users rarely advertise the fact and the few who do suffer adverse effects in their lives become readily apparent to the media, the police, treatment facilities, and ordinary citizens, it is easy to overestimate the proportion of users who suffer any ill effects (Beyerstein and Hadaway, 1990). In a political climate such as that in North America, where psychoactive drug use tends to be automatically equated with abuse, it is politically hazardous to argue that not all use is necessarily abuse. There are excesses on both sides of the debate, however. Responsible users sometimes underplay certain real dangers, or their actual incidence, but they quite rightly decry the hypocrisy that permits some drugs to be legal and promoted while others, arguably less harmful ones, are prohibited – often with penalties out of all proportion to the realistic levels of risk.

It is only to be expected that anything that touches as many of a culture's philosophical "hot buttons" as psychoactive drug use does will engender strong attitudes, misunderstandings, fears, passions, and not infrequently, vilification and persecution. Throughout history, dominant cultures have tended to demonize the socially condoned drugs of other societies or subcultures while tolerating, even applauding, their own which may be objectively worse. Which ones are in or out of the approved category has flip-flopped repeatedly throughout history. Attitudes toward tobacco are a classic example.

The annals of psychopharmacology are replete with examples of drugs that, unpredictably and in amazingly short order, swing from deification, to benign indifference, to demonization, and back again (Brecher, 1972). The emotions generated in this clash of values and political aims are capable of clouding judgement and setting decent and intelligent people onto immoral and (foreseeably) unworkable crusades to eradicate exaggerated threats (Beyerstein and Hadaway, 1990; Buckley *et al*, 1996). This is not to deny that drugs can have adverse effects. It is a question, rather, of how best to deal with these problems without creating problems worse than the ones we set out to remedy – and without doing violence to our democratic values.

As we have seen, even extreme brutalities on the part of colonial powers never succeeded in eliminating the psychoactive drug use of the aboriginal population. For similar reasons, critics charge, the current "war on drugs" in the industrialized nations has failed in its attempt to create a drug-free society – one might as well try to ban sex, they say. While legal prohibition has incurred a tremendous waste of resources and lives, exacerbated racial tensions, and destabilized economies, drugs are cheaper and more available than ever (Duke and Gross, 1993; Buckley *et al.*, 1996). Reformers such as Alexander (1990) argue that as long as societies fail to deal with the social conditions that promote escapist drug use, people will continue to use them whether the authorities approve or not, regardless of the cost. As William F. Buckley (1996) summarized the situation in the pages of the conservative periodical, *The National Review*:

> [Some may say] . . . that the *National Review* favors drugs. We don't; we deplore their use; we urge the stiffest possible sentences against anyone convicted of selling a drug to a minor. But that said, it is our judgement that the war on drugs has failed, that it is diverting intelligent effort away from how to deal with the problem of addiction, that it is wasting our resources, and that it is encouraging civil, judicial, and penal procedures associated with police states. We all [i.e., Buckley and his fellow contributors to the piece] agree on movement toward legalization, even though we may differ on just how far.

In his contribution to the same collection, Joseph McNamara, a former police chief in several large American cities, and currently a fellow at Stanford's Hoover Institution, also emphasized the serious consequences such as the inevitable corruption of the police, customs agents, and the criminal justice system, brought about by the unwinnable war on drugs. The enrichment of organized crime has been another of its unintended consequences. Reformers such as Arnold Trebach join Buckley, McNamara, and former US Secretary of State George Schultz in urging the abandonment of prohibition, while stalwarts such as James Inciardi and James Q. Wilson maintain more effort and resources are all that is needed for it to succeed (Trebach and Inciardi, 1993). Goldberg (1999) has edited a volume of essays by supporters of both sides of the decriminalization debate. Even those who advocate removal of criminal sanctions for simple possession of psychoactive drugs support the need for criminal penalties for those who act in antisocial ways when intoxicated (Nadelmann, 1998). As with alcohol, the police will still have a role to play.

If legal sanctions, including the death penalty in some times and places (Brecher, 1972), cannot eliminate psychoactive drug use, what then? Erickson *et al.* (1997) and Marlatt (1998) have compiled articles by experts from around the world, demonstrating the benefits of the "harm reduction" alternative to legal prohibition. This has long been the official position in the Netherlands and more recently in Germany and Switzerland. Other countries such as Australia, Britain, and Canada have also begun to move in this direction. The harm reduction approach advocates public policies aimed at reducing the personal and social harms associated with drugs (many of which are more the results of attempted prohibition than of the drugs *per se* [Beyerstein and Hadaway, 1990]) rather than prohibition. Harm reduction advocates try to deal with drug abuse rather than drug use, within a comprehensive social services and medical system (Smith, 1995). Countries that have moved in this direction generally have fewer serious social problems with drugs than those that have maintained criminal sanctions for use. Having seen the improvements the harm reduction approach had achieved with the country's heroin problem, the citizens of Switzerland recently rejected by a two-thirds majority, a national referendum aimed at returning to the criminal justice model.

Of course, we all rightly fear what ramifications drug reforms might have for the children we naturally wish to protect. Marsha Rosenbaum (1996) has approached this problem head-on, arguing that prohibition makes children less safe rather than more, and offering useful, humane advice, based on harm reduction principles, to parents concerned, as we all should be, with protecting our children. As Rosenbaum shows, current drug education programs not only don't work, they may even promote usage.

Zinberg (1984) has argued persuasively that, just as in the pre-industrial societies discussed above, the real measures that have always controlled drug use are the ingrained mores and customs of the people. Without a philosophical consensus and willing self-policing, no law enforcement effort is likely to succeed in the long run – gambling, prostitution, and abortion supply ample cases in point. Where drug use becomes problematical is usually when another culture's socially approved drug is parachuted into a society that has not had the time and experience to adapt its informal controls to handle it. If the new drug is associated with marginalized people, such as racial minorities, the dominant culture is all the more likely to react with fear and repression. Problems are most likely to arise when the newly exposed users lack models for safe usage that they can emulate, because the society has not had time to evolve its unwritten cultural controls that encourage responsible behavior. If these traditional means of regulating social interactions are breaking down in general, as in a time of cultural decimation, problematic usage, such as widespread escape into chemical oblivion, is all the more likely. Witness the terrible devastation wrought by the introduction of alcohol in aboriginal communities in North America.

The early history of LSD use in our own culture teaches a similar lesson. When the users were stable, well-socialized adults with a stake in society, and (a) they took the drug in pure form, (b) were well-prepared as to what to expect, and (c) took it in supportive environments with experienced users who were not also on the drug at the same time, the incidence of adverse reactions was remarkably low. When observance of each of these precautions eroded as the drug spilled out onto the streets in socially tumultuous times, predictably, the number of adverse reactions rose.

If it is true, as George Santayana warned, that those who ignore history are condemned to relive it, we can expect more strife as modern societies try to come to grips with the allure of psychoactive drugs. We can expect that, as our ancient ancestors showed, some people will continue to seek ways to alter consciousness regardless of cost, and others will continue to disapprove of certain kinds of usage for moralistic reasons. The problem will intensify as modern neuroscience learns more about the chemistry of the brain and provides more precise tools for manipulating it. The ability to produce drugs for such purposes, outside of officially sanctioned channels, will grow as the equipment and technical know-how becomes more widespread. If we hope to mitigate the harm that this trend could produce, we could do far worse than looking to the kinds of cultural controls discussed in this chapter.

In this chapter we have sampled the enormous literature devoted to this endlessly fascinating topic. It has taken us from archeology to botany to neuroscience to psychology to sociology to religion and the law. In a pluralistic society anything carrying as much metaphysical and social baggage as this field does is unlikely ever to achieve a universal consensus. We can only hope that by dispelling myths on all sides of the debate, calming irrational fears, promoting humane values and tolerance of diversity, and by agreeing that none of us has the unfettered right to engage in anything that harms our fellow citizens, we can reach what Arnold Trebach has called a "middle-level" solution – one that involves compromise on all sides and affords reasonable protections while respecting democratic values. We must promote the understanding that these mind-altering chemicals are a two-edged sword. They are neither inherently evil nor inevitably harmful, but that they can do harm as well if used unwisely. These drugs are capable of both life-enhancing and socially destructive results. Which consequences we see depends on the responsibility of the user and the setting of usage. History and psychology teach us that these substances are unlikely to go away and the lessons of the past as to how best to deal with them may help guide us in the future. If this commonsense approach can prevail, we will all be the better for it.

REFERENCES

Aghajanian, G. and Marek, G. J. (1999) *Neuropsychopharmacology*, 21, pp. 16s–23s.

Alexander, B. K. (1990) *Peaceful Measures: Canada's Way out of the War on Drugs*, Toronto, Ontario: University of Toronto Press.

Allegro, J. (1970) *The Sacred Mushroom and the Cross,* Garden City, NY: Doubleday and Co.

Ashton, H. and Stepney, R. (1982) *Smoking: Psychology and Pharmacology*, London, England: Tavistock Publications.

Barron, F., Jarvik, M. and Bunnell, S. (1964) The hallucinogenic drugs, *Scientific American*, April 1964, New York: W. H. Freeman, *Scientific American* reprint No. 483.

Beyerstein, B. L. (1996a) Visions and hallucinations, in G. Stein (ed.), *Encyclopedia of the Paranormal*, Amherst, NY: Prometheus Books, pp. 789–797.

Beyerstein, B. L. (1996b) Altered states of consciousness, in G. Stein (ed.), *Encyclopedia of the Paranormal*, Amherst, NY: Prometheus Books, pp. 8–16.

Beyerstein, B. L. (1996c) Dissociation, possession, and exorcism, in G. Stein (ed.), *Encyclopedia of the Paranormal*, Amherst, NY: Prometheus Books. pp. 544–552.

Beyerstein, B.L. and Downie, S. (1998) Naturopathy, *The Scientific Review of Alternative Medicine*, 1, pp. 10–18.

Beyerstein, B. L. and Hadaway, P. F. (1990), *J. Drug Issues*, 20, pp. 689–700.

Brecher, E. M. (1972) *Licit and Illicit Drugs,* Boston, MA: Little, Brown and Co.

Brunvand, J. H. (1993) Drug horror stories, in *The Baby Train and Other Lusty Urban Legends,* New York: W. W. Norton and Co., pp. 109–112,

Buckley, W. F., Nadelmann, E., Schmoke, K., McNamara, J., Sweet, R., Szasz, T. and Duke, S. (1996) The war on drugs is lost, *The National Review,* 1 July 1996.

Chafetz, M. (1965) *Liquor: The Servant of Mankind*, Boston, Mass.: Little, Brown and Co.

Clark, W. H. (1970) in B. Aaronson and H. Osmond (eds), *Psychedelics*, New York: Doubleday.

Davis, W. (1988) *Passage of Darkness: The Ethnobotany of the Haitian Zombie,* Chapel Hill, NC: University of North Carolina Press.

Davis, W. (1996) *One River: Explorations and Discoveries in the Amazon Rain Forest,* New York: Simon and Schuster.

deSmet, A. G. M. and Rivier, L. (1989) *J. Ethnopharm*, 25, pp. 127–138.

Devereux, P. (1997) *The Long Trip: A Prehistory of Psychedelia,* Harmondsworth, England: Penguin Books.

Duke, S. B. and Gross, A. C. (1993) *America's Longest War: Rethinking our Tragic Crusade Against Drugs,* New York: Tarcher/Putnam.

Efron, D., Holmstedt, B. and Kline, N. S. (eds) (1967) *Ethopharmacologic Search for Psychoactive Drugs,* Public Health Service Publication No. 1645, Washington, DC: US Government Printing Office.

Emboden, W. (1979) *Narcotic Plants* (revised edn), New York: Collier Books.

Erickson, P. G., Riley, D. M., Cheung, Y. W. and O'Hare, P. (eds) (1997) *Harm Reduction: A New Direction for Drug Policies and Programs,* Toronto, Ontario: University of Toronto Press.

Fernandez, J. W. (1972) in P. Furst (ed.), *Flesh of the Gods: The Ritual Use of Hallucinogens,* New York: Praeger, pp. 237–260.

Festi, F. and Bianchi, A. (1985) *Amanita muscaria*: Mycopharmacological outline and personal experiences, *Psychedelic Monographs and Essays,* Vol. 5, available online at: http:www.erowid.org/plants/amanitas writings4.shtml

Foo, H. and Lemon, J. (1997) Acute effects of kava, alone or in combination with alcohol, on subjective measures of impairment and intoxication and on performance, *Drug and Alcohol Review*, 16, pp. 147–155.

Frazier, J. G. (1890/1993) *The Golden Bough: The Roots of Religion and Folklore,* New York: Gramercy Books.

Furst, P. (ed.) (1972a) *Flesh of the Gods: The Ritual Use of Hallucinogens,* New York: Praeger.

Furst, P. (1972b) To find our life: Peyote among the Huichol Indians of Mexico, in P. Furst (ed.), *Flesh of the Gods: The Ritual Use of Hallucinogens*, New York: Praeger, pp. 136–184.

Furst, P. (1976) *Hallucinogens and Culture*, Novato, CA: Chandler and Sharp.

Garlaschelli, L. (2001) *La Chimica e L'industria*, 7, p. 71.

Goldberg, R. (ed.) (1999) *Taking Sides: Clashing Views on Controversial Issues in Drugs and Society* (fourth edn), New York: McGraw Hill/Dushkin.

Grinspoon, L. and Bakalar, J. B. (1997) Ch. 4, *Psychedelic Drugs Reconsidered* (revised edition), New York: The Lindesmith Foundation.

Grob, C. S., Mckenna, D., Callaway, J., Brito, G., Neves, E., Oberlaender, G., Saide, O., Labigalini, E., Tacla, C., Miranda, C., Strassman, R. and Boone, K. (1996) *J. Nervous and Mental Disease*, 184, pp. 86–94.

Halpern, J. H. and Pope, H. G. Jr (1999) *Drug and Alcohol Dependence*, 53, pp. 247–256.

Hayden, B. (1987) *J. Scientific Study of Religion*, 26, pp. 81–91.

Hofmann, A. (1983) *LSD: My Problem Child: Reflections on Sacred Drugs, Mysticism, and Science*, trans. J. Ott, Los Angeles, CA: Jeremy Tarcher Inc.

Hollister, L. E. (1962) Drug-induced psychoses and schizophrenic reactions: A critical comparison, *Annals of the NY Academy of Sciences*, 96, pp. 80–88.

Houghton, P. J. and Bisset, N. G. (1985) Drugs of ethno-origin, in D. C. Horwell (ed.), *Drugs in Central Nervous System Disorders*, New York: Marcel Dekker, pp. 283–332.

Huxley, A. (1956) *The Doors of Perception: Heaven and Hell,* New York: Harper and Row.

Javitt, D. C. and Zukin, S. R. (1991) *Am. J. Psychiatry*, 148, pp. 1301–1308.

Johanson, C. (1988) in *The Encyclopedia of Psychoactive Drugs: Cocaine*, London, England: Burke Publishing.

Joy, J. E., Watson, S. J. Jr and Benson, J. A. Jr (eds) (1999) *Marijuana and Medicine: Assessing the Science Base*, Washington, DC: Institute of Medicine/National Academy Press.

Julien, R. M. (2001) *A Primer of Drug Action* (ninth edn), New York: Worth Publishers.

Kalix, P. (1990) *Therapeutics*, 48, pp. 397–416.

Kobler, J. (1974) *Ardent Spirits: The Rise and Fall of Prohibition*, London, England: Michael Joseph Ltd.

LaBarre, W. (1972) Hallucinations and the shamanistic origins of religion, in P. Furst (ed.), *Flesh of the Gods: The Ritual Use of Hallucinogens*, New York: Praeger, pp. 261–278.

Lawless, R. (1989) *Latin American Anthropology Review*, 1(1), pp. 5–6.

Leavitt, F. (1995) *Drugs and Behavior* (third edn), Thousand Oaks, CA: Sage Publications.

Lee, M. A. and Schlain, B. (1985) *Acid Dreams: The CIA, LSD, and the Sixties,* New York: Grove Press.

Lewin, L. (1931) *Phantastica, Narcotic and Stimulating Plants,* London, England: Routledge and Kegan Paul. (Translation of original 1924 German edition).

Lewis, W. H. and Elvin-Lewis, M. P. F. (1977) in *Medical Botany*, New York: Wiley-Interscience.

Lucia, S. P. (ed.) (1963) *Alcohol and Civilization*, New York: McGraw-Hill.

Lyttle, T. (1988) *J. Drug Issues*, 18, pp. 271–284.

Marks, J. (1979) The Search for the "Manchurian Candidate": The CIA and Mind Control, New York: Dell Books.

Markus, D. M. and Grollman, A. P. (2002) *New Engl. J. Med.*, 347(25), pp. 2073–2075.

Marlatt, G. A. (ed.) (1998) *Harm Reduction: Pragmatic Strategies for Managing High-risk Behaviors*, New York: The Lindesmith Center.

Martinetz, D. and Lohs, K. (1987) *Poison: Sorcery and Science; Friend and Foe*, Leipzig, Germany: Edition Leipzig.

Matossian (1989) *Poisons of the Past: Molds, Epidemics, and History*, New Haven, CT: Yale University Press.

McKenna, D. J. and Towers, G. H. N. (1984) *J. Psychoactive Drugs*, 16, pp. 347–358.

McKim, W. A. (2000) *Drugs and Behavior: An Introduction to Behavioral Pharmacology*, Upper Saddle, NJ: Prentice-Hall.

Montgomery, C. (2001) High Tea: Letter from the Amazon, *The Vancouver Sun*, 10 February 2001, pp. H1–H2.

Mortimer, W. G. (1901, 1979) *History of Coca: The "Divine Plant" of the Incas*, San Francisco, CA: AND/OR Press.

Nadelmann, E. (1998) Commonsense drug policy, *Foreign Affairs*, 77, pp. 111–126.

Nesse, R. M. and Berridge, K. C. (1997) *Science*, 278, pp. 63–66.

Norman, J. (1977) *National Geographic*, 151, pp. 832–853.

Ott, J. (1993) *Pharmacotheon: Entheogenic Drugs – Their Plant Sources and History*, Kennewick, WA: Natural Products Co.

Ott, J. and Bigwood, J. (1978) *Teonanacatl: Hallucinogenic Mushrooms of North America*, Seattle, WA: Madrona Publishers.

Reichel-Dolmatoff, G. (1972) The cultural context of an aboriginal hallucinogen: *Banisteriopsis caapi*, in P. Furst (ed.), *Flesh of the Gods: The Ritual Use of Hallucinogen*, New York: Praeger, pp. 84–113.

Robinson, T. (1974) *Science*, 184, pp. 430–435.

Rosenbaum, M. (1996) *Kids, Drugs, and Drug Education: A Harm Reduction Approach*, New York: The Lindesmith Center.

Sagan, C. (1995) *The Demon-Haunted World – Science as a Candle in the Dark*, New York: Random House.

Satina, S. and Avery, A. G. (1959) A review of the taxonomic history of *Datura*, in A. G. Avery, S. Satina and J. Rietsma (eds), *Blakeslee: The Genus Datura*, New York: The Ronald Press, pp. 16–38.

Schultes, R. E. (1969) *Science*, 163, pp. 245–252.

Schultes, R. E. (1972) Hallucinogens in the Western Hemisphere, in P. Furst (ed.), *Flesh of the Gods: The Ritual Use of Hallucinogens*, New York: Praeger, pp. 3—54.

Schultes, R. E. and Hofmann, A. (1973) *The Botany and Chemistry of Hallucinogens*, Springfield, IL: Charles Thomas.

Siegel, R. K. (1989) *Intoxication: Life in Pursuit of Artificial Paradise*, New York: Pocket Books.

Smith, H. (1964) Do drugs have religious import?, *Journal of Philosophy*, 61(18), reprinted in D. Solomon (ed.), *LSD: The Consciousness-Expanding Drug*, New York: Putnam, 1968, pp. 155–169.

Smith, R. (1995) *The British Medical Journal* (editorial), 311, pp. 23–30.

Snyder, S. H. (1986) *Drugs and the Brain*, New York: Scientific American Books.

Snyder, S. H. (1989) *Brainstorming: The Science and Politics of Opiate Research*, Cambridge, MA: Harvard University Press.

Strassman, R. J. (1984) *J. Nervous and Mental Disease*, 172, pp. 577–595.

Strassman, R. J. (1996) *Behav. Brain Res.*, 73, pp. 121–124.

Strassman, R. J. (2001) *DMT: The Spirit Molecule*, Rochester, NY: Park Street Press.

Strauss, S. (1994) Taking their own medicine, *The Globe and Mail* (Toronto), 3 June 1994, B4.

Swiss Federal Commission for Drug Issues (1999) Cannabis Report, Basel: Swiss Federal Office of Public Health, Federal Printing and Supplies Office.

Szasz, T. (1975) *Ceremonial Chemistry: The Ritual Persecution of Drugs, Addicts, and Pushers*, Garden City, NY: Anchor Press/Doubleday.

Talalay, P. and Talalay, P. (2001) *Academic Med.,* 76, pp. 238–247.

Trebach, A. (1982) *The Heroin Solution*, New Haven, CT: Yale University Press.

Trebach, A. and Inciardi, J. (1993) *Legalize It?: Debating American Drug Policy*, Washington, DC: American University Press.

Tyler, V. (1993) *The Honest Herbal: A Sensible Guide to the Use of Herbs and Related Remedie*, third edn, New York: Pharmaceutical Products Press/Hayworth.

Vollenweider, F .X., Vontobel, P., Hell, D. and Leenders, K. L. (1999) *Neuropsychopharm.*, 20, pp. 424–433.

Wall, C. H. (1927) *Four Thousand Years of Pharmacy*, Philadelphia, PA: J. B. Lippincott.

Wasson, R. G. (1968) *Soma: The Divine Mushroom of Immortality*, New York: Harcourt, Brace, Jovanovich.

Wasson, R. G. (1972a) The divine mushroom of immortality?, in P. Furst (ed.), *Flesh of the Gods: The Ritual Use of Hallucinogens*, New York: Praeger, 1972, pp. 185–200.

Wasson, R. G. (1972b) What was the soma of the Aryans?, in P. Furst (ed.), *Flesh of the Gods: The Ritual Use of Hallucinogens*, New York: Praeger, 1972, pp. 201–213.

Wasson, R. G., Ruck, C. and Hofmann, A. (1978/1999) *The Road to Eleusis*, William Daily Rare Books.

Wilbert, J. (1972) Tobacco and shamanistic ecstacy among the Warao Indians of Venezuela, in P. Furst (ed.), *Flesh of the Gods: The Ritual Use of Hallucinogens,* New York: Praeger, 1972, pp. 55–83.

Witters, W., Venturelli, P. and Hanson, G. (1992) *Drugs and Society* ,third edn, Boston, MA: Jones and Bartlett.

Zimmer, L. and Morgan, J. (1997) *Marijuana Myths, Marijuana Facts*, New York: The Lindesmith Center.

Zinberg, N. (1984) *Drug, Set, and Setting,* New Haven, CT: Yale University Press.

CHAPTER 2

OCCURRENCES AND FORMS OF THE HALLUCINOGENS

Terry A. Dal Cason
Edward S. Franzosa

2.0 INTRODUCTION

Illicit drugs may be synthesized in clandestine laboratories or may be the result of biosynthesis by plants, or even animals. Drugs produced by plants are often consumed as a dried plant material in capsules, smoked, or brewed to make a "tea." In some instances the dried plant may be ground to a fine powder for inhalation or insufflation. Only rarely are the hallucinogenic drugs of plant origin extracted and processed to provide a relatively pure material. Those drugs which are made in clandestine laboratories are usually basic compounds that are converted into one of their water soluble salts and subsequently cut (i.e. diluted) and packaged as free powders in metal foil or paper bindles, or small plastic bags. Clandestine laboratory operators may place the powder in hard gelatin capsules in order to form pre-measured dosage units (DU) or, in more sophisticated clandestine lab operations, compress it into tablet form with suitable equipment. Occasionally the material may be marketed as a solution in various small containers or large vessels. For some hallucinogens, the solution containing the drug is placed on a substrate such as candy or paper for ingestion. In particular cases, the drug is directly incorporated into a hard gelatin matrix.

PART I: LSD

2.1 LSD AND RELATED ANALOGUES

The best known, and the most ubiquitous, of the hallucination-inducing drugs is d-lysergic acid diethylamide (LSD, LSD-25). The acronym LSD-25 was derived from the German spelling for the compound, *l*yserg*s*aure*d*iethylamid, while the often omitted number, 25, has been attributed to the date of first synthesis by Dr Albert Hofmann (2 May 1938) at the Sandoz laboratory in Basel, Switzerland (Hopes, 1968). Hofmann did not note the effects of the drug until several years later. On 16 April 1943, Dr Hofmann experienced disorientation while working with the substance and attributed his disorientation to accidental absorption of the material through the skin of his fingers (Ray and Ksir, 1990). Subsequently, Hofmann experimented with a 0.5% aqueous LSD tartrate solution, orally administering a 250µg dose. A minimally effective dose for inducement of hallucinations is now believed to be 25–30µg. Hofmann's ingestion of LSD as an aqueous solution represented the first intentional human administration of the drug. Since that time, LSD (as both the free base and the tartrate) has been illegally distributed for use in many physical forms, each of which has a variety of subdivisions within that form:

1. tablets
2. blotter papers
3. aqueous and alcoholic solutions
4. free powders and powders in capsules
5. gelatin matrixes
6. miscellaneous forms unique to the distribution of LSD.

Typical concentrations of LSD circa 1960 were between 250–300µg/DU. The current (1999–2002) dosage concentrations are in the range of 20-80µg/DU.

Lysergic acid amide (LSA) is one of two analogues of LSD which have been reported in illicit use but that are rarely encountered. LSA is virtually always obtained from plants rather than through illicit clandestine laboratory synthesis. The seeds of the Hawaiian baby wood rose and morning glory seeds of the varieties "Heavenly Blues" or "Pearly Gates" (Ray and Ksir, 1990) appear to be favorite LSA sources. Microgram (Anonymous, 1968) reports "dosage units" of 4 to 12 seeds for Hawaiian Baby Wood Rose and approximately 5g of seeds (LSA content from 25 to 213µg) for morning glorys (Hopes, 1968). The seeds of either plant may be ground and then chewed or else put into gelatin capsules for direct ingestion. This method of achieving a hallucinatory experience currently appears unpopular, with only one analysis of morning glory seeds occurring in Federal drug analysis laboratories in the last ten years.

To date, only one analogue of LSD has been reported in a clandestine laboratory synthesis: lysergic acid sec-butylamide. A single occurrence of this LSD analogue was uncovered during the seizure of a clandestine laboratory operating in suburban Vancouver, Canada (Laing, 1998). Although concern had been aroused in the forensic community about the possibility of clandestine laboratory synthesis of lysergic acid methyl propylamide (LAMPA) and the potential for misidentification because of the similarity of mass spectra for these isomers, it appears to have been unwarranted. LAMPA is not known to have been illicitly produced and marketed. Additionally, the current widespread use of capillary columns in gas chromatographs interfaced with mass spectrometers has essentially eliminated the possibility of an error in the identification of these two isomeric compounds or their sec-butylamide isomer. Although the mass spectra for these isomers are similar, the variance in retention time on commonly used capillary column coatings will allow differentiation among these compounds.

2.2. FORMS OF THE DRUG

2.2.1 TABLETS

One of the earliest illicit dosage forms for LSD was the tablet. More than 200 types of LSD tablet have been identified since 1969. Tablets containing LSD have been found in a wide range of colors including yellow, red, orange, green, white, tan, black, purple, and green. Most often, tablets were a uniform solid color, but the occurrence of speckled or mottled tablets was not uncommon in the 1970s and 1980s. One type of light gray tablet even contained flecks of brass imbedded in the dosage unit, while a second type of tablet had a black exterior and a white interior. Over a period of time, tablet dimensions (diameter × height), weight, shape, and concentration of LSD evolved from large (4.5–8.1mm × 2.3–4.1mm), heavyweight (≥150mg), round, high concentration (90–350μg/tab) dosage units (Figure 2.1) to small (2.0–3.5mm diameter), lightweight (as low as 4.7mg/tab), variously shaped, lower concentration (12–85μg/tab, average range 30–40μg/tab) dosage units. LSD tablets have been acquired in a number of shapes including those resembling cylinders, cones, five-pointed stars, mushrooms, the Mercury spacecraft, pumpkin seeds, hearts (Love Drug), and the ubiquitous round, biconvex, unscored "microdots." "Microdots" (Figure 2.2), so named because of their extremely small tablet size, and cylindrical shaped tablets, often called "barrels"

Figure 2.1
Large LSD tablets

Figure 2.2
"Microdot" LSD tablets

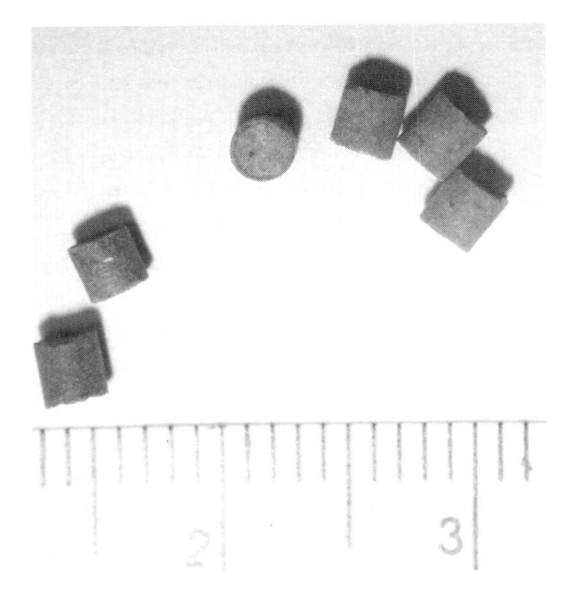

Figure 2.3
"Barrel" LSD tablets

(Figure 2.3), were two of the most prevalent tablet types. The three most popular colors for microdots were yellow and purple, followed by red. Because of the small size of even the larger types of LSD tablets, markings or engravings on tablets are very unusual. A single instance of a "logo" on a tablet was the LSD "Peace Pill" with an engraved "peace symbol" on the tablet surface. However, the name "*Peace Pill*" has more often been associated with, and should not be confused with, the hallucinogen phencyclidine (PCP) that was initially illicitly distributed as a free flowing powder, or a powder in gelatin capsules, rather than as the description "pill" implies. Several other varieties of tablets acquired "street names" based primarily on their color (Anonymous, 1971):

- "Blue moons," blue, flat, round, 3mm × 5mm (height × diameter)
- "Green mescaline," light green, flat, round, 4.1mm × 8.1mm
- "Purple haze", light purple, flat, round, 2.9mm × 8.1mm
- "Pink passion," red-pink, flat, round, 3.4mm × 6.5mm
- "Orange sunshine," light orange, flat, round, 2.0mm × 4.9mm
- "Black beauties," charcoal gray, flat, round, 3.0mm × 5mm
- "Black bombers," black, biconvex, round, 3.0mm × 5mm

2.2.2. BLOTTER PAPER

"Blotter paper LSD" represents the largest classification of LSD dosage units. Since 1975 more than 350 paper designs have been classified. Papers of various types, dimensions and colors have been used as a "vehicle" in addition to the "true" blotter papers. Blotter papers have been prepared by "spotting" LSD on each individual dosage unit (Figure 2.4), but a more frequently used method is to dip the entire pre-printed sheet into the LSD/water/alcohol solution. Occasionally, Vodka or "Everclear" grain alcohol has been used to make the LSD/solvent solution. However, haphazard dipping, *per se*, is not the most common way of applying LSD/solvent to paper. Such ungoverned dipping has two drawbacks: first, valuable LSD solution is lost from paper when too much liquid is applied since it often drips off, and second, there is a real possibility of an accidental overdose for the person manipulating the paper. Most large-scale blotting operations attempt to "calibrate" the paper they are using. This is accomplished through experimentation in order to determine how much solution is required to wet the paper up to the edges through the phenomenon of capillary action. The paper is then "laid" out in the orientation in which it will be dried. Next, the "calibrated" amount of LSD solution is applied. Often a large syringe or pipette, with or without a hose on the end to direct the liquid stream, is used to both measure the solution and apply the "calibrated" amount

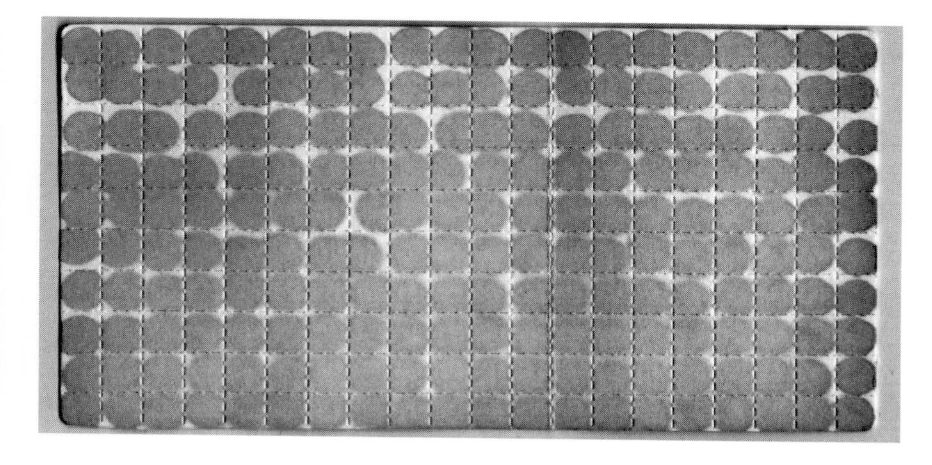

Figure 2.4
"Spotted" LSD blotter paper

to the paper in a spiral pattern intended to help speed uniform soaking. Checks are then made to see that the paper is "wet" to the edges and subsequently, the product is allowed to dry.

Artwork and design

These "dipped" blotter papers occur in a wide assortment of designs with single color (Figure 2.5) or multicolor imprints (Figure 2.6). Artwork on blotter paper is usually commercially printed using inks that do not dissolve in water or alcohol solutions. If soluble inks are used, the solvent vehicle applying the LSD makes the ink "run" throughout the sheet of paper. In multicolor ink designs, the red (or magenta) ink is mostly likely to be affected. Occasionally the imprints are applied with hand stamps. Hand-stamped designs are usually applied after the paper has been treated with LSD since the rubber stamp inks are almost always soluble in water and/or alcohol. Designs can be a single picture covering the entire blotter paper sheet (Figure 2.7), or may be multiples of the same design on each multisquare section of the sheet (Figure 2.8). The design can be repetitive from square to square (Figure 2.9) or may sequentially repeat a uniform number of designs (Figure 2.10).

When LSD blotter paper was first sold in 1972 it was not a popular product on the street market. This was the heyday of the microdot tablets. Most of the various tablet types sold on the street were made by clandestinely manufactured tablet presses rather than by commercial tablet presses diverted from the legitimate pharmaceutical industry. These illicitly manufactured presses were designed to be compact, portable, powered by 110v AC power, and easily

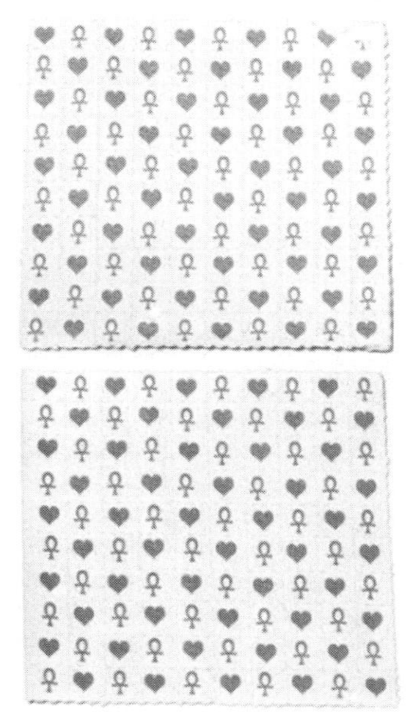

Figure 2.5

Single color LSD blotter paper designs

Figure 2.6

Multicolor LSD blotter paper design

Figure 2.7

Single design for entire LSD blotter paper sheet

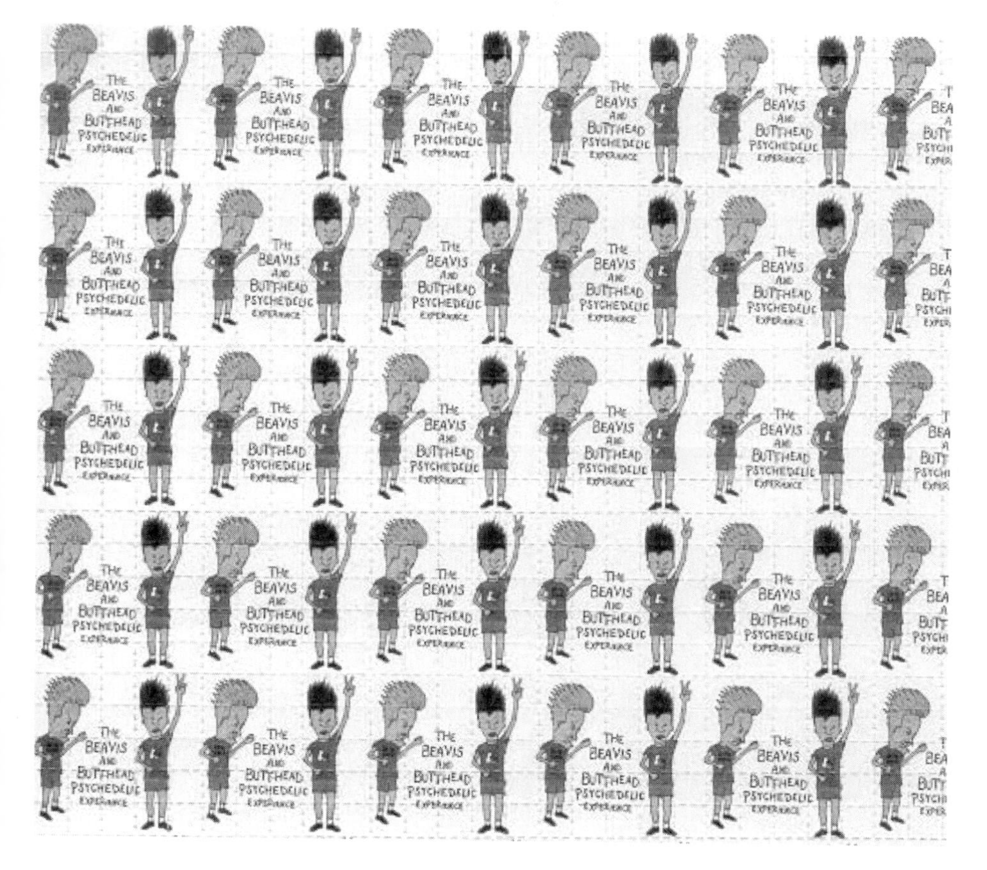

Figure 2.8

Multiples of a single design on an LSD blotter paper sheet

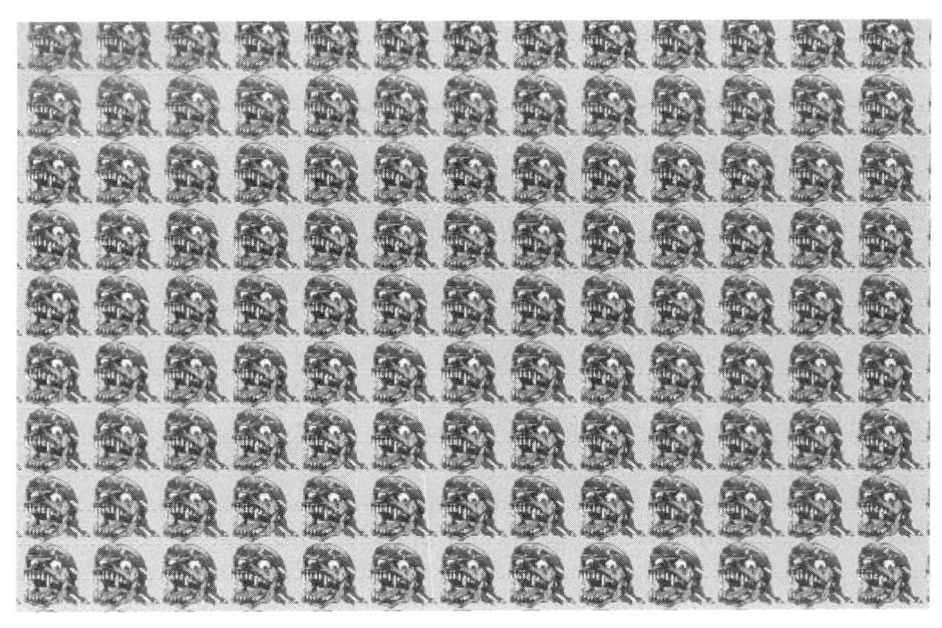

Figure 2.9

Uniform design on each dosage unit of the LSD blotter paper sheet

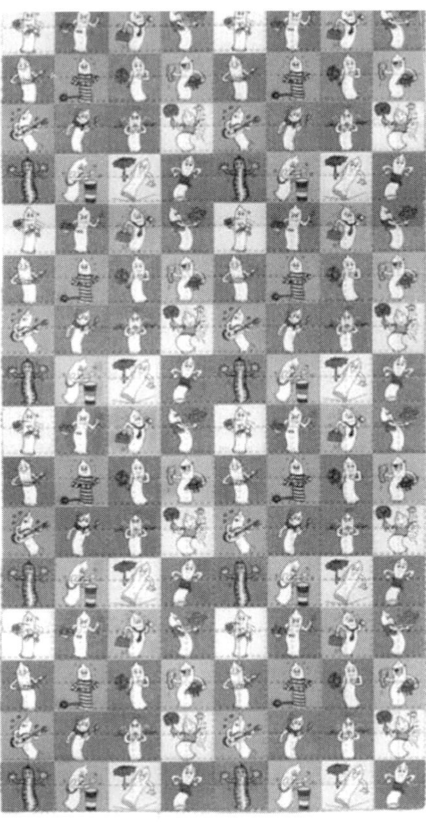

Figure 2.10

Individual non-repeating design on each dosage unit of the LSD blotter paper sheet

broken down into smaller components for storage in large safe deposit boxes. These presses were made from aluminum instead of steel (except for the tablet punches and dies themselves) and wore out with repeated use. The machinist responsible for making these presses died in the late 1970s and clandestine tablet manufacturing groups did not want to be exposed to law enforcement through purchases of commercial equipment to replace their increasingly decrepit presses.

By 1979 several illicit LSD manufacturing groups had reintroduced LSD in blotter form and, within two to three years, blotter paper LSD dominated the market. The first LSD of this period was not perforated. The dimensions of the design determined dosage units and a square grid of lines became the most popular. Later, perforated paper with 0.5 inch by 0.5 inch squares appeared on the street. Perforating the paper made the tearing off of "uniform" individual dosage units so much easier that most LSD blotter makers soon followed suit. Both square and rectangular dosage units were seen in various sizes within each of the geometries. By 1985, 0.25 inch by 0.25 inch-square dosage units in the United States and 8mm by 8mm squares in metric countries were the most common sizes. The kinds of paper used varied quite considerably and total LSD extraction for analysis could present analytical problems (Heagy *et al.*, 1995). The weight of an average single LSD blotter "tab" or dosage unit was approximately 8mg. Coincidentally, this was the same as the average weight of the most popular size microdot tablet. It was easy to apply 25 to 100μg of LSD to a piece of paper weighing 8mg and to be confident that the consumer would receive most of the product. Although blotter paper as a vehicle for drug distribution is almost exclusively the domain of LSD, 4-bromo-2,5-dimethoxyamphetamine (DOB) is potent enough so that when applied to 0.5 inch squares, a usable quantity of the drug is supplied to the consumer.

Whether on blotter paper or as a powder, LSD is degraded by three routes: 1) UV light, 2) heat, and 3) free radicals, especially ozone. Clandestinely manufactured LSD can be a mixture of LSD, iso-LSD, and various precursors and "breakdown" components as a final product. LSD in solution and on blotter paper can undergo even further degradation. In a cascade of events, the powder or solution received by an LSD blotter paper manufacturer has already started to decompose to some unknown degree. Storage conditions, prior to application of LSD on the paper, contribute to additional degradation of the drug. Exposure to UV light and free radicals after the LSD has been applied to the paper furthers the decay. By the time a law enforcement laboratory analyzes a blotter paper sample, all of these factors have been at work. In addition to the decomposition noted above, two further complications for the clandestine laboratory chemist in the preparation of dosage units of a known and uniform concentration are the uneven distribution of the LSD when applying the solution to the paper and the subsequent migration of the solution by capillary action while the paper is drying. Some older LSD blotter paper samples (originally on white paper) have turned quite brown, thereby giving visual signs of the degradation process. These exhibits illustrate, through the distribution of brown color, the manner in which the paper was physically supported during the drying process. If a portion of the paper is not exposed to air directly, then evaporation at that point is restricted, and capillary action carries the LSD solution to the nearest area where solvent is evaporating, thus enhancing the local concentration of LSD. Eventually the paper shows a darker brown color where the LSD concentration was highest. Color patterns show that some blotter papers were dried while lying on large bottle caps, while other papers were laid upon plastic supports with many small columns in a square grid array. These plastic grids are sold as inserts for the vegetable drawers in refrigerators in order to keep fruits and vegetables from direct contact with cold porcelain surfaces. The use of these plastic grids is the most popular method of drying LSD blotter paper. Typically, LSD blotter paper exhibits show a range of 20–80μg of LSD per dosage unit.

OTHER FORMS

Liquids

Solutions of LSD have been found in a number of commercially marketed product containers, either admixed with the legitimate product or replacing the product in the original container. Small (0.25oz.) food coloring containers, in both the older style "rectangular" glass or the newer plastic "tear-drop" shaped bottles, have been used to contain red, yellow, green, blue and orange colored liquids which were found to contain LSD. In one instance, the quantitation of this type of solution showed an LSD concentration of 720μg/milliliter (mL). Several instances of LSD solutions (1000μg/mL) repackaged in "Murine" brand eyedrop containers have also been encountered. In 1971, admixtures of LSD and the commercial perfume, "Shantung," were analyzed by the Australian authorities and were found to have concentrations in the range of 1,490μg/mL to 21,120μg/mL. A variety of "breath-freshener" containers have also been used to conceal LSD solutions. "Ice Drops" and "Binaca" brand bottles, containing brown solutions, as well as "Sweet Breath" bottles, with clear liquids, all have been used for packaging the hallucinogen.

One of the principal purposes of selling or distributing LSD in aqueous solutions is to facilitate the "blotting" or soaking of LSD paper (see below). These small containers are purported to have enough LSD to properly soak from 100 to 1000 dosage units. Dealers intent on manufacturing their own blotter paper will experiment to determine how much solvent is required to wet a piece of paper equal to the desired number of dosage units. The LSD solution in the small bottle is then diluted to this "experimental" volume and applied to the paper. Using this process LSD is distributed more or less uniformly over the whole sheet of paper. After the solvent(s) have evaporated, the LSD particles adhere to all the fibers of the sheet of paper making a durable form for sale or distribution.

Powders

LSD base and tartrate have each been found as pure (~99%) powders with colors typically varying from off-white through yellow-brown to tan. In some instances, glass vials with capacities of 1, 5, and 10 grams were used to contain the drug. Pure LSD has also been "cut" with a variety of powdery agents. Material sold as "chocolate mescaline" was found to be a mixture of LSD, cocoa, and lactose. A pinkish-colored powder encountered under the name "Strawberry Fields" was a mixture of LSD and a strawberry flavoring product intended to be mixed with water to provide a sweetened beverage. In one of the first LSD exhibits to be analyzed by the federal government in 1967, the drug had been mixed with wheat flour and packaged in gelatin capsules. Powdered gelatin itself has been used as a cut for LSD.

Gelatin matrices

One of the most unusual forms of drug distribution is the LSD/gelatin matrix composed of a mixture of LSD, gelatin and a food-coloring agent. The earliest encounter with this formulation occurred in London, England in mid-1970. Approximately 10 000 dosage units in the shape of small squares were seized. These squares varied in size from 3mm × 3mm to 6mm × 6mm with a thickness of 0.38mm and had an average weight of 1.065 mg. The LSD concentration for these squares was approximately 120µg. This type of LSD dosage unit acquired names such as "clear lights," "window glass" and, more frequently, "window panes" (Figure 2.11). Both larger (1cm × 1cm × 0.5mm) and smaller (2.0mm × 2.0mm × 0.2mm) squares, and occasionally rectangles, have been found. Large quantities of the smaller version of the window pane were often found packaged in

Figure 2.11
"Window pane" LSD gelatin squares

100 dosage unit quantities in small plastic or glass vials with 40 of the plastic vials placed in small (3.5 inch × 2.5 inch × 1.5 inch) brown wooden boxes. A round form of the LSD/gelatin matrix dosage unit referred to as "contact lenses" or "soapers" contained LSD concentrations between 140–180μg/DU for the "lenses" and 50μg/DU for the "soapers." An even more unusual version of the "window pane" was found as a clear/brown bi-level configuration with a raised waffle pattern. The overall dosage unit measured 6.5mm × 6.5mm × 1.2mm, but only the clear layer contained LSD. A similar type dosage unit, consisting of LSD paper laminated with a plastic layer on top of the paper, was seen in a sample from France in the early 1980s. The design was the head of the Pink Panther on each dosage unit and the plastic laminate was applied over the printed side of the paper. The LSD was impregnated in the paper. The plastic laminate was quite tough and may not have been biodegradable.

Another distinct type of LSD/gelatin dosage unit was prepared by using clear or opaque plastic fluorescent light fixture covers as a mold or template. Aqueous mixtures of gelatin and LSD were prepared, poured into these "molds," allowed to harden, and subsequently removed as a multidosage sheet. These sheets of multiple dosage units have been found in both a pyramidal (Figure 2.12) and a half bubble (or "egg crate") shape in several different transparent colors. Pyramidal dosage units measured 3.3mm at the base and were 1.4mm high with a range of LSD concentrations from 52μg/DU to 126μg/DU.

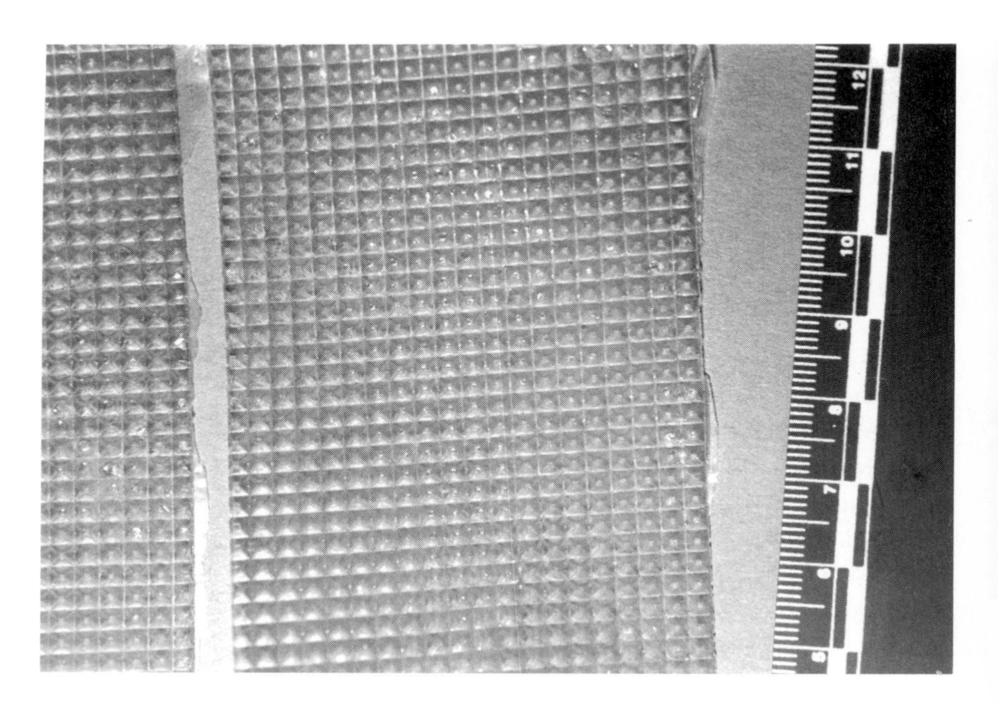

Figure 2.12

Uncut pyramidal hard gelatin LSD squares

Miscellaneous forms

LSD has been encountered in a number of dosage forms that don't fall into a particular category. In general these dosage units have been created by dropping or spotting an LSD solution onto a solid substrate, or, alternatively, by dipping the substrate into an LSD solution. In this manner, a variety of dosage forms have been manufactured including sugar cubes, animal cookies (also known as "animal crackers"), saccharin tablets, gumdrops, licorice, "Smarties" brand candy, plastic pellets, and chewing gum. "Hakerol" brand yellow colored pastilles, of Swedish manufacture, have been spotted with LSD to give a dosage of 58μg/lozenge. "Bullseyes," a caramel candy with a soft white center, have been acquired that have had single tablets pressed into and hidden within the soft center. "PEZ" brand candy tablets proved very popular vehicles because the oblong tablet has a concave "well" on each side that facilitates the application of a drop of LSD solution. This "well" helps to prevent loss of any drug while the solution evaporates. Both cooked mushrooms and small star-shaped pasta have been found containing LSD. In the case of small star-shaped pasta, the LSD was mixed with food coloring dyes. The pasta stars were dropped into the LSD solution and then screened out. Dry pasta is not very permeable and most of the LSD was found at or near the surface, thus giving a product that lost potency with rough handling. Both round (Figure 2.13) and flat toothpicks (5μg/toothpick) have been dipped in LSD solutions to form "dosage units." Finally, individual matches have been torn from matchbooks, dipped in LSD solutions, and re-inserted to form a carrier for the drug.

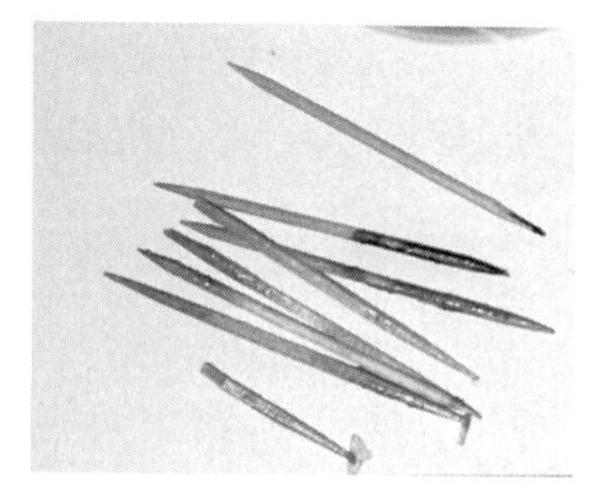

Figure 2.13

Round toothpicks with LSD coating

LSD mixtures

In addition to being encountered as the sole illicit component of a dosage unit, LSD has been found mixed with a variety of additional drugs or drug containing substances. Several of these additional drugs are also hallucinogenic substances. Among these additional drug combinations are 4-bromo-2,5-dimethoxyamphetamine (DOB), phencyclidine (PCP), 3,4-methylenedioxy-amphetamine (MDA), benactyzine, and psilocybin. LSD has also been mixed with several other non-hallucinogenic drugs including cocaine, amphetamine, strychnine, tetrahydrocannabinol (THC), and an ephedrine/methamphetamine/N-ethyl-3,4-methylenedioxyamphetamine (MDEA) mixture. The last two of these mixtures were found as blotter papers rather than tablets. LSD has been found mixed with substances, such as nutmeg, that have hallucinogenic properties of their own when ingested in large quantities.

PART II: INDOLALKYLAMINES

2.3 PSILOCIN, BUFOTENINE AND OTHER SUBSTITUTED TRYPTAMINES

Among the best known of this group of hallucinogens are psilosin (4-hydroxy-N,N-dimethyltryptamine) and its phosphate ester, psilocybin (4-phosphoryl-N,N-dimethyltryptamine). These two compounds occur as biosynthesized components in a number of different mushrooms, primarily of the genus *Psilocybe.* Bufotenine (5-hydroxy-N,N-dimethyltryptamine) is a positional isomer of psilocin and has gained a degree of notoriety from reports of "toad smoking" (Chamakura, 1994). It is found in several plants and their seeds or bark, a few species of mushrooms, and the secretions of several varieties of toads (Chamakura, 1994). Recently, several drug exhibits were analyzed and found to contain 5-methoxy-N,N-diisopropyltryptamine, N,N-dipropyltryptamine, and N-methyltryptamine. Clandestinely synthesized N,N-dimethyltryptamine (DMT) and N,N-diethyltryptamine (DET), although popular during the late 1960s and 1970s, are infrequently encountered in the Federal system of forensic drug laboratories.

2.4 FORMS OF THE DRUGS

Bufotenine was first isolated in 1893 from the secretions of the parotoid gland of the toad, *Bufo vulgaris* Laur. (Saxton, 1965). Apparently, the only known source of illicitly marketed bufotenine is from the venom of certain species of toads. This material, sold under a variety of names including "Chinese Love Stones," "Black Stone," "Stones," "Rock Hard," "Rock," and "Stud 100," has been encountered as "reddish brown cubes/pieces weigh(ing) 0.5g to 2.0g, packaged in small Ziploc bags or cellophane packages (Chamakura, 1998). Bufotenine must be injected or inhaled since it is inactive by oral administration (Chamakura, 1998) unless accompanied with a monoamineoxidase inhibitor (MAOI). In the last ten years, federal laboratories have only identified bufotenine on three occasions.

The United Nations Laboratory reported (Anonymous, 1989a) that there are 140 species of *Psilocybe* mushrooms and that at least 80 of these contain psychotropic substances. Use of these hallucinogenic mushrooms by the Aztecs of Central America dates back several thousand years. The religious importance the Aztecs ascribed to these mushrooms is evident in the name given to them, "Teonanacatl," which translates into "God's flesh" or "sacred mushrooms" (Hofmann, 1971). Although these mushrooms are principally found in temperate zones, their climatic habitat can range from the tropics to the arctic

(Anonymous, 1989a). Three of the more important species, from the aspect of drug abuse, are *Psilocybe semilanceata* (FR.) QUEL., *Psilocybe cubensis (*EARLE) SINGER, and *Psilocybe mexicana* HEIM (Anonymous, 1989b). Stamets (1996) gives previously reported percentage composition ranges of psilocybin:psilocyn as follows: 0.50:0.25%; 0.63:0.11%; 0.15:0.11% (a Mexican strain); 0.15–0.33% (an Amazonian strain); and 1.3:0.35% (extraordinarily high). One sample of the species *Psilocybe azurescens*, was found to contain 1.78% psilocybin, 0.38% psilocin, and 0.35% of a commonly encountered third hallucinogenic component, baeocystin. In addition to the genus *Psilocybe*, the following genra each have members which contain psilocybin/psilocin*: Panaeolus, Conocybe, Gymnopilus, Inocybe* and *Pluteus* (Stamets, 1996).

Psilocin is a degradation product of psilocybin and may be formed enzymatically in the live plant or as a result of poor drying and storage techniques. Often, in psilocin-containing mushrooms, a blue coloration can be seen in the stems or caps. According to Stamets (1996), the chemical identity of the blue compound is unknown; however, its appearance seems to parallel the degradation of psilocin within the mushroom, thus indicating a decrease in potency. The number of mushrooms ingested depends upon the concentration of the hallucinogenic components, typically an unknown parameter, since factors such as growing conditions, age and species play a role in biosynthesis of these components. Mushrooms may be consumed by chewing either the raw or (more likely) the dried plant (Figure 2.14); by grinding the dried material into a powder for encapsulation and direct administration; or by brewing into a "tea." Although psilocin is reported to be 1.4 times as potent as psilocybin (Hofmann, 1971), dephosphorylation of psilocybin into psilocin occurs during digestion (Stamets, 1996) so that, on a molar basis, the two compounds are equipotent (Hofmann, 1971). Since 1990, nearly 400 exhibits have been

Figure 2.14

Dried "Psilocybe" mushrooms

reported by Federal (US) authorities having either psilocin or psilocybin as the active ingredient.

The parent compound for bufotenine, psilocin and psilocybin is N,N-dimethyltryptamine (DMT). DMT is a hallucinogenic component of Cohoba snuff and, unless taken with an MAO inhibitor to prevent rapid enzymatic destruction, must be smoked, inhaled or injected to produce an effect (Ray and Ksir, 1990). When encountered as a product of clandestine laboratory synthesis in the 1970s, the material was often a dark brown/black, sometimes gummy, solid with a disagreeable odor. By the 1990s, reports of illicit DMT (and DET) synthesis were rare. Microgram (Anonymous, 1993a) reports only nine submissions of the compound to the DEA laboratory system in the ten-year period from 1983 to 1992. Several instances of clandestinely manufactured 5-methoxy-N,N-diisopropyltryptamine; N,N-dipropyltryptamine; and N-methyltryptamine have been reported in the mid-western United States and more recently in tablet form in Canada (Figures 2.15, 2.16).

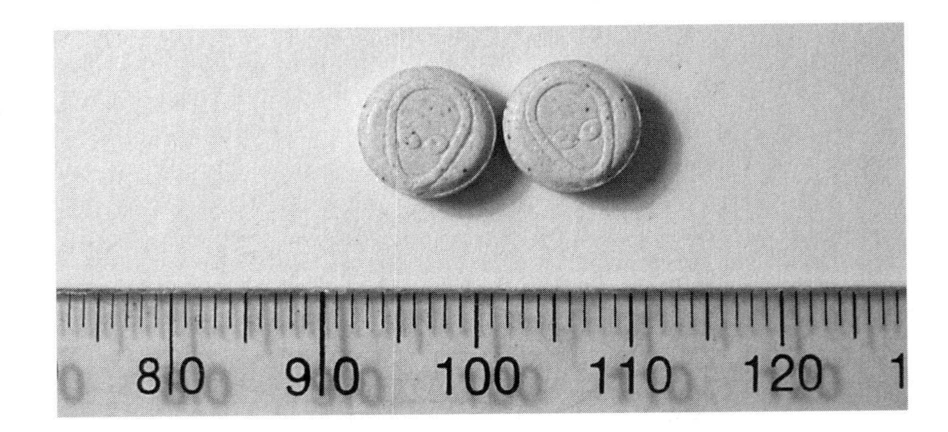

Figure 2.15
"Purple Alien" 5-OMe-N,N-diisopropyl-tryptamine

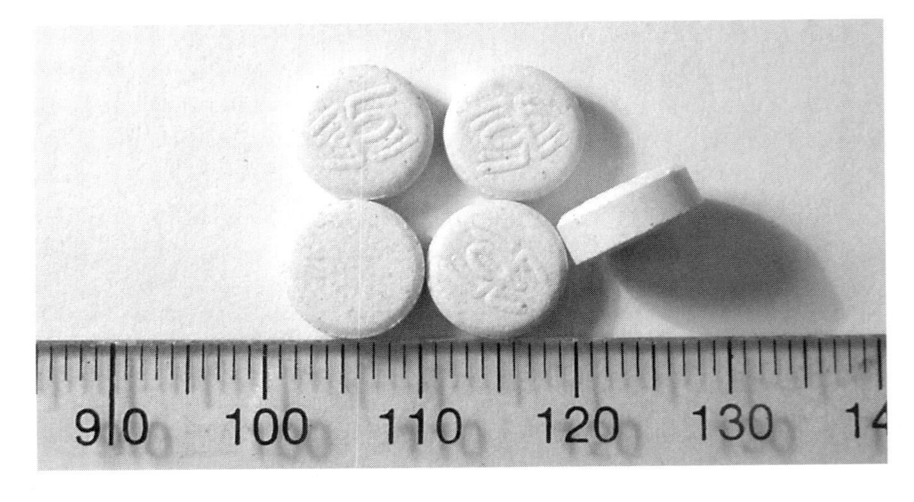

Figure 2.16
"Pink Spider" 5-OMe-N,N-diisopropyl-tryptamine and dextromethorphan

PART III: HALLUCINOGENIC PHENETHYLAMINES

2.5 MESCALINE/PEYOTE

Two of the best-known illicit phenethylamines are mescaline (3,4,5-trimethoxy-β-phenethylamine) and MDMA (N-methyl-3,4-methylenedioxyamphetamine, Ecstasy). Mescaline may be considered the prototypical compound for the hallucinogenic phenethylamines (Shulgin, 1978), and is the primary hallucinogenic alkaloid of peyote (Hardman *et al.*, 1973), which is described as "a small spineless carrot-shaped cactus, *Lophophora williamsii* Lemaire" (Ray and Ksir, 1990). Peyote (Figure 2.17) was used by the Aztecs as a ritual hallucinogen and was given the name "Peyotl" meaning "furrything" in reference to the small thick, tufts of off-white colored hairs on the mature cactus (Anonymous, 1989a). Mescaline content is between 0.5% and 1.5% of the weight of the dried cactus (Anonymous, 1989b). Clandestinely synthesized mescaline is usually obtained as a powder or crystalline material, often in hard gelatin capsules, as the sulfate salt. It is probably encountered more frequently in the biosynthetic form as "mescal buttons," the dried, hard upper section, or "crown" of the cactus. In this form, the drug is usually consumed by chewing and swallowing the buttons, by grinding the buttons to a powder and placing the powder in gelatin capsules, or by making into a "tea" and drinking. Nausea, vomiting, and headaches are often the accompanying side effects (Shulgin and Shulgin, 1991; Ray and Ksir, 1990) that precede the vivid colorful hallucinations

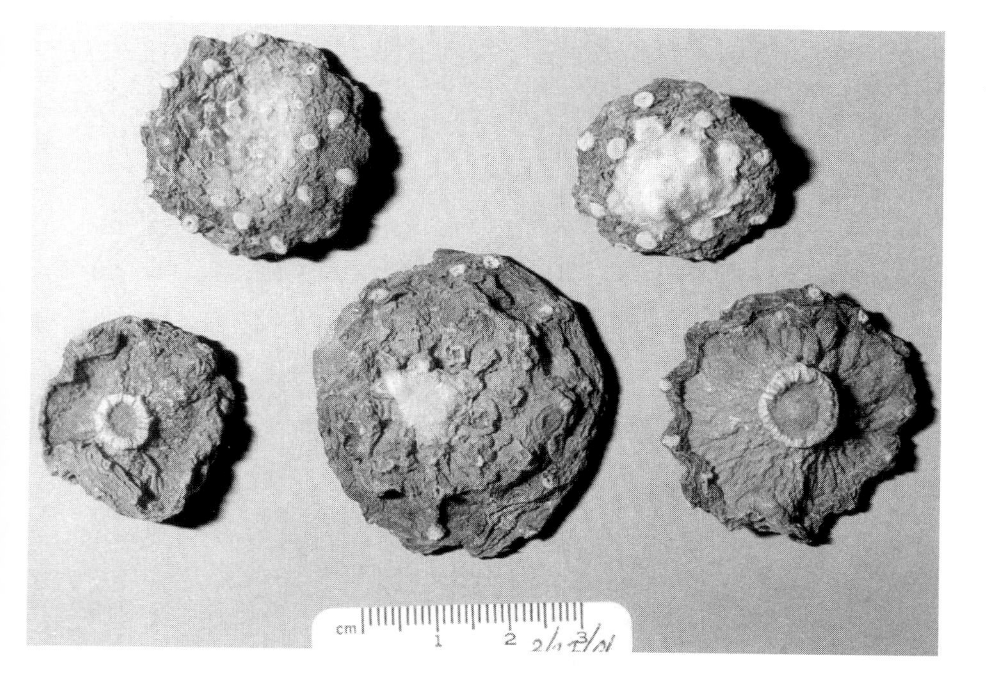

Figure 2.17

Dried peyote "buttons"

induced by mescaline. Mescal buttons contain approximately 6% mescaline (Hopes, 1968) along with other alkaloids. Doses in the range of 3mg/kg are reported as euphoric, while at the 5mg/kg range hallucinations occur (Ray and Ksir, 1990). In the form of the sulfate salt, mescaline is active at the 200–400mg level (Shulgin and Shulgin, 1991). Mescaline also serves as a relative quantitative standard for measuring the activity of other hallucinogenic compounds (Shulgin and Shulgin, 1991). This measure of hallucinogenic activity is the mescaline unit (MU), essentially defined as a level of 3.75mg of mescaline per kg of body weight (Shulgin *et al.*, 1969), or 300mg for an 80kg (170lb) individual. Peyote buttons are apparently more difficult to obtain than are *Psilocybe* mushrooms, with an average of only one Federal (US) case in each of the last ten years.

2.6 MDA, MDMA

The compounds 3,4-methylenedioxyamphetamine (MDA) and 3,4-methylene-dioxymethamphetamine (MDMA), are each unusual but for substantially different reasons. Because of a chiral carbon in its side chain, MDA can exist in two distinct absolute configurations designated as the [R] configuration and the [S] configuration. MDA appears to be unique in that the [R] configuration is hallucinogenic while the [S] configuration is analeptic (Glennon and Young, 1984). All reported clandestine synthesis of MDA has resulted in an equal mixture of the two configurations, designated by [R,S], thus imparting the unique combination of both stimulant and hallucinogenic properties to the same compound. Although MDA has been acquired in paper or metal foil packets as a loose powder, more often it is packaged in hard gelatin capsules or compressed into tablets (Figure 2.18). MDA tablets have been acquired with a variety of imprinted designs and a range of concentrations, a sampling of which includes:

Figure 2.18

Black "2H" MDA hydrochloride tablets

1. Off-white, round tablets with a butterfly design; 85mg/tablet.
2. White, round tablets with a "Z" monogram; 70–74mg/tablet.
3. White, round tablets with a flying dove design; 29–40mg/tablet.
4. Black, round tablets with a "2HH" monogram on both sides; 116mg/tablet.
5. White, round tablets with either a "G" or "H" monogram; 148mg/tablet.
6. Off-white, round tablets with a star design; 36mg MDA and 22mg MDMA/tablet.

Powders, loose or in capsules, were generally some shade of white with a single report of a rose red colored powder. MDA is usually encountered as the hydrochloride salt, but the sulfate, phosphate, tartrate, and acetate have also been seen.

MDMA, and its N-ethyl homologue MDEA, seem to be devoid of hallucinogenic activity and are usually described as entactogens (Nichols and Oberlender, 1989; Nichols *et al.*, 1986; Kovar, 1998) rather than hallucinogens. Nonetheless, MDMA will be discussed here in some length due to its current popularity, especially at all-night dance parties referred to as "raves," and because MDMA tablets comprise the largest component in the realm of clandestinely manufactured illicit drug-containing tablets.

E. Merck, a German pharmaceutical company, first synthesized MDMA in 1912. Popular rumor insists that the drug was initially manufactured and tested as an appetite suppressant; however, there seems to be little or no evidence (i.e., Merck's archives) to support this assumption. Merck's patent, issued in 1914, does not indicate any therapeutic category or purpose. A second patent was granted to Merck in March 1921, but this patent is also strictly a chemical synthesis item. In the 1950s the CIA and the US Army tested MDMA on animals (and perhaps human subjects, however no records have been released) at Edgewood Chemical Warfare Services in Maryland. As one of eight mescaline-related substances, MDMA was examined and functional effects in animals were observed and assessments on its toxicity were evaluated.

The first identified recreational use of MDMA occurred in 1972. In the next 12 years only four clandestine lab seizures (out of some 2400) produced MDMA exhibits. Starting in 1980 Dr George Greer, a psychiatrist, treated approximately 75 patients with MDMA (Greer and Tolbert, 1986). While most mainstream psychiatrists did not advocate the use of MDMA in the treatment of patients, a vocal minority did use MDMA as a psychotherapeutic drug and published favorable reports.

MDMA was first put into Schedule I of the Controlled Substances Act (CSA), under emergency provisions, in July 1985, extended in July 1986 and made permanently Schedule I on 23 March 1988. Prior to the final scheduling action taking effect, there was considerable testimony before Congress by various psychiatrists in North America and Western Europe who had worked with MDMA as a psychotherapeutic drug. The considerable publicity during the CSA scheduling hearings alerted the illicit market that there was another potential money-making drug awaiting their collective attention.

Control of MDMA ushered in the first wave of illicit "Ecstasy" popularity. This popularity was spurred by the decreasing availability of another illicit drug, methaqualone, usually marketed in the form of counterfeit Rorer or Lemmon "714" tablets. By 1985 counterfeit methaqualone tablets, often referred to by the trade names "Quaalude" and "Sopor" or the street names "Ludes" and "Sopors," had become very difficult to purchase. Generated by the Quaalude era, there was an enormous market for a drug that would facilitate and/or promote sexual relations. Young persons, 15–30 years old, of both sexes, had made "Ludes" almost as popular as marijuana in terms of numbers of users/abusers.

In the last half of the 1980s (1985–1990), Ecstasy was widely available in the gay community. This was the era of the "large" MDMA tablets (7/16 inch and half-inch diameters) usually made in Mexico, just south of the United States border, particularly in Tijuana. Outside of Mexico, one large international MDMA manufacturing group used the German firm of Imhausen to perform the synthesis, while the tableting operation was conducted by a second German company, Dragnopharm. This operation resulted in tons of pharmaceutical grade MDMA being converted to five different tablet types and sold throughout the world. Ecstasy rapidly developed popularity outside the gay club scene as a "safe love drug."

Most of the Ecstasy tablets sold in the late 1980s were white or off-white, round, biconvex or flat/beveled and had either no scoremark or a single score mark. At that time monograms were rare. MDA, MDEA, and N-hydroxy-MDA were seen in combination with MDMA in tablet exhibits.

The end of the 1980s saw the growth of the "Rave" music scene in the United Kingdom and Western Europe. Huge gatherings of teenagers and/or young adults in their twenties became ready markets for Ecstasy tablets. At the same time, the unavailability of MDMA precursors in North America shifted the synthesis operations, first to Amsterdam and then to surrounding areas of Holland, Belgium, Luxemburg, and northwestern Germany. This began the current era of Ecstasy tablets in metric diameters (6, 7, and 8mm diameters). European clandestine laboratories capable of manufacturing MDMA in multi-hundred kilogram lots have been seized. In one such seizure, Dutch Police dismantled a lab with 750kg of finished MDMA and enough precursors for another 750kg. The operators of this laboratory claimed they could make at least a metric ton (1000kg) per month. As the RAVE party phenomenon immigrated to the United States, the smuggling and sale of European Ecstasy tablets followed closely behind.

The second wave of Ecstasy tablets showed much more variety than the first wave. Tablets have been made in a wide range of colors, although white, off-white and tan were the most common. Tablets have also been encountered in a variety of shapes other than the ubiquitous round geometry, including oval, oblong, triangular (Figure 2.19), square, pentagonal (Figure 2.20), hexagonal, heart, shamrock, and diamond. Single-scored tablets are the most common. Monograms of various types have been identified: letters ("D&G," "E-mail," "SEX," "DEA"); trademarks (Mitsubishi, Rolex, Nike, Motorola, Warner Brothers); animals (dolphins, elephants, birds); cartoon characters (Fred Flintstone, Barney Rubble, Dino, Pokemon, Tweety Bird, Fido-Dido, Pink Panther); and diverse symbols (omega, stars, arrows, Buddhas, Batman, Superman,

Figure 2.19
Triangular Ecstasy tablets

Figure 2.20
Pentagonal Ecstasy tablets

lightning bolts, Christmas trees, anchors, butterflys, cupid, skull-and-crossbones, smiley faces). The DEA has identified over 200 different logos, and Dutch police authorities have found over 600.

The number of Ecstasy samples seen in DEA's Source Determination Program (SDP) showed an early peak in the late 1980s, then lower activity in the first half of the 1990s, when production had shifted from Mexico to Europe, and then a steady, fast growth in the late 1990s. The quantity of Ecstasy tablets currently available is beginning to approach the number of methaqualone tablets found in the early parts of the Quaalude era. The MDMA annual consumption in the United States is now above 100 million dosage units. Quaaludes at their peak exceeded 1000 million tablets per year. The demand for and availability of Ecstasy seems to be headed in the same direction.

A recent review of Ecstasy samples shows that 80% of all exhibits examined by the DEA laboratory system contain MDMA. Of that percentage, 70% contain only MDMA while 10% contain additional drugs, usually amphetamine or methamphetamine. In 12% of the tablets, only amphetamine and/or methamphetamine were found. In the remaining tablets, less than 3% have MDEA, MBDB, chloro-MDMA, 2-CB, Ketamine, etc., and the other 5% have non-controlled drugs such as ephedrine, pseudoephedrine, phenylpropanolamine, caffeine or acetaminophen.

Any discussion of specific MDMA tablet monograms must take into consideration the multiple clandestine origins of the most popular logos. One of the best known of these monograms is the "three diamonds" (or Mitsubishi trademark) logo. Nine separate Mitsubishi logos from different clandestine sources have been identified. Seven of these tablet types contained MDMA while two contained p-methoxyamphetamine (PMA) and p-methoxymethamphetamine (PMMA). Police officials from Holland report the identification of 26 different Mitsubishi logos! At this time, three different sitting Buddha logos; four four-leaf clovers; six butterflies; four crowns; three Nike trademarks; six Ying-Yang symbols, and two Motorola trademarks have been differentiated, with more likely to occur. As soon as any monogram becomes popular on the street other clandestine sources often market copies.

2.7 NEXUS (2-CB), STP, DOB, TMA

One of the earliest of the synthetic hallucinogens was 4-methyl-2,5-dimethoxyamphetamine (STP, DOM), popular between the spring of 1967 and the end of 1972. This material was primarily encountered in the form of tablets, and occasionally as capsules. Orange, yellow, and white tablets were reported with approximate dimensions of 4.8–4.9mm (d) × 3.1–3.3mm (h) weighing 75–80mg. Potency varied between 9.2 and 10.2mg/tablet. Capsules encountered were of the No. 3 hard gelatin type and contained 183mg of green powder that assayed at 3.2mg/capsule. By mid-1975, this compound had been supplanted by a more potent analogue, 4-bromo-2,5-dimethoxyamphetamine (DOB). This compound had a number of interesting associations with LSD. Rose/pink colored tablets [5mm × 2–4mm; weight 18mg] of DOB were encountered mixed with LSD and made from the same set of punches previously identified as those used to tablet only LSD. Besides LSD, DOB is the only hallucinogen frequently occurring in the blotter paper dosage form. Unlike LSD, however, DOB is usually spotted on blotter paper without designs leaving a noticeable, and often colored, spot. There are, however, two instances where designs have been recorded. One detailed an eagle on a flaming pyre with the word "Rebirth" across the top of the pyre. The printing was in blue ink on orange blotter paper. The second design consisted of a red colored lightning bolt on plain white paper. Although white is the predominant color for blotter paper, yellow, dark green and purple blotter papers have also been acquired. Spots on the blotter paper have occurred as brown, green or blue roughly shaped circles. In one instance, 16mm × 18mm squares had 10mm diameter purple stains with an outer concentric pink stain of 1mm width. Blotter paper squares usually were found in a 1cm × 1cm dosage unit format. Concentration of the DOB varied from 0.8mg/square up to 1.8mg/square. A 30–35mg dose of DOB was reported to have resulted in an overdose death. The deceased may have snorted the drug believing it was cocaine due to its method of packaging typical of cocaine sales: a folded waxy paper with the impression of a seal and the word "Sno-Seal" inscribed.

DOB retained its popularity through 1985. By 1993, however, 4-bromo-2,5-dimethoxyphenethylamine (Nexus, 2-CB), a homologue of DOB, had supplanted DOB. Originally identified in a 1979 Texas exhibit, the drug was obtained (1993–1997) as a white powder in clear colorless or opaque yellow hard gelatin capsules. Tablets containing 2-CB first appeared in 1997, analyzed by the Laboratory of Drugs in the Balearic Islands. Tablets were of two types: white inscribed with a "sun"; or speckled beige inscribed with "rays." White, round, flat, beveled half scored tablets with a weight of 250mg containing 9.4mg of 2-CB were analyzed in the US in the same time period. Additional tablets have been quantitated with a concentration of 2-CB between 8–10mg. Typical oral ingestion is between 10–20mg (Anonymous, 1993b). US Federal authorities have only reported the identification of DOB in five instances since 1990, while in that same time period, 2-CB has been reported 24 times.

Trimethoxyamphetamine (TMA) is a generic term representative of a group of six compounds. With five positions available for substitution on the phenyl ring of amphetamine, and three methoxy-substituents to be attached, a series of six distinct compounds can be made with the following numbers representing the location of the methoxy- groups on the ring: 3, 4, 5; 2, 4, 5; 2, 3, 4; 2, 3, 5; 2, 3, 6; and 2, 4, 6. Shulgin and Shulgin (1991) refer to theses compounds, respectively, as TMA, and TMA-2 through TMA-6. The two most interesting of this group are TMA, with a ring structure identical to mescaline, and TMA-2, the most potent compound of the series with hallucinatory experiences at doses as low as 20mg. Although all six of the isomers are active, TMA and TMA-2 are the only two of the six compounds, which have been confirmed in Federally (US) analyzed drug exhibits. The most recent encounters with TMA occurred in 1993 while TMA-2 was reported for the first time in 2000. The TMA drug exhibits analyzed between 1972 and 1975 were described as light brown/tan colored powders and were concealed either in metal foil or paper packets.

PART IV: PCP, PCP ANALOGUES, AND KETAMINE

2.8 PCP AND KETAMINE

Phencyclidine (1-[1-*P*henyl*c*yclohexyl])*p*iperidine, PCP) has had a substantial history of legitimate pharmaceutical use as a short-acting analgesic and a general anesthetic under the trade name "Sernyl" (Munch, 1974). Human use as an anesthetic was eventually abandoned when 10–20% of the patients experienced post-surgical hallucinations (Analine and Pitts, 1982). Under the name "Sernylan," the drug continued to be used as an animal tranquilizer until 1979 (Analine and Pitts, 1982). The hallucinations that made PCP medically unacceptable made it desirable on the street where it became available under a variety of names including Angel Dust, Dead on Arrival (DOA), *P*ea*c*e *P*ill, and Elephant Tranquillizer. PCP and cyclohexamine (N-ethyl-1-phenylcyclohexylamine), in red and white mottled tablets, have each been referred to as "Rocket Fuel." PCP, an easily synthesized drug, first appeared on the drug scene in 1966 (Schnoll, 1980). Cheap precursor chemicals, easily conducted syntheses, no necessity for chemical glassware or equipment and a resulting high yield all made PCP the favorite of clandestine drug labs for years. The drug is powerful so that only 2–3mg is needed for the average adult person to get high (standard dosage unit is 5mg). Side-effects of PCP ingestion, in addition to hallucinations, may be manifest in a combination of unmanageable, hostile, and aggressive behaviors (Analine and Pitts, 1982) that are sometimes accompanied by displays of extraordinary strength. In one incident, a 14-year-old girl successfully resisted arrest by four burly state troopers.

Because of its legitimate medical usage, phencyclidine was originally placed in Schedule III of the Controlled Substances Act (CSA). In 1978 it was moved to Schedule II, reflecting reduced medical applications and increased popularity on the street markets. PCP is often sold as powder wrapped in metal foil or plastic packets, or as powder sprayed onto leafy materials such as marijuana, parsley or mint. Some drug exhibits of PCP have been obtained as tablets or capsules. The most common dimensions for tablets were 5mm × 2–3mm, in a round, unscored, flat/beveled shape with various colors including red, orange, yellow, and purple most frequently seen. Additionally, cylindrical blue colored tablets measuring 7mm × 2mm weighing 93mg have also been found. The concentration of PCP in various tablets ranged from 0.8mg to 11.5mg.

Primarily during the 1970s, PCP was occasionally encountered mixed with additional illicit drugs. In most cases, these mixtures were found as powders in plastic bags, capsules, or metal foil packets. Two notable exceptions to this were the occurrence of biconvex, tan-pink colored LSD/PCP tablets (6.4mm × 3.2mm; 160mg) with LSD concentrations in the range of 280–480mg and PCP levels between 3–5mg and a solution of PCP/methamphetamine. Other unusual combinations of PCP with a controlled substance include heroin, meperidine, phenobarbital, phentermine, MDA (4% + 20% PCP), mephentermine (19% + 21% PCP), and N-methyl-3-piperidylbenzilate (JB-336, LBJ; 0.74–1.8mg + 0.80–3.32mg PCP) as brown or white powders in various size capsules. Multiple controlled substance mixtures with PCP include cocaine/phendimetrazine as a light brown powder and phenobarbital/phentermine as a green powder referred to as "green snow."

By 1991 PCP tablets had become rare, although the drug stayed popular throughout the early 1990s. The preferred methods for ingestion of PCP are through snorting powdered PCP hydrochloride or, more often, through smoking. Dipping cigarettes or "cigarillos" in ether solutions of phencyclidine produces a popular form of the drug for smoking, often referred to as "Shermans." Transportation and sales of phencyclidine in bulk quantities often occurs in 5-gallon containers as a solution of PCP base in diethyl ether. Although PCP has lost some of its early popularity, it still remains a major drug of abuse with over 3000 exhibits of the substance analyzed Federally (US) since 1990.

Phencyclidine has given rise to several analogues that, although never as popular or prevalent as PCP, have been clandestinely synthesized and introduced on the "streets." Among these analogues are those produced by substitution of the piperidine ring with ethylamine, n-propylamine, morpholine, pyrrolidine or 4-methylpiperidine to give, respectively, N-ethyl-1-phenylcyclohexylamine, (N-n-propyl-1-phenylcyclohexylamine), 1-(1-phenylcyclohexyl)-morpholine (PCM), 1-(1-phenylcyclohexyl)pyrrolidine (PCPy) and 1-(1-phenylcyclohexyl)-4-methylpiperidine (4-methyl-PCP). Substitution of the phenyl ring with 2-thiophene produced a popular product, TCP (1-[1-{2-

thienyl}cyclohexyl]piperidine). Substitution of the phenyl ring with either a benzyl or 4-methylphenyl ring resulted in two additional analogues that were identified in Quebec, Canada (Lodge *et al.*, 1992). These analogues have primarily been found as white or off-white powders, or as a coating on various plant materials including marijuana, parsley, and mint. A wide variety of substitutions in the phencyclidine molecule have been presented by Maddox *et al.*, (1965) and Kalir *et al.* (1969) including replacement of both the phenyl and piperidine rings within the same molecule. One analogue that falls into this category that has been analyzed in a clandestinely produced sample is 1-[1-(2-thienyl)cyclohexyl]pyrolidine (TCPy). Not all of the analogues prepared in the Maddox and Kalir papers have a potency approaching PCP, with some having little or no PCP-like effect. They are all easily synthesized, however, by substitution of one or two reactants in the formulas used to make PCP. The lack of CNS activity in these other analogues has probably precluded their clandestine synthesis.

One popular compound that is related to PCP, but is synthesized in a different manner, is the drug ketamine or "Special K." Structurally, this compound is 2-(o-chlorophenyl)-2-methylaminocyclohexanone. Originally marketed by Parke-Davis as an anesthetic for surgical procedures under the name Ketalar (or Ketavet by Bristol Laboratories for veterinary use), it was available as injectable solutions with concentrations of 10, 50, and 100 mg/mL. Most of the ketamine available on the "streets" is diverted from legitimately manufactured material, primarily from veterinary supplies, and is available as a solution or as fine, fluffy white crystals resembling pharmaceutical cocaine hydrochloride. The compound is snorted, injected, used in alcoholic beverages, or smoked on marijuana and has become one of a number of drugs that are popular at "rave parties." As a powder, ketamine is usually sold in small plastic bags, folded paper or aluminum foil, or capsules holding 70mg or 200mg of the substance. Seventy milligrams up to 500mg are usually snorted, with the higher dose resulting in a state referred to as "K-hole" and described as an "out-of-body, near death" experience (Anonymous, 1997).

Ketamine is only rarely seen in tablets. DEA's Source Determination Program (SDP) has reported Ecstasy (MDMA) tablets that contain ketamine. The first of these were encountered as round, off-white tablets with monogram of "XTC" and "BLD" from Peru. This type of tablet assayed at 39mg ketamine, 97mg MDMA, and 13mg methamphetamine per tablet. A second exhibit occurred as green tablets with a "Y2K" monogram and contained 26mg MDMA/tablet plus methamphetamine, ketamine and caffeine (none of which were quantitated.) Off-white tablets containing a mixture of amphetamine and ketamine with the imprint of a dolphin were reported in Alabama and single score white tablets, with the image of a bird, were found by Spanish authorities to contain ketamine and ephedrine.

PART V: OTHERS

2.9 β-CARBOLINES

The two best known of the Harmala alkaloids are, perhaps, the β-carbolines harmaline and harmine, both of which are monoamineoxidase inhibitors. Only two reports have been noted on the occurrence of either compound in a forensic drug laboratory. In one instance, two red colored hard gelatin capsules were found to contain harmaline, and in the other, approximately 4g of ground vegetable material was found to contain harmine. Although both compounds may be mild hallucinogens (neither one is currently a controlled substance), their surreptitious use is probably related to their MAOI action rather than their hallucinogenic activity. Orally inactive hallucinogens, such as DMT and bufotenine, can be made active by this route of administration if taken in conjunction with harmaline or harmine. These compounds may also be taken with orally active drugs in order to prolong and/or enhance the action or potency of the hallucinogen. When taken in combination, effects may be experienced which are noticeably different from either compound by itself. Shulgin and Shulgin (1997) has extensively reported on these compounds and their plant (and animal) sources. Both of these compounds, as well as additional alkaloids, are found in the plants *Banisteriopsis caapi* and *Peganum harmala* (Syrian rue). *Banisteriopsis caapi,* when taken in conjunction with a second plant, *Psychotria viridis,* and brewed as a "tea," is known as ayahuasca (hoasca), a hallucination-inducing drink widely used in the Amazon area of South America (Shulgin, 1997).

ACKNOWLEDGEMENT

Microgram is a publication of the Drug Enforcement Administration, Office of Forensic Sciences, with both a national and international distribution restricted to "forensic scientists serving law enforcement agencies." A significant portion of the information for this chapter was culled from various issues of *Microgram* that had been submitted for publication in the form of general intelligence information rather than as the work of a particular author. A second valuable source of information was provided directly through the DEA's Source Determination Program (SDP). This program, available to the national and international forensic science community, provides comparisons of tablets and blotter papers to determine if various submissions to the program have a common source.

REFERENCES

Analine, O. and Pitts, F. N. (Jr) (1982) *CRC Critical Reviews in Toxicology*, Boca Raton, FL: CRC Press, pp. 145–177.

Anonymous (1968), *Microgram*, 1(10), p. 1.

Anonymous (1971) *Microgram*, 4(9), pp. 122–123.

Anonymous, (1989a) *Recommended methods for testing Peyote Cactus (Mescal Buttons)/ Mescaline and Psilocybe Mushrooms/Psilocybin*, New York: United Nations, pp. 4–12.

Anonymous, (1989b) *Recommended Methods for Testing Peyote Cactus (Mescal Buttons)/Mescaline and Psilocybe Mushrooms/Psilocybin*, New York: United Nations, pp. 20–32.

Anonymous (1993a) *Microgram* 26(10), p. 219.

Anonymous (1993b) *Microgram* 26(12), p. 271.

Anonymous (1997) *Substance Reports: Ketamine Abuse Increasing*, US Dept. of Justice, DEA Diversion Control Program, pp. 1–2.

Chamakura, R. P. (1994) *Microgram*, 27(9), pp. 316–329.

Chamakura, R. P. (1998) *Microgram*, 31(5), pp. 127–149.

Glennon, R. A. and Young, R. (1984) *Euro. J. Pharm.*, 99, pp. 249–250.

Greer, G. and Tolbert, R. (1986) *J. Psycho. Drugs*, 18(4), pp. 319–327.

Hardman, H. F., Haavik, C. O. and Seevers, M. H. (1973) *Tox. and App. Pharm.*, 25, pp. 299–309.

Heagy, J. A., Moriwaki, W. M. and Chan, K. T. (1995) *Microgram*, 28(3), pp. 85–90.

Hofmann, A. (1971) *Bul. Narco.*, 23(1), pp. 3–14.

Hopes, T. M. (1968) *Microgram*, 1(4), pp. 2–8.

Kalir, A., Edery, H., Pelah, Z., Balderman, D. and Porath, G. (1969) *J. Med. Chem.*, 12, pp. 473–477.

Kovar, K.-A. (1998) *Pharmacopsychiat.*, 31 (Supplement), pp. 69–72.

Laing, R. R. (1998) *Journal of the Clandestine Laboratory Investigating Chemists Association*, 8(3), pp. 15–16.

Lodge, B. A., Duhaime, R., Zamecnik, J., MacMurray, P. and Brousseau, R. (1992) *For.. Sci. Int.*, 55, pp. 13–26.

Maddox, H. V., Godefroi, E. F. and Parcell, R. F. (1965) *J. Med. Chem.*, 8, pp. 230–235.

Munch, J. C. (1974) *Bul. Narco.*, 26(4), pp. 9–17.

Nichols, D. E., Hoffman, A. J. Oberlender, R. A., Jacob III, P. and Shulgin, A. T. (1986). *J. Med. Chem.*, 29, pp. 2009–2015.

Nichols, D. E. and Oberlander, R. A. (1989) Pharmacology and toxicology of amphetamine and related designer drugs, in K. Ashghar and E. De Sousa (eds), NIDA Research Monograph 95, Washington, DC: Superintendent of Government Documents, US Government Printing Office, pp. 1–29.

Ray, O. and Ksir, C. (1990) *Drugs, Society and Human Behavior*, St Louis: Times Mirror/Mosby.

Saxton, J. E. (1965) *The Alkaloids, Vol. 8: The Indole Alkaloids*, R. H. F. Manske (ed.), New York: Academic Press, pp. 16–19.

Schnoll, S. H. (1980) *J. Psych. Drugs,* 12(3–4), pp. 229–223.

Shulgin, A. T. (1978) *Handbook of Psychopharmacology*, Vol. 11, L. L. Iverson, S. D. Iverson and S. H. Snyder (eds), New York: Plenum Publishing Co., pp. 243–333.

Shulgin, A. T., Sargent, T. and Naranjo, C. (1969) *Nature*, 221, pp. 537–541.

Shulgin, A. and Shulgin, A. (1991) *PIHKAL: A Chemical Love Story*, Berkeley, CA: Transform Press.

Shulgin, A. and Shulgin, A. (1997) *TIHKAL: The Continuation*, Berkeley, CA: Transform Press, pp. 247–284, 285–309.

Stamets, P. (1996) *Psilocybin Mushrooms of the World*, Berkeley, CA: Ten Speed Press.

CHAPTER 3

BASIC PHARMACOLOGY AND EFFECTS

Alexander T. Shulgin

3.0 INTRODUCTION: WHAT ARE THE HALLUCINOGENIC DRUGS?

This chapter is intended to present an overview of the current knowledge of the hallucinogenic drugs, in a format that will be of value to a readership composed of forensic scientists, prosecution and defense lawyers, law enforcement, pharmacologists, chemists, toxicologists, and other scientists. This review will cover a very narrow area, a small sliver of the immense world of psychopharmacological agents.

A possible alternative title for this chapter could have been "The Psychedelic Drugs," but it will be another 20 years before that word will become acceptable in the text of a technical volume directed towards the scientific community. To many, this word conjures up an image of the 1960s, with its Vietnam war protests, its love-ins, its wildly colored posters and raucous music, and of the drugs that were widely associated with that era. Even today, a third of a century later, the editorial advisers of scientific journals and government granting agencies are loath to use that terminology.

There is a yet older alternative term, "Psychotomimetic Drugs," which is discounted for a different reason. Its etymological origin, the imitation of psychosis, gives an inaccurate description of the effects that are produced. These states may be mysterious, instructive, fascinating, and to some quite frightening, but they are not "crazy" states as one would infer from that name. The nature of the effects of all these drugs was considered pathological a few decades ago, and the transient changes they effected in the experimental subject were seen as disruptive and negative. It was believed at that time that the major value to be obtained from them, as research tools in medicine, was that they allowed the creation of an authentic state of psychosis (of limited duration) in an otherwise normal individual. Their medical research value was, according to the philosophy of those years, that interning psychiatrists, those who would be called upon to treat madness, could get insight into the schizophrenic state.

Some researchers came up with another interpretation of the actions of these materials, the previously unacknowledged potential for personal inquiry and self-discovery, and what might be called the mystical or spiritual nature of the drug experience. It became apparent to some researchers that these were not simply agents of mental disruption but, rather, potentially valuable research and therapy tools. The psychotomimetic concept gradually disappeared from the literature and is today almost unknown.

There are several newly proposed euphemisms for these drugs, such as "entheogens," "empathogens" or "entactogens" (the finding of God within us, the experience of empathy for others, or the touching of our inner psychic nature); all of these have both rational justifications and avid supporters, but to some extent they are no more than further efforts to avoid the term "psychedelic."

The term "hallucinogen" is, of course, a euphemism as well, but it is one that is, at the moment, medically acceptable. Quite literally, this term should only be used to define drugs that give rise to hallucinations, states wherein the subject is unaware of the fact that what he or she is experiencing is unreal. With rare exceptions, there are no hallucinations observed or reported in the use of the drugs in this review. There are some plants in the Solanaceae family that can produce a mental state that is characterized by a dream-like hypnogogia with an accompanying syndrome of confusion and amnesia, plants and compounds that could be classified with some validity as hallucinogenic agents. Datura, Belladonna, Mandrake, Henbane – all of these are rich in the atropine/scopolamine family of alkaloids that could more rightly be called the drugs of hallucination, and in which amnesia is a major symptom. However, these compounds are not included as part of our coverage. Nor are the anesthetic parasympatholytics such as PCP or ketamine which can consistently produce a delusional, out-of-body state. The word "hallucinogen" is also not correct for the description of this area of pharmacology.

The psychotropic agents described in this chapter lack a proper name that would define a unique collection of drugs that produce an alteration of one's state of consciousness in a generally predictable manner, for a reasonably short period of time, and without any lasting sequelae in most people. No such term exists at the present time. For the purposes of this review, these drugs will be referred to as hallucinogens.

The diversity of pharmacological effects resulting from the use of these drugs has never been acknowledged as deserving the attention of the scientific or medical world. Historically, the Western cultures have viewed these particular forms of intoxication as being somehow demonic. They were expressions of some toxic psychosis produced by the use of a so-called magical or sacred plant, or of some toxin derived from it. A large body of human experience was dismissed as simply being a form of poisoning, either from the plant, or an active isolate of it, or a synthetic modification of that isolate. What was accepted as a mystical or religious experience by some people was dismissed by others as being some form of blasphemy. Our present medical ethic is that any agents capable of producing such changes of states of consciousness are deemed dangerous to the physical and mental well-being of individuals given the many social and economic side-effects of drug proliferation and use. As such, currently any extensive research for possible scientific and medical benefit is largely not practiced or not held in high regard.

A number of different properties of the hallucinogens have been used in organizing reviews. There is their relationship to specific neurotransmitters, or the subjective nature of the human response to them, or the molecular structure. The neurotransmitter argument is appealing but not completely satisfactory. With the clear and obvious division of the hallucinogenic drugs into two large classes, the phenethylamines and the tryptamines, it is most tempting to view them in relation to the two major neurotransmitters, dopamine and serotonin. These two biochemicals have the structures of a phenethylamine and a tryptamine, respectively. It would be much simpler if those drugs that were phenethylamines invoked the dopaminergic systems, and those that were tryptamines were active through some serotonergic mechanism. The explanation would not only be short but simple in concept. However, there is a terribly complex crossover between drug classes and neuroreceptors, and no easily described relationship is yet apparent. There is certainly no one-to-one correspondence.

Most of the hallucinogenic drugs, be they phenethylamines or tryptamines, are intimately involved as ligands of the serotonin (5-HT) receptors (currently 5-HT_2 is the preferred favorite) in some balance of agonist/antagonist interplay, but the dopamine receptors are involved with many of them as well. Efforts to group them by the known receptor relationship have made no sense when the nature of action or the potency were considered. Arrangements by potency have given a listing that was more like a check-list than a logical progression. Animal behavior ranking, or toxicity measures, were intrinsically faulted in that the experimental animal is not the human subject, and the principal reason for any study of these compounds is to investigate subjective responses reported by people.

All efforts to organize drugs for correlative purposes by these very subjective responses have been unsatisfactory. The qualitative nature of the induced responses are extremely variable from subject to subject, and they have never been brought into a framework that is broadly acceptable. Each person's change of consciousness is unique to him or her, and the efforts to find threads of commonality have been unsuccessful. It is the quantitative measure of response that shows greater consistency.

So, the molecular structures are usually accepted as the basis of organization, and the potency of a drug in humans as the single measure of its efficacy. This review is based upon these two properties. Nine drugs are used as archetypes and shall be the reference starting points for the discussion of the potency and activity of a large collection of related drugs. In short, this is a structure/activity relationship (SAR) review.

The usual presentation format that is seen as an SAR review is a series of tables, with a parent skeleton substituted with numbered Rs, then a listing of the things that the R represents. This is the structure half of the concept of SAR. The activity half is a listing of numerical values such as potency, toxicity, receptor binding kinetics, orbital computations – something that the experimental scientist can use on the "other axis" for plotting his or her own findings with these compounds. The pharmacologist can make an easy correlation between his or her own experimental animal findings and human activity. The neurochemist can evaluate the receptor binding, the effectiveness of a compound as agonist or antagonist, or relate brain regional distribution to central activity.

So the one axis of this correlation is easily defined: it is the chemical structure modified by a cascade of substituents. The other axis is the effectiveness reported in human clinical studies. This is based on a complex set of qualitative actions, which must somehow be reduced to a number that represents potency. There are three major dimensions of quality of action that contribute to this end-number. The nature of the altered state of consciousness can be described as involving intoxication, introspection, and escape. Let me try to give form to each of these parameters of drug effect.

Intoxication can be equated to "getting high." This property brings into play the disinhibition and diffuseness that is familiar to anyone who has become inebriated from consuming too much alcohol. Often, superimposed on this "high" is an amplification of sensory inputs that can be expressed in any of several ways. Visual amplification is often seen, colors may be brighter, or they may be completely new. Changes in the delineation of shapes or edges of objects are often commented upon. Movement of things immobile can be distracting. Unexpected interpretations of the significance of familiar things are common. One may report lights flashing, being enveloped by darkness, or feeling surrounded by chaos. Changes in sound interpretation can occur. These are all part of the intoxication character of the hallucinogenic drug.

Introspection is the phenomenon of "going inside." It is the viewing of one's own psyche as opposed to viewing one's relationship with the external world. The Jungians might refer to it as a dialog with the unconscious. Here, one can have access to early memories, down to and including (according to some analysts) the birth trauma and, according to others, earlier lives. Some of these compounds can be valuable adjuncts to psychotherapy sessions by allowing the subject to circumvent the denial and fear, allowing a deep acceptance of oneself. Relationships with one's contemporaries, with one's trusted lover, with one's parents or children, are potentially exposed.

Escape is "being out there." The immediate external world and the immediate internal world are both abandoned. A person may feel as if he or she were on some "astral plane" or at some "cosmic level" that makes the acknowledgement of the real world quite unnecessary. This aspect of the state of consciousness change may be perceived as the ultimate escape by the user. Some philosophies call it ecstasy, and others call it death.

Any attempt to categorize the effects of a hallucinogenic drug must reckon with each of these areas: the intoxication, the introspection, and the escape. The need to put a quantitative number on the potency of such a drug is an exercise in weighing and measuring, in an objective way, values that are totally subjective. Any SAR will always be firm and objective as to the "S" (structure), but be necessarily loose and subjective as to the "A" (activity). This chapter is a presentation of the "R" (relationship) between the "S" and "A" properties, using the nine classic hallucinogens and their immediate analogues.

3.1 THE PHENETHYLAMINES

The first of the two major groups of the hallucinogens is called the phenethylamines. These are compounds of extremely simple structure, involving a molecular combination that constitutes a benzene ring (the "phen" composed of carbon and hydrogens atoms), a two carbon chain (the "ethyl," also carbon and hydrogen atoms), and a new atom known as a nitrogen (the amine). So, the phenethylamine system is a ring, attached to a chain, attached to a nitrogen atom (see Figure 3.1).

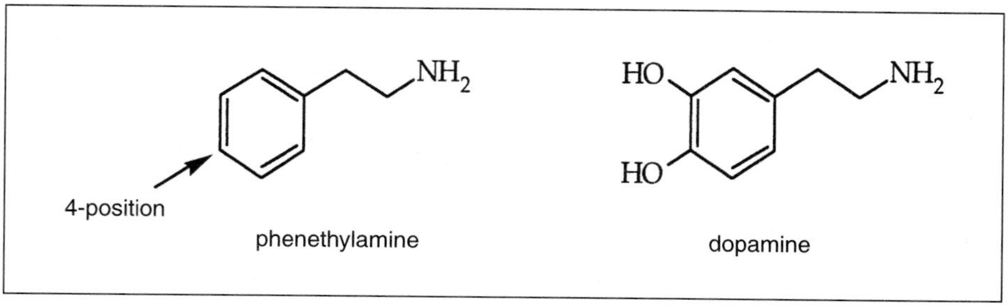

Figure 3.1
Chemical structures of phenethylamine and dopamine

The parent phenethylamine is closely associated with the neurotransmitter dopamine, a vital brain neurotransmitter. It is a phenethylamine with two hydroxyl substituents at the benzene 3- and 4-positions.

3.1.1 MESCALINE ANALOGUES

The first of these classic hallucinogens is indeed a classic in every sense, being the first example of an isolate from a sacred plant that was assigned a chemical structure, and was found to be responsible for the activity of the plant itself. This is the compound mescaline, 3,4,5-trimethoxyphenethylamine seen in Figure 3.2.

Figure 3.2

Chemical structure of mescaline

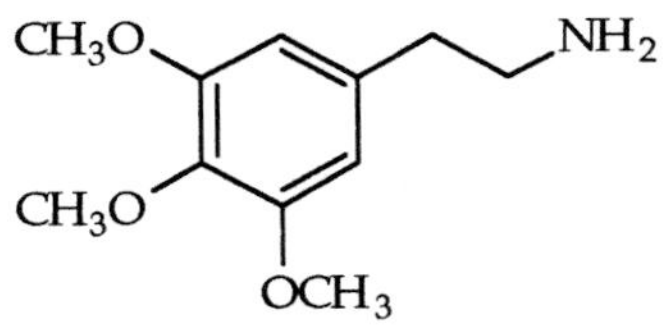

CH_3O NH_2 CH_3O OCH_3

#1 MESCALINE

The plant from which it was first isolated is a small cactus found in the southwest of the United States and the north of Mexico. Its common name is Peyote, and its botanical binomial is *Anhalonium lewinii* or *Lophophora williamsii.* Its human pharmacology was described in 1896 and its structure was verified by synthesis 23 years later. Although a large number of psychoactive plants have been known since antiquity, this isolated compound remained the only structurally identified hallucinogen until after World War II. The second structurally characterized hallucinogen (LSD) was a synthetic derivative of an ergot alkaloid and was described in the 1940s. In the years since then, there has been a flood of new information; some of it has come from the identification and characterization of plant products, and much of it has come from the methodical synthetic variations of these isolates.

Most of the structural variations of mescaline that have been studied in a clinical environment have had modifications on one or more of the three substituents on the benzene ring. These are organized into two groups, those with variations at the 4-position, and those with other variations.

The 4-position-modified compounds are shown below (Table 3.1) with the nature of the 4-substituent, the common name, the code names, the potency (orally, in man) in milligrams, and the potency relative to mescaline (called mescaline units, or M.U.). In some of these entries (as with other listing later in this chapter), the greater than symbol (>) indicates the highest level that had been evaluated, and that there had been no pharmacological action noted at that level. There is a corresponding (<) usually found in the potency relative to mescaline column, since there can be no placement of relative activity when there is no known activity. This symbol does not imply that the compound

Table 3.1

Mescaline analogues substituted at the 4-position

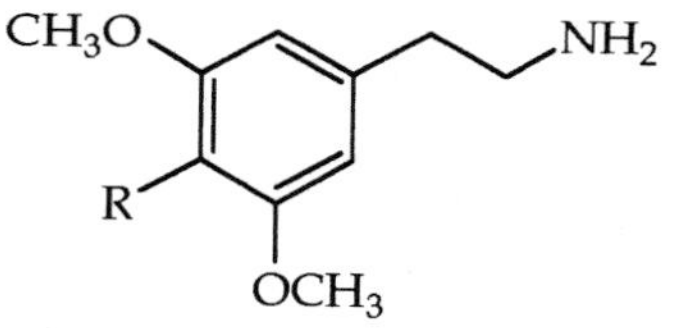

CH_3O NH_2 R OCH_3

mescaline analogues substituted at the 4-position

R =	common name	code	potency (mg)	M.U.
CH_3O-	mescaline	M	200–400	1
CH_3CH_2O-	escaline	E	40–60	6
$CH_3(CH_2)_2O$-	proscaline	P	30–60	7
$(CH_3)_2CHO$-	isoproscaline	IP	40–80	5
$CH_3(CH_2)_3O$-	buscaline	B	>150	<1
$(CH_2CH_2)CHCH_2O$-	cyclopropylmethyl	CPM	60–80	5
$CH_2{=}CHCH_2O$-	allyloxy	AL	20–35	10
$CH_2{=}C(CH_3)CH_2O$-	methallyloxy	MAL	40–65	6
$CH{\equiv}CCH_2O$-	propynyloxy	PROPYNYL	>80	<2
CH_3-	4-desoxymescaline	DESOXY	40–120	4
$C_6H_5CH_2CH_2O$-	phenescaline	PE	>150	<1

either is or is not active. This relationship to the action of mescaline has been occasionally called the Mescaline Unit measure within the substituted phenethylamine hallucinogenic drugs, following the fact that mescaline was the earliest known example of this family and the least potent example. Thus, for correlation purposes, all activity values are 1 or greater.

A generality is apparent; there is an increase in potency following an increase in chain length of the aliphatic group attached to the oxygen atom at the 4-position of mescaline, at least to a total of three atoms. The n-butyl homologue has not been reported to be active and no higher homologues have been clinically assayed. The introduction of certain structural parameters into this chain (branching, a small ring, or a double bond) maintains the compound's capability of provoking central activity, but the effect of a triple bond is uncertain. The removal of the oxygen atom from the 4-methoxy group of mescaline yields the compound DESOXY, which proves to have an activity that is difficult to classify. The removal of additional oxygen atoms yields 3,5-dimethyl-4-methoxyphenethylamine which produces a potent rage response in the cat but which has not been evaluated in man and thus has not been entered in Table 3.1.

Retaining all three of the oxygen atoms in the 3,4,5-positions that are defined by mescaline, but varying the substituent on other than just the 4-position has led to compounds that are no more potent than mescaline, if active at all. This was designed as an indirect way of exploring any pharmacological or metabolic sensitivity that might be displayed by changes to neuroreceptor specificity by simple changes in the steric bulk at these geographic locations. The compound lophophine is not yet known as a natural product despite the trivial name. It is, however, a logical bioprecursor of the several 6-methoxy-7,8-methylenedioxy-1,2,3,4-tetrahydroisoquinoline alkaloids, such as anhalonine, lophophorine and peyophorine, which are natural components of the peyote cactus, *Lophophora williamsii.* (See Table 3.2.)

Table 3.2

Other alkoxylated mescaline homologues

R^1O, R^2O, OR^3, NH_2

other alkoxylated mescaline homologues

R^1=	R^2=	R^3=	common name	code	potency (mg)	M.U.
CH_3O-	- OCH_2O -		lophophine	L	>200	<1
CH_3O-	CH_3O-	CH_3CH_2O-	metaescaline	ME	200–350	1
CH_3O-	CH_3O-	$CH_3(CH_2)_2O$-	metaproscaline	MP	>240	<1
CH_3O-	CH_3CH_2O-	CH_3CH_2O-	asymbescaline	ASB	200–280	1
CH_3CH_2O-	CH_3O-	CH_3CH_2O-	symbescaline	SB	>240	<1
CH_3CH_2O-	CH_3CH_2O-	CH_3CH_2O-	trescaline	TRIS	>240	<1

Some isotopic analogues of mescaline should be mentioned here. One labelled form of mescaline that has been studied in man, is the radioactive isomer with ^{14}C- at the benzylic position. It has been studied as a metabolic tracer. A study was made in humans using three different quantities of mescaline (350mg, 4mg, and 60µg) but with a constant amount of radiocarbon label. This measured the stochiometry of the amine-oxidase enzyme system that is responsible for the disposition of mescaline in humans. The ratio of the deamination metabolite 3,4,5-trimethoxyphenylacetic acid to unchanged mescaline was a measure of this enzyme's participation capability in this route of metabolism. Two of the five stable deuterated analogues of mescaline have also been studied in humans. The α,α-dideutero mescaline would be compromised by this conversion to the phenylacetic acid, but still could be valuable as a measure of the chiral position sensitivity of metabolism as the separate R and S isomers, but the β,β-dideutero analogue of mescaline has been made and evaluated. Also, the 4-trideuteromescaline (4-D) has been explored as a separate and new drug. The question asked here is whether any of these hydrogen atom positions represent reaction sites that might contribute to the understanding of the mechanism of action of mescaline. In both of these analogues, the observed psychopharmacological activity was in the 200–400mg range in humans, indistinguishable from mescaline itself. The three possible remaining deutero-analogues (the 3,5-dimethoxyl group hexadeuteromescaline, the ring 2,6-dideuteromescaline and the di-alpha-deuteromescaline) are unexplored.

One additional isomer of mescaline and one dimethoxyphenethylamine have been studied in humans. The mescaline isomer is isomescaline (IM) which maintains the three methoxyl groups of mescaline but in effect has the side chain relocated next to one of them (see Figure 3.3).

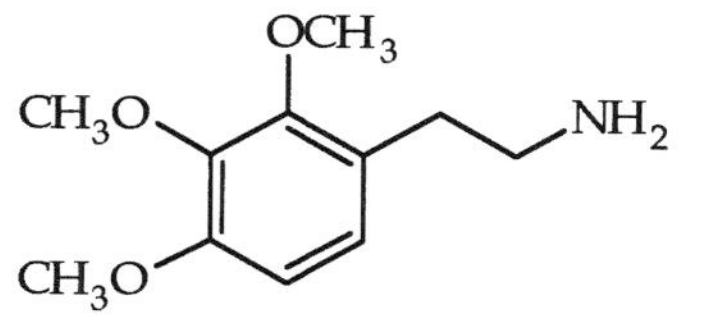

OCH_3 CH_3O NH_2 CH_3O

isomescaline (reciprocal mescaline)
inactive at 400mg orally

Figure 3.3

Chemical structure of isomescaline

This compound was perhaps the first synthetic mescaline analogue to be studied in humans. A report was published In 1936 stating that it was not active at dosages where mescaline would be fully effective, but that in schizophrenic patients it would intensify the psychotic symptoms that they experienced. To my knowledge, these studies have never been confirmed.

3,4-Dimethoxyphenethylamine (DMPEA) shown in Figure 3.4 is of particular interest in that it has been reported to be present in the urine of schizophrenic patients.

The wide publicity given this so-called "pink spot" inspired a great deal of speculation as to some possible biochemical significance of this compound in

Figure 3.4

Chemical structure of O,O-dimethyldopamine

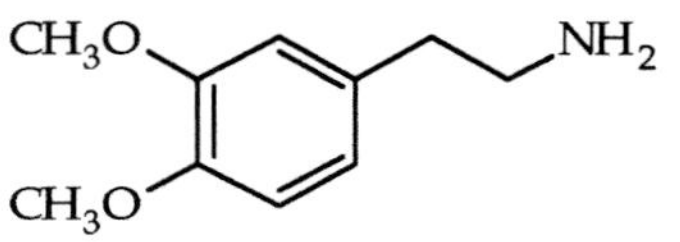

CH_3O NH_2 CH_3O

DMPEA, O,O-dimethyldopamine
inactive at 1000mg orally
inactive at 10mg iv

relation to mental illness. Some people could, and some could not confirm its presence in the urine of schizophrenic patients. Some researchers reported the presence of this compound in the urine of normals as well, and other researchers couldn't find it anywhere. Might DMPEA be some metabolic endogenous psychotomimetic, or might it be simply some dietary component? Its close structural resemblance to the neurotransmitter dopamine lent appeal to the causality association. The answer came from the direct assay of the chemical in humans as a possible psychoactive agent; it is not active. Neither is the 3,4-methylenedioxyphenethylamine counterpart, MDPEA, at least at dosages of up to 300mg orally. The 4-ethoxy homologue of DMPEA (MEPEA) has been assayed to 300mg orally, and it also has little if any psychopharmacological activity. The totally unsubstituted analogue with the backbone skeleton that defines this entire section on phenethylamines, is phenethylamine (PEA) itself. It is a natural component of the human nervous system and has received much lay press as a component of chocolate and as a "love drug," but the pharmacological truth is that it is not an active compound in humans. Studies with oral administrations of 1600mg and i.v. injections of up to 50mg have been without observed response.

3.1.2 THIOMESCALINE ANALOGUES

The replacement of the 4-position oxygen of mescaline with a sulfur atom provides the archetype of a second class of hallucinogenic drugs, 4-thiomescaline shown in Figure 3.5. From a purist's point of view, this is a true analogue, in that the relationship between oxygen and sulfur is one of their being neighbors vertically in the periodic table.

Figure 3.5

Chemical structure of thiomescaline

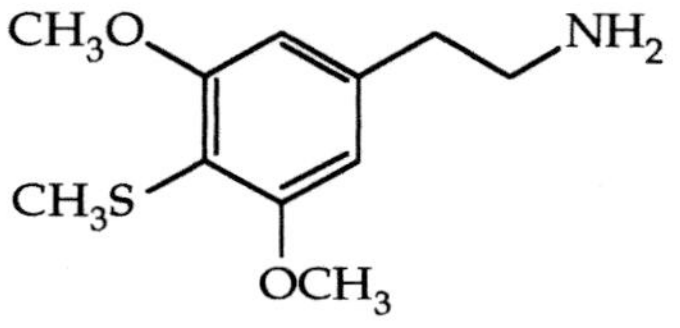

CH_3O NH_2 CH_3S OCH_3

#2 THIOMESCALINE

Although thiomescaline (or 4-thiomescaline, 4-TM) is a relatively recent synthetic compound (first prepared and with activity discovered in 1977), its unexpectedly high potency inspired the synthesis and evaluation of perhaps 20 analogues of which nearly half are more potent than mescaline.

The highest levels of activity have been associated with compounds that are homologated at this 4-position. This theme will be played again and again in the discussion of the comparative activities of both the substituted phenethylamines and (later) the substituted amphetamines. It is the structural manipulation of this 4-position, the position opposite to the aliphatic chain, that provides the greatest swings of biological activity as well as affording the highest potency compounds.

Homologation to the four-carbon n-butyl chain maintained a relatively high potency. However, the compounds that represent a further extension of this series to the amyl or higher bid fair to be active but have not yet been reported in the synthetic literature. (See Table 3.3.)

A completely parallel set of compounds has been made with the meta-oxygen of mescaline, rather than the para-oxygen, having been replaced with a sulfur atom. In general this series of isomers is of a somewhat reduced potency when compared with the 4-thio counterparts. An added complexity is the fact that when both the sulfur atom and the ethoxy group are meta to the aliphatic

Table 3.3

4-Thiomescaline homologues

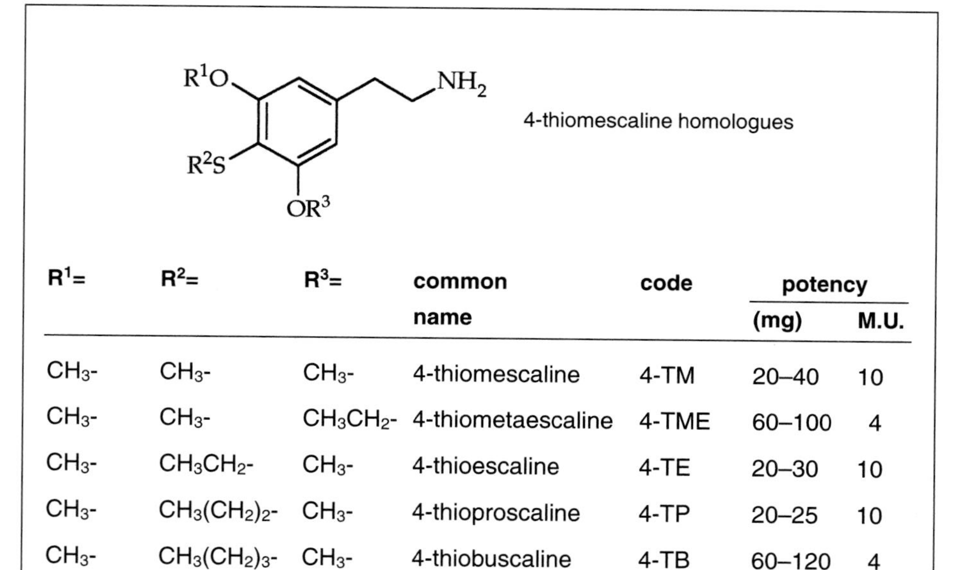

R^1=	R^2=	R^3=	common name	code	potency (mg)	potency M.U.
CH_3-	CH_3-	CH_3-	4-thiomescaline	4-TM	20–40	10
CH_3-	CH_3-	CH_3CH_2-	4-thiometaescaline	4-TME	60–100	4
CH_3-	CH_3CH_2-	CH_3-	4-thioescaline	4-TE	20–30	10
CH_3-	$CH_3(CH_2)_2$-	CH_3-	4-thioproscaline	4-TP	20–25	10
CH_3-	$CH_3(CH_2)_3$-	CH_3-	4-thiobuscaline	4-TB	60–120	4
CH_3-	CH_3CH_2-	CH_3CH_2-	4-thioasymbescaline	4-TASB	60–100	4
CH_3CH_2-	CH_3-	CH_3CH_2-	4-thiosymbescaline	4-TSB	>240	<1
CH_3CH_2-	CH_3CH_2-	CH_3CH_2-	4-thiotrescaline	4-T-TRIS	>200	<1

chain, there are two possible isomers that can exist. All such isomers have been made, and the findings of their evaluation are shown below. One additional note: two of the given weights are preceded by an approximate symbol (~). This is not an indication that the weight of material administered is unknown, but that there were indeed effects produced by these compounds but they could not correctly be classified as hallucinogenic. The reports of 3-TASB indicate a strong adrenergic component with extensive stimulation. Those of 5-TASB suggest some neurological hyperactivity and extended physical malaise.

Table 3.4

3-Thiomescaline homologues

3-thiomescaline homologues

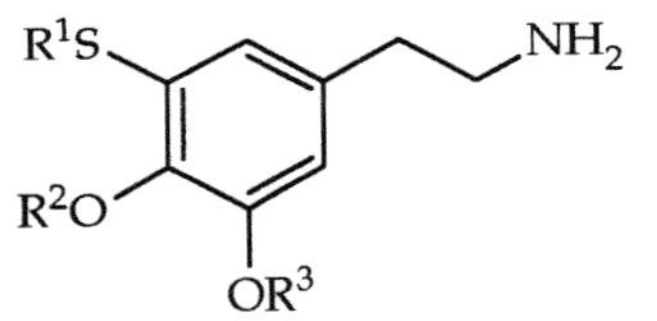

R^1=	R^2=	R^3=	common name	code	potency	
					(mg)	M.U.
CH_3-	CH_3-	CH_3-	3>thiomescaline	3-TM	60–100	4
CH_3-	CH_3CH_2-	CH_3-	3-thioescaline	3-TE	60–80	5
CH_3CH_2-	CH_3-	CH_3-	3-thiometaescaline	3-TME	60-100	4
CH_3-	CH_3-	CH_3CH_2-	5-thiometaescaline	5-TME	>200	<1
CH_3CH_2-	CH_3-	CH_3CH_2-	3-thiosymbescaline	3-TSB	>200	<1
CH_3CH_2-	CH_3CH_2-	CH_3-	3-thioasymbescaline	3-TASB	~160	<1
CH_3-	CH_3CH_2-	CH_3CH_2-	5-thioasymbescaline	5-TASB	~160	<1
CH_3CH_2-	CH_3CH_2-	CH_3CH_2-	3-thiotrescaline	3-T-TRIS	>160	<1

As neither of these effects are mescaline-related, the potency × mescaline column is proper at <1.

There are three possible sulfur analogues of isomescaline and all three have been synthesized and assayed. None have the action of mescaline, which is surprising considering the several-fold enhancement of potency that usually accompanies the replacement of an oxygen atom with a sulfur atom. (See Table 3.5.)

This lends additional weight to the intrinsic inactivity of the parent, isomescaline.

Table 3.5

Thioisomescaline isomers

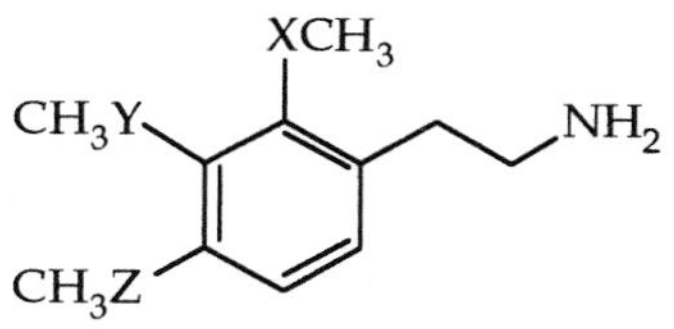

thioisomescaline isomers

X=	Y=	Z=	common name	code	potency (mg)	potency M.U.
S	O	O	2-thioisomescaline	2-TIM	>240	<1
O	S	O	3-thioisomescaline	3-TIM	>240	<1
O	O	S	4-thioisomescaline	4-TIM	>160	<1

3.1.3 2,5-DIMETHOXY-4-METHYLPHENETHYLAMINE (2C-D) ANALOGUES

All of the above described compounds have the basic ring orientation of three vicinal oxygens (or their sulfur analogues). Mescaline has this adjacency pattern, and it is one of the two patterns that have been widely studied as hallucinogens. The second is the so-called 2,4,5-trisubstitution arrangement. The prototype for this orientation is the 2-carbon homologue of DOM shown in Figure 3.6 and discussed later under 3.1.6).

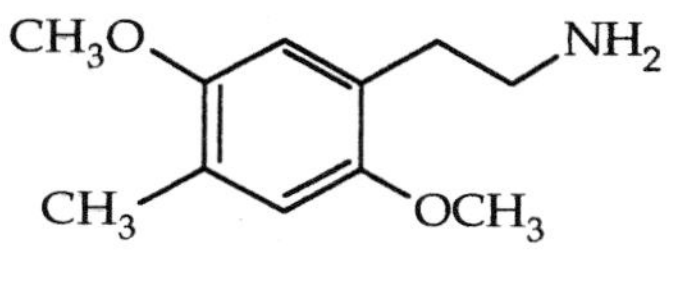

#3 2C-D (LE-25)

Figure 3.6

Chemical structure of 2,5-dimethoxy-4-methylphenethylamine

This is the first of a large number of 2C compounds, so named for the fact that they have the 2-carbon chain of the phenethylamine. The following letter has generally been chosen to reflect either the substituent at the 4-position, or the parent from which it had been derived.

This reference compound, 2C-D, is the simplest of the entire series. It is of relatively low potency relative to many examples in this group, but it has two properties that call for special comment.

Table 3.6

2C-D and its 4-alkyl homologues

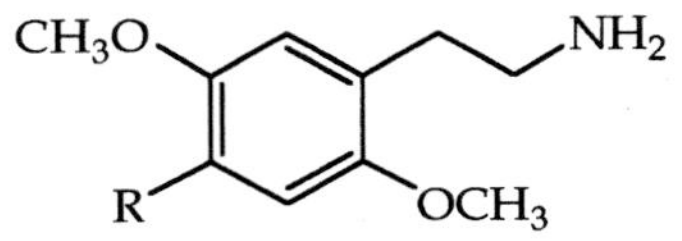

2C-D and its 4-alkyl homologues

R=	common name	code	potency (mg)	M.U.
H-	2,5-DMPEA	2C-H	?	
CH_3-	4-methyl-2,5-DMPEA	2C-D (LE-25)	20–60	8
CH_3CH_2-	4-ethyl-2,5-DMPEA	2C-E	10–15	24
$CH_3CH_2CH_2$-	4-propyl-2,5-DMPEA	2C-P	6–10	40

It has been very successfully explored in clinical trials in Germany under the name of LE-25, as an adjunct to psychotherapy. Some of the dosage regimens there were quite high (100 or more milligrams) but in its evaluation as a hallucinogenic drug, the lower range was published. The proteo-homologue, 2,5-dimethoxyphenethylamine (2,5-DMPEA), remains unassayed. Its major interest is the fact that it can play the role of a starting material (both synthetically and logically) for many of these surprisingly potent phenethylamines. It is generally believed that it would have little activity orally, due to its potential deamination by the hepatic monoamine oxidase systems, but this is difficult to rationalize in light of the extraordinarily high potency of many of the 4-substituted analogues which should be equally effective substrates.

As the chain lengthens, the potency increases but there is a change in the qualitative nature of the induced effects. The ethyl homologue, 2C-E, is a most extraordinary drug, evoking for many subjects a dramatic recall of forgotten events, with a remarkable ability to interpret and analyze them. The alteration of one's state of consciousness is dramatic, often quite frightening, and yet consistently allowing a complete recall. The extension by just a single carbon atom more, to give 2C-P, produces a yet more potent compound but one which is physically uncomfortable and considered to be unpleasant. No further homologation has been studied.

A completely separate type of homologation has been reported, with two groups located between the methoxyl groups of 2,5-DMPEA. This family has been explored along with the 3-carbon homologue amphetamines, and they have proven to be exceptionally potent compounds.

Table 3.7

3,4-Dialkyl homologues of 2C-D

3,4-dialkyl homologues of 2C-D

R	R	common name	code	potency (mg)	potency M.U.
CH_3-	CH_3-	3,4-dimethyl-2,5-DMPEA	2C-G	20–35	10
$-CH_2CH_2CH_2-$		3,4-trimethylene-2,5-DMPEA	2C-G-3	16–25	14
$-CH_2(CH_2)_2CH_2-$		3,4-tetramethylene-2,5-DMPEA	2C-G-4	?	
$-CH(CH_2)_2CH-$ (bridged by CH_2)		norbornyl-2,5-DMPEA	2C-G-5	10–16	24
-CH=CH-CH=CH-		1,4-dimethoxynaphthyl-2-ethylamine	2C-G-N	20–40	10

The tetramethylene homologue is entered here because it has been synthesized but the clinical study of it has not yet been completed. A great number of fascinating analogues are suggested by this collection, such as the 3,4-diethyl (or the two distinct methyl ethyl isomers) or multiple ring systems of the norbornyl type with heteroatoms in them. Synthetically, these should present few problems.

An exceptionally rich family of compounds has come from the substitution of groups at the 4-position of 2C-D which are not simple alkyl homologues. There seems to be little if any correlation between either the size or the electronegativity of the group, and the potency of the resulting phenethylamine. The iodine-analogue, one of the most potent of the series, has proven to be a valuable ligand for positron emission tomography, as it has been substituted with radio iodine and, as is so rarely the case in receptor-site studies of labelled ligands, the heavy atom here is intrinsic to the activity of the molecule. An enigma is 2,4,5-trimethoxyphenethylamine, a positional isomer of mescaline (the 3,4,5-counterpart). It is devoid of activity even at doses that with mescaline would be fully effective. (See Table 3.8.)

And yet, the addition of an alpha-methyl group to mescaline (a move that presumably protects it from oxidative deamination) only doubles the potency, whereas the same protective modification of this "inactive" isomer (to give the

Table 3.8

4-Substituted analogues of 2C-D

CH_3O / NH_2 / R / OCH_3

4-substituted analogues of 2C-D

R	common name	code	potency (mg)	potency M.U.
F-	4-fluoro-2.5-DMPEA	2C-F	>250	<1
Cl-	4-chloro-2,5-DMPEA	2C-C	20–40	10
Br-	4-bromo-2,5-DMPEA	2C-B	12–24	16
I-	4-iodo-2,5-DMPEA	2C-I	14–22	16
CH_3O-	2,4,5-trimethoxy-PEA	TMPEA	>300	<1
NO_2-	4-nitro-2,5-DMPEA	2C-N	100–150	2
$(CH_3)_2CHO$-	4-isopropoxy-DMPEA	2C-O-4	>60	?
CH_3S-	4-methylthio-DMPEA	2C-T	60–100	4
CH_3Se-	4-methylseleno-DMPEA	2C-SE	~100	~3

compound TMA-2), there is an increase of more than an order of magnitude. The 4-methylthio analogue (2C-T) is included here for comparative purposes, but it is the starting point for a large family to be discussed below.

Another location for structural variation is the β-position on the aliphatic side chain where one might place a methoxyl group. This is the position of the hydroxyl group on the neurotransmitters norepinephrine and epinephrine (noradrenaline and adrenaline) as well as on the natural phenethylamine ephedrine. This collection has been referred to as the BOX family. (See Table 3.9.)

Three O-demethyl compounds in this BOX group have been clinically studied in humans. The β-hydroxy analogue of BOD is 4-methyl-2,5-dimethoxyphenyl-β-ethanolamine (BOHD). It has a side-effect of causing a precipitous drop of both systolic and diastolic blood pressure. At an oral dosage of 50mg a drop of 36mm was observed and no further assays were conducted with this compound. There were no indications of hallucinogenic activity. The analogous derivative of BOH is 3,4-methylenedioxyphenyl-β-ethanolamine (BOHH). In the early literature it is occasionally referred to by the name MDE but this code has now been exclusively assigned to the N-ethyl

Table 3.9

Phenethylamines substituted with a β-methoxy group

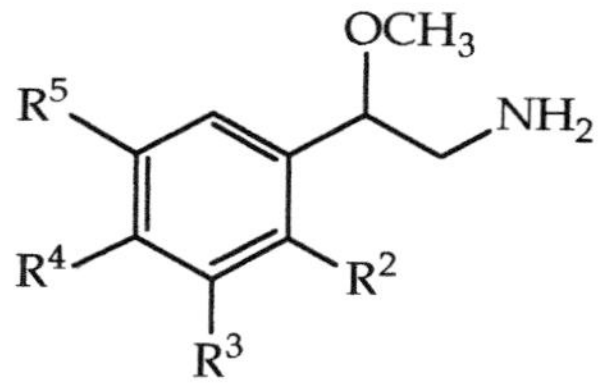

phenethylamines substituted with a β-methoxy group

R^2	R^3	R^4	R^5	common name	code	potency	
						(mg)	M.U.
CH_3O-	H-	Br-	CH_3O-	4-bromo-2,5-dimethoxyphenyl-M[a]	BOB	10–20	20
CH_3O-	H-	CH_3	CH_3O-	4-methyl-2,5-dimethoxyphenyl-M	BOD	15–25	15
H-	-OCH_2O-		H-	3,4-methylenedioxyphenyl-M	BOH	80–120	3
H-	CH_3O-	CH_3O-	CH_3O-	3,4,5-trimethoxyphenyl-M	BOM	>200	<1

[a] "M" stands for β-methoxyethylamine

homologue of MDMA. It is inactive at an oral dosage of 100mg. The corresponding dimethoxy compound (3,4-dimethoxyphenyl-β-ethanolamine, DME) has been studied at doses of up to 115mg and is without psychoactivity.

3.1.4 2,5-DIMETHOXY-4-METHYLTHIOPHENETHYLAMINE (2C-T) ANALOGUES

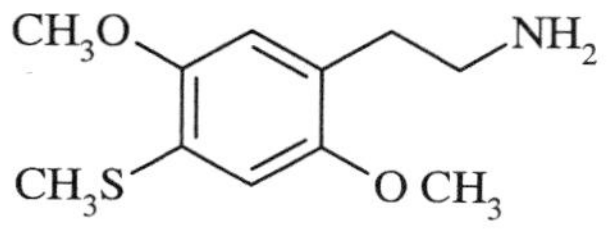

#4 2C-T

Figure 3.7

Chemical structure of 2,5-dimethoxy-4-methylthiophenethylamine

This modestly active methylthio-analogue of 2C-D has proven to be a rewarding starting point for an extensive study of sulfur-contained homologues and analogues. These are presented in a format of increasing mass and complexity of the substituents; the number suffix of the code name simply reflects the sequence of synthesis. (See Table 3.10.)

The optimum alkyl substitution seems to be two to three carbons, with 2C-T-2, 2C-T-4 and 2C-T-7 being both potent and similar in effect to LSD. The >30mg dosage indicated for 2C-T-15 indicates that this level showed some suggestion of activity, but the final active range had not yet been determined.

Table 3.10

Chemical structures of 2C-T and its alkyl and heteroalkyl analogues

CH_3O / NH_2 / RS / OCH_3

alkyl homologues and heteroalkyl analogues of 2C-T

R	common name	code	potency	
			(mg)	M.U.
CH_3-	4-methylthio-2,5-DMPEA	2C-T	60–100	4
CH_3CH_2-	4-ethylthio-2,5-DMPEA	2C-T-2	12–25	16
$CH_3(CH_2)_2$-	4-n-propylthio-2,5-DMPEA	2C-T-7	10–30	15
$(CH_3)_2CH$-	4-i-propylthio-2,5-DMPEA	2C-T-4	8–20	20
s-C_4H_9-	4-s-butylthio-2,5-DMPEA	2C-T-17	60–100	4
t-C_4H_9-	4-t-butylthio-2,5-DMPEA	2C-T-9	60–100	4
c-C_3H_5-	4-cyclopropylthio-2,5-DMPEA	2C-T-15	>30	?
c-$C_3H_5CH_2$-	4-cyclopropylmethylthio-2,5-DMPEA	2C-T-8	30–50	8
$CH_3O(CH_2)_2$-	4-(2-methoxyethyl)thio-2,5-DMPEA	2C-T-13	25–40	10
FCH_2CH_2-	4-(2-fluoroethyl)thio-2,5-DMPEA	2C-T-21	8–12	30
	— with an N-hydroxy substitution —			
CH_3CH_2-	4-ethylthio-N-hydroxy-DMPEA	HOT-2	10–18	24
$CH_3(CH_2)_2$-	4-n-propylthio-N-hydroxy-DMPEA	HOT-7	15–25	14
s-C_4H_9-	4-s-butyl-N-hydroxy-DMPEA	HOT-17	70–120	3

The final compound of the first group on this list deserves some additional comment. One, it is a novelty, being the first hallucinogenic material with six different elements in it (C, H, N, O, S and F), and as the hydrochloride salt, that number becomes seven. Secondly, it carries an ideal atom for radiolabelling the material for PET scanning. The introduction of ^{18}F should be readily achieved, and the drug might well be a potent ligand for the serotonin receptors involved with the action of the hallucinogenic drugs. And lastly, the presence of the fluorine atom at the end of the aliphatic chain suggests the possibility of lipophobic and hydrophobic chain terminals such as β,β,β-trifluoroethylthio-2,5-DMPEA. These are discussed below with the amphetamine homologues, and are presently completely unexplored.

The small second group on this list is from a small study of psychoactive hydroxylamines. The generic family name is HOT for "N-HydrOxyThio" and the assigned serial number corresponds to those of the corresponding 2C-T compound. The hallucinogenic potency of these analogues is very similar to the primary -NH2 counterparts. The 2,4,6-substituted positional isomer of 2C-T-4, is, 4-isopropylthio-2,6-dimethoxyphenethylamine (ψ-2C-T-4). In preliminary evaluations, it appears to be an active compound in the 10–20 milligrams range, but more studies are needed to firmly establish the dosage and the nature of the intoxication produced.

I learned of a remarkable study that had been methodically carried out but never published, in which the methoxy groups of a number of the 2C- family were replaced with ethoxy groups. As there are two methoxy groups, there are three possible homologues: the 2-methoxy-, the 5-methoxy- or both methoxys can be lengthened. These compounds were called the TWEETIO compounds in a humorous way of pronouncing the simplest of them, a compound with an ethoxy (or EtO-) group at the 2-position, vis., 2-EtO-something. (See Table 3.11.)

Table 3.11

The TWEETIO homologues of several 2C- compounds

R^5O — NH_2 — R — OR^2

the TWEETIO homologues of several 2C- compounds

R	dosage (parent)	R^2	R^5	code	trial dose (mg)	comments of trial
Br	12–24	Et-	Me-	2CB-2-ETO	15	30–50 prolongs action
		Et-	Et-	2CB-DI-ETO	>55	restless sleep, only
CH_3-	20–60	Et-	Me-	2CD-2-ETO	60	intimate, no intox, 4 hrs
		Me-	Et-	2CD-5-ETO	50	12 hour duration
		Et-	Et-	2CD-DI-ETO	>55	mild, 4 hr duration
C_2H_5-	10–15	Me-	Et-	2CE-5-ETO	15	16–24 hr duration
I	14–22	Et-	Me-	2CI-2-ETO	5	50 mg longer duration only
CH_3S-	60–100	Et-	Me-	2CT-2-ETO	50	mild, 4 hr duration
		Me-	Et-	2CT-5-ETO	30	15 hr duration
EtS-	12–25	Et-	Me-	2CT2-2-ETO	50	9 hr duration
		Me-	Et-	2CT2-5-ETO	20	16 hr duration
		Et-	Et-	2CT2-DI-ETO	50	only longer with higher dose
iPrS-	8–20	Et-	Me-	2CT4-2-ETO	25	dosage affects duration only
PrS-	10–30	Et-	Me-	2CT7-2-ETO	20	fast, out at 5 hrs

In most of these listings the human studies involved only two subjects, never more than six. For this reason, ranges are not given, only typical dosages, and the short statement of effect is simply a terse abstract of the reports that were written. Some generalities are possible. Homologation at the 2-position appears to lead to a shorter duration and a decrease in potency. Homologation at the 5-position can enhance the potency and increase the duration to a remarkable degree. With 2CE-5-ETO for example, the administration of Valium or Halcion allowed sleep, but did not abort the long duration of action. Longer ether chains such as the propoxy groups have not been synthesized as yet.

3.1.5 3,4,5-TRIMETHOXYAMPHETAMINE (TMA) ANALOGUES

The first synthetic modification of the mescaline molecule that was found to be hallucinogenic was the alpha-methyl homologue, 3,4,5-trimethoxyamphetamine. This simple structural manipulation, the adding of a methyl group adjacent to the amine function, is a well documented change that in general protects the amine from metabolic deamination.

Figure 3.8

Chemical structure of TMA

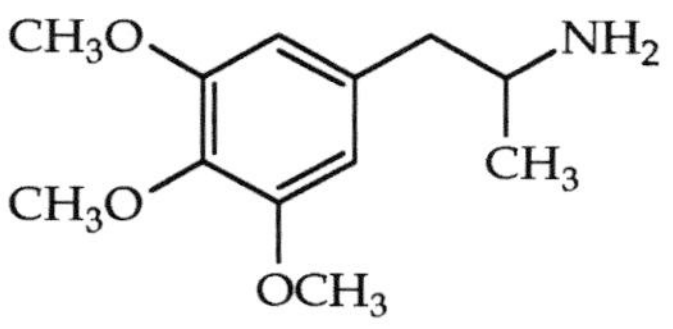

CH_3O NH_2 CH_3 CH_3O OCH_3

#5 TMA

The prototype for this change was the design of amphetamine itself as a metabolically stable stimulant from the extremely labile compound phenethylamine. With very few exceptions, the alpha-methyl homologues of the above-described phenethylamines are more potent and longer-lived in the human subject. From the historic view, the amphetamine derivative was discovered first in most cases, but the more logical structure-activity presentation in this review calls for the progression from the two-carbon phenethylamine to the three-carbon amphetamine. In the case of TMA, a continuous extension of this alpha-group led to the alpha-ethyl homologue alpha-ethylmescaline (AEM) which showed no human activity at over 200mg, and the longer-chained homologues were not evaluated. This alpha-homologation is discussed below for some of the 2,4,5-trisubstituted analogues.

A large number of positional variations have been explored, based on this prototype. In general, as the number of methoxy groups increases, the character of central action of the drug progresses from that of a stimulant to that of a hallucinogen, with a trisubstitution pattern being the most effective example of this latter action. And, in general, the most effective locations for the three substituents are in the 2,4,5- or the 2,4,6-orientations. Special attention should be paid to the first entry in the above table, for 4-methoxyamphetamine (4-MA). The earliest clinical studies with this compound were inspired by the frequency with which patients were seen displaying a schizophrenic-like syndrome, that was in fact a consequence of chronic excessive amphetamine use. One of the metabolites of amphetamine in humans is 4-hydroxyamphetamine. (See Table 3.12.)

Table 3.12

Alkoxy analogues of TMA

alkoxy analogues of TMA

R^2	R^3	R^4	R^5	R^6	common name	code	potency (mg)	potency M.U.
H-	H-	[a]MeO-	H-	H-	4-methoxy-A[b]	4-MA (PMA)	50–80	5
MeO-	H-	MeO-	H-	H-	2,4-dimethoxy-A	2,4-DMA	>60	?
MeO-	H-	H-	MeO-	H-	2,5-dimethoxy-A	2,5-DMA	80–160	2.5
H-	MeO-	MeO-	H-	H-	3,4-dimethoxy-A	3,4-DMA	in 100's	<1
H-	MeO-	MeO-	MeO-	H-	3,4,5-trimethoxy-A	TMA	100–250	1.7
H-	MeO-	[c]EtO-	MeO-	H-	4-ethoxy-3.5-dimethoxy-A	3-CE	30–60	7
H-	MeO-	[d]BzO-	MeO-	H-	4-benzyloxy-3.5-dimethoxy-A	3C-BZ	100–180	2
MeO-	H-	MeO-	MeO-	H-	2,4,5-trimethoxy-A	TMA-2	20–40	10
EtO-	H-	MeO-	MeO-	H-	2-ethoxy-4.5-dimethoxy-A	EMM	>50	?
MeO-	H-	EtO-	MeO-	H-	2,5-dimethoxy-4-ethoxy-A	MEM	20–50	10
MeO-	H-	MeO-	EtO-	H-	2,4-dimethoxy-5-ethoxy-A	MME	~60	~5
MeO-	H-	OEt-	OEt-	H-	4,5-diethoxy-2-methoxy-A	MEE	nt	
EtO-	H-	OMe-	OEt-	H-	2,5-diethoxy-4-methoxy-A	EME	nt	
EtO-	H-	EtO-	MeO-	H-	2,4-diethoxy-5-methoxy-A	EEM	nt	
EtO-	H-	EtO-	EtO-	H-	2,4,5-triethoxy-A	EEE	nt	
MeO-	H-	[e]PrO-	MeO-	H-	2,5-dimethoxy-4-(n)-propoxy-A	MPM	>30	?
MeO-	MeO-	MeO-	H-	H-	2,3,4-trimethoxy-A	TMA-3	>100	?
MeO-	MeO-	H-	MeO-	H-	2,3,5-trimethoxy-A	TMA-4	>80	?
MeO-	MeO-	H-	H-	MeO-	2,3,6-trimethoxy-A	TMA-5	~30	~10
MeO-	H-	MeO-	H-	MeO-	2,4,6-trimethoxy-A	TMA-6	25–50	8
MeO-	MeO-	MeO-	MeO-	H-	2,3,4,5-tetramethoxy-A	TA	>50	?

[a] "Me" represents a methyl group
[b] "A" represents the word amphetamine
[c] "Et" represents an ethyl group
[d] "Bz" represents a benzyl group
[e] "Pr" represents an n-propyl group
"nt" not yet clinically tested in humans

The question was asked that, if 4-MA were metabolized by O-demethylation to this same metabolite, might it be the causative agent for the dementia? 4-MA is indeed metabolized to this hydroxy derivative but there is no evidence that it is psychoactive. But an entirely separate issue has come up, in that there is a considerable amount of 4-MA (called PMA but distributed as "Ecstasy") is being distributed in the current rave scene, and a number of fatalities have followed its use. At the level of about 1mg/kg (in humans) it is a light hallucinogen with some pressor side-effects. At twice this dosage, there can be a considerable cardiovascular disturbance, and with several of these tablets taken at once (not uncommon in the rave scene) the user can face a lethal crisis.

N-Methylation of primary amines became the structural challenge. If MDA went to MDMA with such beauty, maybe other primary amines would become magical with a similar N-methylamine.

A common analogue of the vicinal dimethoxy structure, found frequently in nature, is the methylenedioxy group. These two substitution patterns are biosynthetically close cousins, and many of the essential oils found in spices and as food fragrances have one or the other of these systems present. Safrole is the cousin of methyleugenol, myristicin of elemicin, and apiole or dill apiole of tetramethoxyallylbenzene. (See Table 3.13.)

Table 3.13

Alkoxy methylenedioxy analogues of TMA

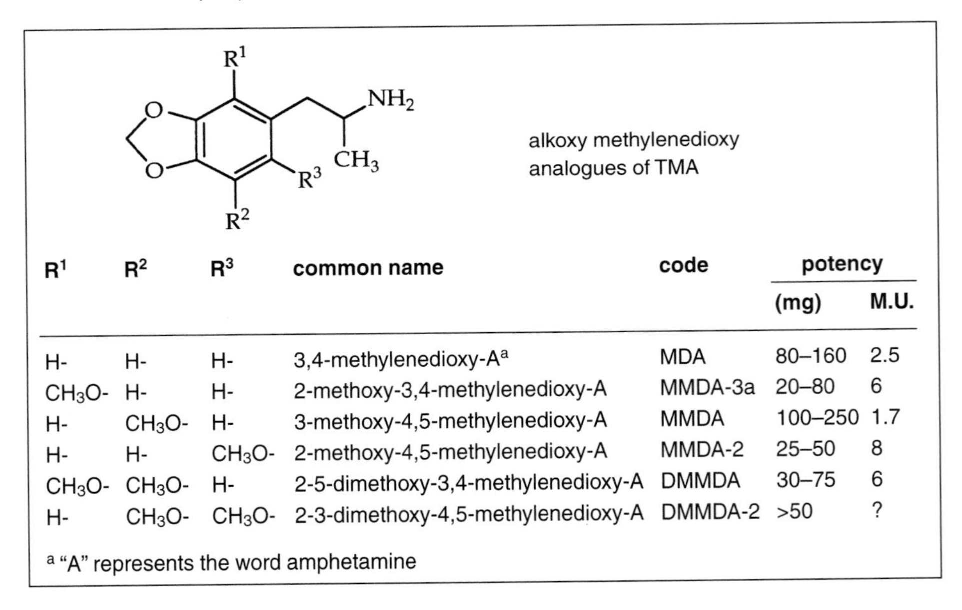

R[1]	R[2]	R[3]	common name	code	potency (mg)	potency M.U.
H-	H-	H-	3,4-methylenedioxy-A[a]	MDA	80–160	2.5
CH_3O-	H-	H-	2-methoxy-3,4-methylenedioxy-A	MMDA-3a	20–80	6
H-	CH_3O-	H-	3-methoxy-4,5-methylenedioxy-A	MMDA	100–250	1.7
H-	H-	CH_3O-	2-methoxy-4,5-methylenedioxy-A	MMDA-2	25–50	8
CH_3O-	CH_3O-	H-	2-5-dimethoxy-3,4-methylenedioxy-A	DMMDA	30–75	6
H-	CH_3O-	CH_3O-	2-3-dimethoxy-4,5-methylenedioxy-A	DMMDA-2	>50	?

[a] "A" represents the word amphetamine

This same parallel in structures may be seen in the alkoxylated amphetamine derivatives usually with a small increment of increased potency given to the methylenedioxy example. Efforts to enlarge the methylenedioxy ring by homologation of MMDA and MDA were synthetically achieved, but the hallucinogenic activity was lost. The products (3-methoxy-4,5-ethylenedioxyamphetamine, MEDA, homologue of MMDA, and 3,4-ethylenedioxy-N-methylamphetamine, EDMA or MDMC, homologue of MDA or, more accurately, of MDMA) were not active at 200mg, and so fall in the less-than-mescaline classification. Similarly, the second possible positional isomer of

MMDA-3a is known (4-methoxy-2,3-methylenedioxyamphetamine, MMDA-3b) and has been clinically explored at up to 50mg without any central effects being noted. With the methylenedioxy in that 2,3-orientation, next to the amphetamine chain, two more isomers are possible. The 2,3,5-arrangement has been named MMDA-4 and the 2,3,6-counterpart is known as MMDA-5; no human activity for either compound has been reported in the literature. There cannot be, of course, a MMDA-6 as the 2,4,6 orientation lacks adjacent oxygen atoms.

A large number of nitrogen-substituted homologues of the MDA-related amphetamines have been synthesized and assayed, with listed potencies again compared to mescaline as being the reference standard, and most of them have been found to be relatively inactive. The N-methyl homologue of MDA is MDMA, which turned out to be not a hallucinogen but rather a mild stimulant that produced a most gentle and friendly altered state of consciousness, one that permitted both easy personal interaction and complete recall. It had seen extensive use as a clinical adjunct in psychotherapy, but with its appearance on the street in 1985 under the name of "Ecstasy" or "X," it was placed in Schedule I of the Controlled Substances Act by the Drug Enforcement Administration. It has been only in the last few years that moves to start human research with it again, mostly abroad, have become successful. The other two active members of this subfamily (MDE and MDOH) are very similar to MDA in their qualitative effects. Interestingly, the ring-methyl homologue of MDMA (2,N-dimethyl-4,5-methylenedioxyamphetamine, MADAM-6) is devoid of either stimulant or hallucinogenic activity in humans, at a dosage of 280mg. The corresponding ring-methoxy analogue (N-methyl-2-methoxy-4,5-methylenedioxyamphetamines, METHYL-MMDA-2) can also be viewed as a mono-substituted analogue of MDMA. It has been assayed in humans up to 70 milligrams orally, and is without activity.

The ethanolamine amide (MDHOET) shows central effects at higher doses that are anesthetic, in the manner of ketamine. (See Table 3.14.)

Of all the compounds of this group that had been assayed for pain suppression in mice, this compound was the most effective. Two phentermine analogues showed some toxicity, and neither had the visual effects of a hallucinogenic drug nor the magic of MDMA. The parent α,α-dimethylhomologue of MDPEA (the α-methyl homologue of MDA) is MDPH and it is of about the same potency as MDA. The N-methyl homologue is MDMP and was virtually inactive at 110mg, and no higher level has yet been explored.

The group of compounds apparent from the ETHYL-J and METHYL-K in the table below speak of a structure activity relation study that was never carried out. The syntheses were straight forward, and the collection of names was clear. The named alkyl group (the METHYL, the ETHYL or the PROPYL) was the

Table 3.14

Nitrogen-substituted homologues and analogues of MDA

nitrogen-substituted homologues and analogues of MDA

R^1	R^2	R^3	common name	code	potency (mg)	potency M.U.
H-	H-	H-	MDPEA[a]	MDPEA	>300	<0.5
H-	H-	Me-	MDA[b]	MDA	80–160	2.5
H-	H-	Et-	α-ethyl-MDPEA	BDB	150–230	1
H-	H-	$(Me)_2$	α,α-dimethyl-MDPEA	MDPH	160–240	1
Me-	H-	Me-	N-methyl-MDA	MDMA[c]	80–150	2.5
Me-	H-	Et-	N-methyl-α-ethyl-MDPEA	MBDB	180–210	1
Me-	H-	Pr-	N-methyl-α-propyl-MDPEA	METHYL-K	>100	<2
Me-	H-	$(Me)_2$	α,α-dimethyl-MDPEA	MDMP	>110	<2
Me-	Me-	Me-	N,N-dimethyl-MDA	MDDM	>150	<1
Me-	HO-	Me-	N-hydroxy-N-methyl-MDA	FLEA	100–160	2
Et-	H-	Me-	N-ethyl-MDA	MDE	100–200	2
Et-	H-	Et-	N,α-diethyl-MDPEA	ETHYL-J	>65	?
Et-	H-	Pr-	N-ethyl-α-propyl-MDPEA	ETHYL-K	>40	?
Pr-	H-	Me-	N-(n)-propyl-MDA	MDPR	>200	<1
iPr-	H-	Me-	N-(i)-propyl-MDA	MDIP	>250	<1
Bu-	H-	Me-	N-(n)-butyl-MDA	MDBU	>40	?
$(C_3H_5)CH_2$-	H-	Me-	N-cyclopropylmethyl-MDA	MDCPM	nt	
Al-	H-	Me-	N-allyl-MDA	MDAL	>180	<1
$CH{\equiv}CCH_2$-	H-	Me-	N-propargyl-MDA	MDPL	>200	<1
$C_6H_5CH_2$-	H-	Me-	N-benzyl-MDA	MDBZ	>150	<1
HO-	H-	Me-	N-hydroxy-MDA	MDOH	100–160	2
MeO-	H-	Me-	N-methoxy-MDA	MDME	>180	<1
$HO(CH_2)_2$-	H-	Me-	N-(β-hydroxyethyl)-MDA	MDHOET	>50	?
$MeO(CH_2)_2$-	H-	Me-	N-(β-methoxyethyl)-MDA	MDMEOET	>180	<1

[a] "MDPEA" represents 3,4-methyleneioxyphenethylamine
[b] "MDA" represents 3,4-methylenedioxyamphetamine
[c] "MDMA" is not hallucinogenic, see text
"nt" Not yet clinically tested in humans.
The BDB and MBDB codes are derived from the exact chemical names for these two compounds, i.e., 1-(1,3-benzodioxol-5-yl)-2-butanamine and N-methyl-1-(1.3-benzodioxol-5-yl)-2-butanamine.
The MDPH and MDMP are codes derived from the commercial names for the molecular skeleton, i.e., 3,4-methylenedioxyphentermine and 3,4-methylenedioxymephentermine.

substituent on the nitrogen atom of the 3,4-methylenedioxyphenethylamine basic skeleton, and the letter employed (the J or the K) representing the α-substituent (the α-H and the α-I were not used, as the indicated molecule already has a trivial code, but the J was the α-ethyl, the K was α-propyl, and so on). All of the synthetic chemistry was easily completed, but the loss of hallucinogenic potency with the heavier substituents quickly discouraged further exploration in people.

Another N-alkyl homologue of these simple amphetamine derivatives merits special mention. PMA was mentioned above as a drug requiring close attention. The N-methyl homologue of this compound has recently been seen in the street trade, first as an impurity in PMA and then as an agent in its own right. This material, called para-methoxy-methamphetamine or PMMA, is active at something over 100 milligrams, but it still produces cardiovascular stimulation without much virtue as a hallucinogen. Its easy availability may be because it can be made directly by the synthetic procedures used for MDMA, but with the use of the rather innocent anethole (anise camphor) in place of isosafrole which is being closely watched by the government. The corresponding analogue of 2,5-DMA (N-methyl-2,5-dimethoxyamphetamine, DMMA or methyl-DMA) is without human activity at 250mg.

3.1.6 2,5-DIMETHOXY-4-METHYLAMPHETAMINE (DOM) ANALOGUES

Another starting point for the viewing of a family of amphetamine homologues of the phenethylamine family is the drug DOM, or STP as it became known on the street in the 1960s. It is an unusually potent amphetamine derivative that was designed as an analogue of TMA-2. (See Figure 3.9.)

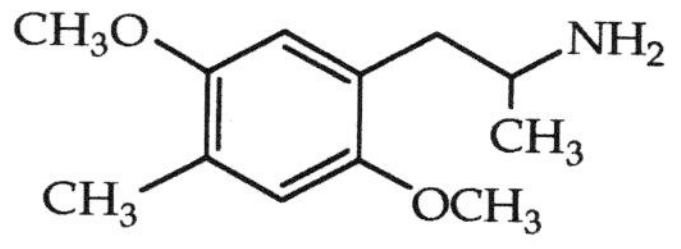

CH_3O NH_2 CH_3 CH_3 OCH_3

#6 DOM (STP)

Figure 3.9

Chemical structure of DOM (STP)

The unusually high potency of TMA-2 was ascribed to the group found at the 4-position, the methoxy group, as indicated by the fact that it was only through homologation at this position that higher potency and hallucinogenic activity were maintained. The replacement of the methoxyl function with a methyl group would interfere with its metabolism, as the methyl group cannot be removed by any obvious hydrolysis scheme. It was believed that if DOM were to act as a receptor agonist and it could not be easily destroyed metabolically, it would be a very potent drug. If it was not intrinsically active but did indeed occupy the receptor, it might well block some endogenous psychotogenic factor, and thus serve as a prophylactic or treatment for spontaneous mental illness. It proved however, to be intrinsically active, and has served as a pilot structure for many analogues.

The first structural variations were changes made on the 4-alkyl group. As it was lengthened, the activity was found to increase, and then it abruptly dropped off. The compounds with longer chain lengths were apparently evaluated simply up to the point where it could be said that the activity was no longer increasing. The one intriguing homologue that lies in the middle of the range of high potencies is the isopropyl isomer, DOIP. In the study that provided the above values, this isomer was investigated. At dosages of 4 and 10mg, there were substantially no effects. Apparently there are "valid changes in the mental state" in the 20 to 30mg range, but these are not described. In any event it appears that the isopropyl analogue does not compete in potency with the ethyl or the propyl counterparts. The β-fluoroethyl isomer is compelling evidence that the ethyl chain is enough for holding on to full activity, and that there is the potential for making extensive modification at the outer end of that ethyl group. (See Table 3.15.)

Table 3.15

4-Alkyl homologues of DOM

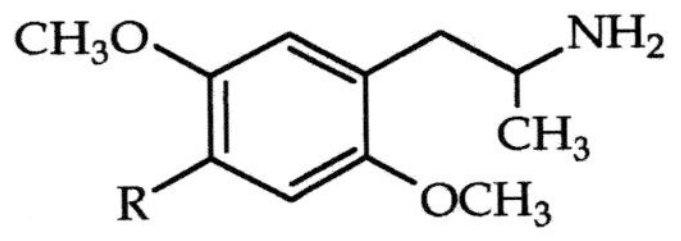

4-alkyl homologues of DOM

R=	common name	code	potency (mg)	potency M.U.
CH_3-	4-methyl-2,5-dimethoxy-A[a]	DOM	3–10	50
CH_3CH_2-	4-ethyl-2,5-dimethoxy-A	DOET	2.0–6.0	80
$CH_3(CH_2)_2$-	4-(n)-propyl-2,5-dimethoxy-A	DOPR	2.5–5.0	80
$(CH_3)_2CH$-	4-(i)-propyl-2,5-dimethoxy-A	DOIP	>10	?
$CH_3(CH_2)_3$-	4-(n)-butyl-2,5-dimethoxy-A	DOBU	>3	?
$(CH_3)_2CHCH_2$-	4-(i)-butyl-2,5-dimethoxy-A	DOIB	>10	?
$CH_3CH_2CH(CH_3)$-	4-(s)-butyl-2,5-dimethoxy-A	DOSB	>25	?
$(CH_3)_3C$-	4-(t)-butyl-2,5-dimethoxy-A	DOTB	>10	?
$CH_3(CH_2)_4$-	4-(n)-amyl-2,5-dimethoxy-A	DOAM	>10	?
FCH_2CH_2-	4-(β-fluoroethyl)-2,5-dimethoxy-A	DOEF	2.0–3.5	100

[a] "A" represents the word amphetamine

Considering this remarkably potent compound along with the expectedly potent β-fluoroethylthio- compound 2C-T-21 mentioned above, one could speculate that a small lipophobic (and hydrophobic) atom at the end of a short aliphatic (and thus lipophilic) chain in the appropriate receptor site makes for a remarkable agonist. If this is valid, what might be the activity of the rather easily made analogues with groups such as CHF_2CH_2- or CF_3CH_2- in these positions?

Two homologues, one ring analogue, and two heterocyclic structural modifications of DOM warrant mention. Neither the N-methyl homologue (N,4-dimethyl-2,5-dimethoxyamphetamine, BEATRICE) nor the 5-ethoxy homologue (5-ethoxy-2-methoxy-4-methylamphetamine, IRIS) have shown any psychopharmacological activity at oral dosages that would be completely effective with DOM itself. The closure of the chain-methyl group with the adjacent benzylic carbon to create a three-member ring produces a compound (2-(2,5-dimethoxy-4-methylphenyl)cyclopropylamine, DMCPA) which is an effective hallucinogen in the 10–20mg range. The incorporation of the oxygen atom at the DOM 5-position into a furan ring has led to two heterocyclic analogues of DOM. The furanyl-2-methyl and 2,2-dimethyl compounds are 6-(2-aminopropyl)-5-methoxy-2-methyl-2,3-dihydrobenzofuran (F-2) and 6-(2-aminopropyl)-2,2-dimethyl-5-methoxy-2,3-dihydrobenzofuran (F-22) and were, as with the two homologues mentioned above, inactive at oral dosages (15mg) that would have been effective for the parent compound, DOM.

Instead of an alkyl group at the sensitive 4-position, there have been a number of compounds synthesized and assayed with other subsituents at that position. (See Table 3.16.)

Table 3.16

4-Substituted analogues of DOM

CH_3O, NH_2, CH_3, R, OCH_3

4-substituted analogues of DOM

R=	common name	code	potency (mg)	potency M.U.
CH_3-	4-methyl-2,5-dimethoxy-A[a]	DOM	3–10	50
Cl-	4-chloro-2,5-dimethoxy-A	DOC	1.5–3.0	150
Br-	4-bromo-2,5-dimethoxy-A	DOB	1.0–3.0	150
I-	4-iodo-2,5-dimethoxy-A	DOI	1.5–3.0	150
NO_2-	4-nitro-2,5-dimethoxy-A	DON	3.0–4.5	80

[a] "A" represents the word amphetamine

The halides that have been tried have been extremely active, and the two heavier ones have enjoyed an extensive popularity as serotonin receptor site ligands (the methyl-bearing DOM is included in this table for comparison). The second of these, DOI, is commercially available, both as the racemate and as the separate optically active isomers. Animal studies on the 4-fluoro analogue (DOF) have suggested that it is several times less potent than DOB and DOI, but no human studies on it have yet been reported. Two positional isomers and two N-alkyl homologues of these halogenated amphetamine derivatives have been clinically explored, and all are less active than their parent prototypes. The interchanging of the 5-methoxy group and the bromine atom of DOB produces META-DOB (5-bromo-2,4-dimethoxyamphetamine) which proved to develop a complex toxic syndrome in the 50 to 100mg range. The relocation of the 2-methoxy group to the 3-position produces 4-Br-3,5-DMA (4-bromo-3,5-dimethoxyamphetamine) which is a remarkably effective anesthetic to skin surfaces at 4 to 10 mg orally. The N-methyl homologue of DOB (4-bromo-2,5-dimethoxy-N-methylamphetamine, METHYL-DOB) produced a long-lasting broad array of physical discomfiture at 8mg and was not explored higher. The N,N-dimethyl homologue of DOI was called IDNNA (2,5-dimethoxy-N,N-dimethyl-4-iodoamphetamine) and, although not active at levels where DOI would be, it led to an extensive series of some 15 mono- and di-N-alkylated amines related to DOI. They were prepared for studies of ^{131}I labelled compounds for rat pharmacology (and eventual ^{122}I PET scanning agents for human studies) but none of them had been clinically explored.

What other hetero-atoms could be brought into that 4-position? The obvious one is suggested by the thiomescaline and 2C-T which call upon the sulfur atom. This is the series of compounds called the "Aleph" group, made with the 4-position sulfur substituted in place of the 4-position oxygen atom. (See Table 3.17.)

Table 3.17

4-Thioalkyl analogues of DOM

CH_3O NH_2 CH_3 RS OCH_3

4-thioalkyl analogues of DOM

R=	common name	code	potency (mg)	potency M.U.
CH_3-	4-methylthio-2,5-dimethoxy-A[a]	Aleph (DOT)	5–10	40
CH_3CH_2-	4-ethylthio-2,5-dimethoxy-A	Aleph-2	4–8	50
$CH_3(CH_2)_2$-	4-(n)-propylthio-2,5-dimethoxy-A	Aleph-7	4–7	50
$(CH_3)_2CH$-	4-(i)-propylthio-2,5-dimethoxy-A	Aleph-4	7–12	30
C_6H_5-	4-phenylthio-2,5-dimethoxy-A	Aleph-6	>40	?

[a] "A" represents the word amphetamine

There are examples of chains up to three carbons in length (the n-propyl group) and the activity gives no indication of beginning to diminish. There is much manipulation possible here, with longer chain alkyls, branched chain alkyls, with small rings, with fluoro-atoms, and it is all still unexplored. One observation is needed. In the broader clinical studies with some of the Aleph compounds, a sizable individual variation was noted, so care must be taken in the exploration of new levels of new compounds. Some people may be unduly sensitive or insensitive.

The determination of the effects of replacing the oxygen atom in active compounds with a sulfur atom has been extended into several families of the hallucinogens.

This substitution has been thoroughly explored with the 4-alkyl-2,5-dimethoxyamphetamines DOM and DOET. With DOM, each of the oxygen atoms separately and both of the oxygen atoms together have been replaced with sulfur, and with one, the sulfoxide was made as well. There is 5-methoxy-4-methyl-2-methylthioamphetamine (2-TOM) which is active at 60–100mg, 2-methoxy-4-methyl-5-methylthioamphetamine (5-TOM) which is active at 30–50mg, and 4-methyl-2,5-bis-(methylthio)amphetamine (BIS-TOM) which shows no activity even at 160mg. The sulfoxide of 5-TOM (2-methoxy-4-methyl-5-methylsulfinylamphetamine, TOMSO) also was without activity at 150mg. The two oxygen atoms in DOET have also been replaced with sulfur. There is 4-ethyl-5-methoxy-2-methylthioamphetamine (2-TOET) which is inactive at 65mg, and 4-ethyl-2-methoxy-5-methylthioamphetamine (5-TOET) which is an active hallucinogenic drug in the 12–25mg range.

The two other possible oxygen substitutions of DOT have been synthesized and evaluated; this completes the three theoretical thio-analogues of TMA-2. 4,5-Dimethoxy-2-methylthioamphetamine and 2,4-dimethoxy-5-methylthioamphetamine (Ortho-DOT and Meta-DOT respectively) were without any central effects at levels of 25mg orally.

Two of the methylenedioxyamphetamine compounds have also been similarly modified. The 2-methoxy group of MMDA-3a is replaced with a methylthio group giving 3,4-methylenedioxy-2-methylthioamphetamine, or 2T-MMDA-3a. It is not active at a dosage of 12mg. The second analogue has one of the oxygens of the methylenedioxy group changed to a sulfur. This is 6-(2-aminopropyl)-5-methoxy-1,3-benzoxathiol (4-T-MMDA-2) and it shows no activity at 25mg orally.

The last group of hallucinogens that has been derived by structural manipulation of DOM is the Ganesha series. This is a collection of 3,4-dialkyl derivatives similar to those that were seen earlier in the 2C-G family, but these are the corresponding amphetamines. Interestingly, the difference between the two families (the normally more potent 3-carbon amphetamine set and the less potent2-carbon phenethylamines) is substantially lost. (See Table 3.18.)

Table 3.18

3,4-Dialkyl homologues of DOM

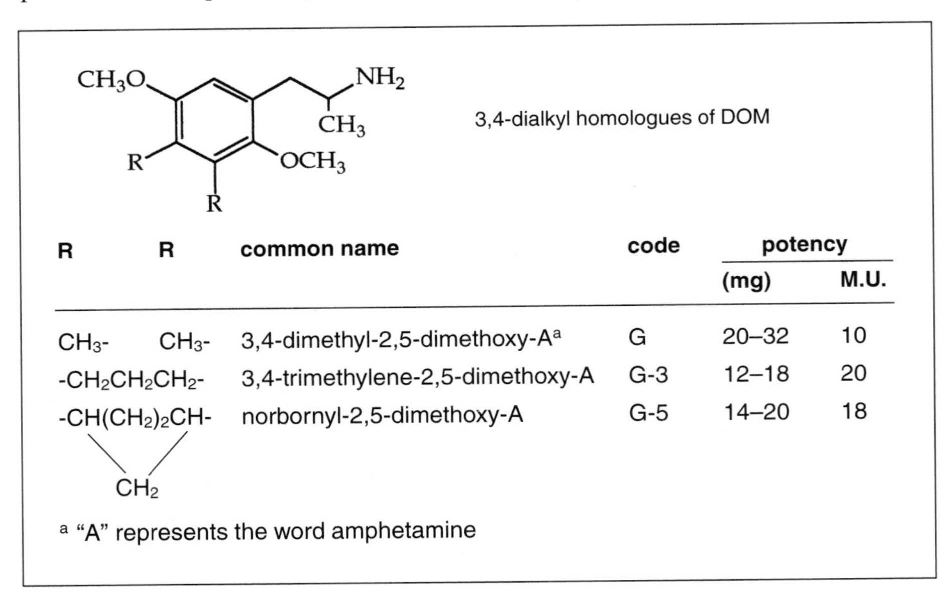

R	R	common name	code	potency (mg)	potency M.U.
CH_3-	CH_3-	3,4-dimethyl-2,5-dimethoxy-A[a]	G	20–32	10
-$CH_2CH_2CH_2$-		3,4-trimethylene-2,5-dimethoxy-A	G-3	12–18	20
-$CH(CH_2)_2CH$- (bridged by CH_2)		norbornyl-2,5-dimethoxy-A	G-5	14–20	18

[a] "A" represents the word amphetamine

The 3,4-dimethyl compounds (G vs. 2C-G) are equipotent. The three-carbon G-3 is half-again more potent than the 2-carbon 2C-G-3. But the norbornyl two-carbon compound 2C-G-5 is believably more active than the three-carbon G-5 listed here. It would seem that within this 3,4-disubstituted-2,5-dimethoxy family, as the substituent become more bulky, the phenethylamine analogue assumes the role of being the more potent. The analogous G-4 and G-N have not been evaluated in humans.

This review to date has considered the relatives of the 3,4,5-trisubstitution ring pattern (modest activity) and the considerably more potent 2,4,5-trisubstitution pattern. As was noted in the comments comparing TMA-2 with TMA-6, the 2,4,6-orientation bids fair to be every bit as important as the 2,4,5-system, although it has as yet been almost unexplored, either chemically or pharmacologically. A nomenclature that has been used to refer to this branch which is parallel to the 2,4,5-group, is to use the code name of the drug and precede it with the Greek letter psi. This was introduced above with the compound Ψ-2C-T-4. Thus, the lead drug of this section (DOM or 2,5-dimethoxy-4-methylamphetamine) becomes Ψ-DOM (2,6-dimethoxy-4-methylamphetamine). Clinical studies have shown it to be active as a hallucinogen in the 15–25mg range, with a mescaline equivalency of 15. There is too little data at the present time to determine any quantitative relationship between the 2,4,5-normal series and the 2,4,6-Ψ-series, but it appears quite possible that the two parallel families are, at least as to their quantitative properties, quite similar.

Just as the addition of a carbon atom in the form of an N-methyl group (converting a primary amine $-NH_2$ to a secondary amine $-NHCH_3$) largely dispels the hallucinogen property, so does the addition of a carbon atom in the form of an extension on the alpha-methyl group of the amphetamine chain (converting an alpha-methyl to an alpha-ethyl group). Again, with DOM as an illustration, the homologue with an alpha-ethyl group is an antidepressant, but not hallucinogenic in any way. This drug, 1-(2,5-dimethoxy-4-methylphenyl)-2-amino-butane, was developed by Bristol Laboratories under the name Dimoxamine, although in the early literature it had the code name Ariadne. A parallel is given in the tryptamine world below, with the comparison of alpha-methyl tryptamine with Monase, the alpha-ethyl homologue, also an antidepressant.

3.2 THE TRYPTAMINES

The drugs to be covered in this section of this review contain a skeleton of two-rings known as an indole. This, with an aminoethyl chain attached to its 3-position, becomes indolethylamine, commonly known as tryptamine. The phenethylamine versus tryptamine balance can be seen in the comparison of the chemical structures of the two major central neurotransmitters, dopamine and serotonin. In these diagrams, the positions of maximum sensitivity for producing hallucinogenic activity, the 4-position in the phenethylamine family, and the 5-position in the tryptamine, are indicated. Each is occupied in the native neurotransmitter by a hydroxyl group. (See Figure 3.10.)

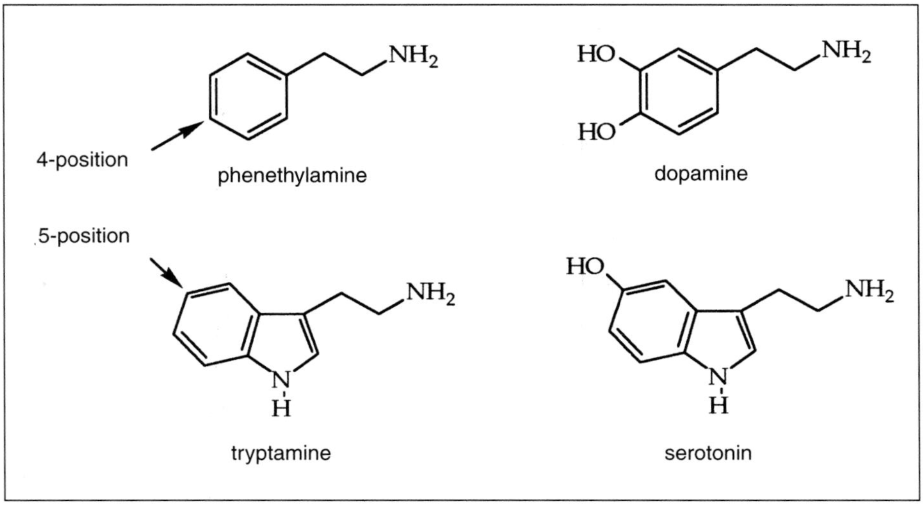

NH2
HO
NH2
HO
4-position
phenethylamine
dopamine
5-position
HO
NH2
NH2
N
H
N
H
tryptamine
serotonin

Figure 3.10

Chemical structures of dopamine and serotonin and their similarity to phenethylamine and tryptamine respectively

The pharmacological universe would be a neat and simple place if the hallucinogens that were phenethylamines were to work via the dopamine receptors and those that were tryptamines were to use the serotonin receptors: a sort of a structural allegiance. The universe is far from simple. All hallucinogens involve one or more of the serotonin subtypes, but the dopamine system is frequently involved as well.

There are structural consistencies between the two families, and differences as well. In both groups, the simplest and most structurally open examples are subject to metabolic oxidative deamination, which can be minimized by placing a methyl group alpha to the basic nitrogen atom. And in both groups, the placement of a non-polar group (as a methoxy group) at the position of the neurotransmitter's binding hydroxy group (4- with the phenethylamines, and 5- with the tryptamines) greatly enhances central activity and versatility of response. In contrast, whereas the substitution of groups on the nitrogen eradicates activity in the phenethylamines, it is an essential structural feature with the tryptamines (unless there is an amphetamine-like alpha-methyl group present).

3.2.1 N,N-DIMETHYLTRYPTAMINE (DMT) ANALOGUES

There is a very wide distribution of the two hallucinogens DMT (discussed here) and 5-methoxy-N,N-dimethyltryptamine (5-MeO-DMT, discussed below) throughout the plant kingdom. But both are also well-established normal constituents in human urine, blood, and cerebral spinal fluid. Efforts to find a relationship between levels of these natural alkaloids and the mental health of humans have been futile. They are always there but at very small levels. When these levels are increased through some parenteral route of self-administration, a short-lived but intense hallucinogenic crisis can occur. DMT is the most convenient compound to use as a base reference point. (See Figure 3.11.)

Figure 3.11

N,N-Dimethyltryptamine (DMT)

N CH_3 CH_3 N H

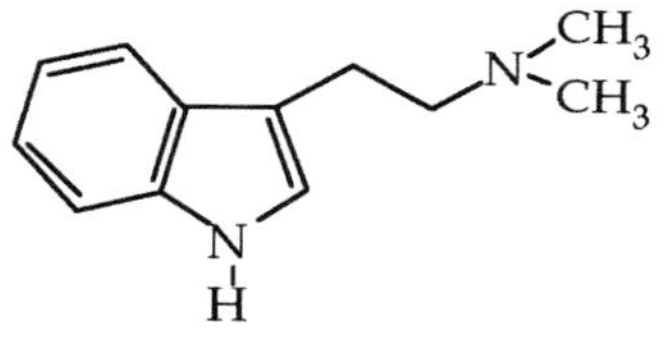

#7 N,N-dimethyltryptamine DMT

It, like mescaline in the phenethylamine group, is one of the oldest, best documented, and least potent of the entire system.

A number of homologues of DMT have been synthesized, and evaluated in human subjects. (See Table 3.19.)

Not all of these chemicals have been rigidly defined by the rigors of the "scientific method," double blind experiments with placebo controls. In fact,

Table 3.19

N,N-Dialkyl homologues of DMT

N,N-dialkyl homologues of DMT

R^1	R^2	common name	code	potency (mg)	x-DMT
H-	H-	tryptamine	T	>100	<1
H-	Me-	monomethyltryptamine	NMT[a]	50–100	1
H-	s-Bu-	mono-(s)-butyltryptamine	NSBT	50–75	1
H-	t-Bu-	mono-(t)-butyltryptamine	NTBT	~20	~4
H-	Am-	mono-(n)-amyltryptamine	NAT	>100	<1
H-	He-	mono-(n)-hexyltryptamine	NHT	>100	<1
Me-	Me-	dimethyltryptamine	DMT	60–100	1
Me-	Et-	methyl-ethyltryptamine	MET	80–100	1
Me-	Pr-	methyl-(n)-propyltryptamine	MPT	>50	?
Me-	iPr-	methyl-isopropyltryptamine	MIPT	10–25	4
Me-	Bu-	methyl-(n)-butyltryptamine	MBM	250–400	0.3
Et-	Et-	diethyltryptamine	DET	60–150	1
Et-	iPr-	ethyl-isopropyltryptamine	EIPT	24–40	3
Pr-	Pr-	di-(n)-propyltryptamine	DPT	~100	1
iPr-	iPr-	diisopropyltryptamine	DIPT[b]	40–100	1
Al-	Al-	diallyltryptamine	DAT	80	1
Bu-	Bu-	di-(n)-butyltryptamine	DBT	>100	<1
—$(CH_2)_4$—		pyrrolidyltryptamine	pyr-T	~100	1

[a] Several of these latter codes employ the first letter "N" for nitrogen rather than "M" for mono-. This latter has been reserved for the methyl group.
[b] Extreme auditory harmonic distortion, with no visual changes at all.

very few of the human studies on any of the hallucinogens have been structured in this way. The simple reason is that all clinical studies, at least with new and previously unexplored compounds that might show some of the subjective properties of this class of drugs, depend upon the awareness of the test subject of the nature of the experiment and the potential range of effects that might be expected. Thus, many studies are single trials in single individuals, with the results commingled in the effort to bring some statistical validity to any conclusions found. Several of the values in the listings above are findings in one or two individuals only.

A logical extension of this family of DMT homologues embraces substitutions on the aromatic ring of the indole. A collection of aromatic-substituted ring-substituted compounds is shown in the Table listing. Analogues that are substituted on the indolic 4-position will be considered separately. (See Table 3.20.)

Table 3.20

Ar, N-Substituted analogues of DMT; nothing at 4-position

Ar, N-substituted analogues of DMT; nothing at 4-position

R^5	R^6	R^7	R^1	R^2	common name	code	potency (mg)	x DMT
HO-	H-	H-	H-	H-	serotonin	5-HT	–	–
HO-	H-	H-	CH_3-	CH_3-	5-hydroxy-N,N-dimethyl-tryptamine	bufotenine	8–16	7[a]
CH_3O-	H-	H-	Ac-	H-	melatonin	melatonin	1–3	[b]
CH_3O-	H-	H-	CH_3-	CH_3-	5-methoxy-N,N-dimethyl-tryptamine	5-MeO-DMT	6–20	7[c]
CH_3O-	H-	H-	CH_3-	iPr-	5-methoxy-N-isopropyl-N-methyl-tryptamine	5-MeO-MIPT	4–6	16
CH_3O-	H-	H-	Et-	Et-	5-methoxy-N,N-diethyl tryptamine	5-MeO-DET	1–3	40
CH_3O-	H-	H-	Pr-	Pr-	5-methoxyl-N,N-dipropyl-tryptamine	5-MeO-DPT	6–10	10
CH_3O-	H-	H-	iPr-	iPr-	5-methoxyl-N,N-diisopropyl-tryptamine	5-MeO-DIPT	8–12	8
CH_3O-	H-	H-	-$(CH_2)_4$-		5-methoxyl-N,N-tetramethylene-tryptamine	5-MeO-pyr-T	0.5–2	[d]
CH_3S-	H-	H-	CH_3-	CH_3-	5-methylthio-N,N-dimethyl-tryptamine	5-MeS-DMT	15–30	4
CH_3O-	CH_3O-	H-	CH_3-	iPr-	5,6-dimethoxy-N-isopropyl-N-methyl-tryptamine	5,6-MeO-MIPT	>75	<1
-OCH_2O-		H-	CH_3-	CH_3-	5,6-methylenedioxy-N,N-dimethyl-tryptamine	5,6-MDO-DMT	?	?[e]
-OCH_2O-		H-	CH_3-	iPr-	5,6-methylenedioxy-N-isopropyl-N-methyl-tryptamine	5,6-MDO-MIPT	>60	<1
-OCH_2O-		H-	iPr-	iPr-	5,6-methylenedioxy-N,N-diisopropyl-tryptamine	5,6-MDO-DIPT	?	?[e]
H-	HO-	H-	CH_3-	CH_3-	6-hydroxy-N,N-dimethyl-tryptamine	6-OH-DMT	>80	<1
H-	CH_3O-	H-	CH_3-	iPr-	6-methoxy-N-isopropyl-N-methyl-tryptamine	6-MeO-MIPT	>50	<1

Table 3.20 (continued)

R^5	R^6	R^7	R^1	R^2	common name	code	potency (mg)	potency x DMT
H-	F-	H-	Et-	Et-	6-fluoro-N,N-diethyl-tryptamine	6-F-DET	>80	<1[f]
H-	H-	CH_3O-	CH_3-	iPr-	7-methoxy-N-isopropyl-N-methyl-tryptamine	7-MeO-MIPT	>70	<1

[a] administered parenterally – see comments below
[b] sedation, dream-state
[c] smoked – not orally active
[d] not hallucinogenic – produces long-lived amnesia and unconsciousness
[e] clinical trials started, not yet completed
[f] with autonomic effects similar to those of DET but without the hallucinogen component, this has found clinical value as an active placebo in human experiments

It is immediately obvious from the representation that substitution in the 5-position is needed for maximum activity (the 4-substituted isomers will be discussed below).

For a number of reasons, some pharmacological and some political, the compound 5-hydroxy-N,N-dimethyltryptamine, bufotenine, deserves special comment. From the pharmacological point of view, the compound is clearly active, but the nature of this activity is difficult to classify. The early studies that report effects in humans followed intravenous administration, and the responses noted (anxiety, panic, visual distortion, intense flushing) have been ascribed to extreme cardiovascular action and possible increases in interocular pressure. No effects have been observed following intranasal or oral administration. Recent studies with snuffs from roasted red seeds of the South American trees of the *Anadenanthera* species have proved highly active and yet careful analysis have shown that the only alkaloid present was bufotenine. Yet there are several reports in the medical literature of human studies where the compound is reported to be without activity.

The political aspects deal largely with the law. In the earliest studies, the comment was made that some of the observed effects were reminiscent of LSD. This prompted the inclusion of bufotenine into the legal structure as a hallucinogen, and it has been referred to in this way ever since. This alkaloid can be found in a number of plants and animals and is, in fact, a normal component of human urine. A recent event involving the smoking of dried toad venom may lead to a legal challenge of consuming a Schedule I drug (bufotenine) and questions such as its presence in the venom, its pyrolytic stability, and its pharmacological nature should be well defined ahead of time.

A number of tryptamines have been studied that imitate the structural pattern of amphetamine, where an alpha-alkyl group effectively protects the molecule from enzymatic deamination. (Table 3.21.)

Table 3.21

α,Ar, N-Substituted tryptamines

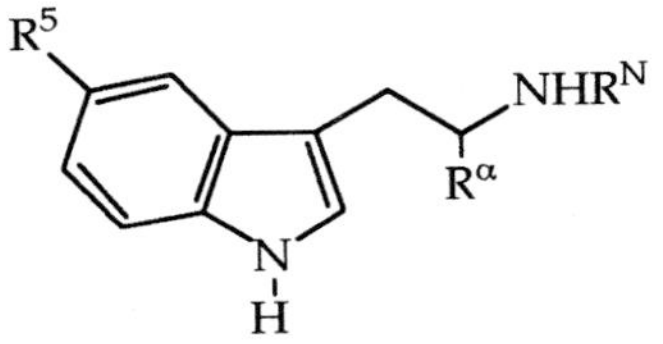

5,α,N-substituted tryptamines

R^5	R^α	R^N	common name	code	potency (mg)	potency x-DMT
H-	CH_3-	H-	α-methyltryptamine	α-MT, IT-290	15–30	3
H-	CH_3-	CH_3-	α,N-dimethyltryptamine	α,N-DMT	50–100	1
H-	Et-	H-	α-ethyltryptamine	α-ET, Monase	100–160	0.5
CH_3O-	CH_3-	H-	α,O-dimethylserotonin	α,O-DMS	3–5	20
CH_3O-	CH_3-	CH_3-	α,N,O-trimethylserotonin	α,N,O-TMS	10–20	5
CH_3O-	Et-	H-	5-methoxy-α-ethyltryptamine	5-MeO-α-ET	~70	1

A close structural analogue of the neurotransmitter serotonin is seen in α,O-DMS. Serotonin, when administered orally, is not available to the brain for two reasons. It is polar (with the 5-hydroxyl group exposed), and is rapidly deaminated by monoamine oxidase attack. A methyl group on the oxygen decreases the hydrophilicity, and a methyl group on the alpha-position affords metabolic protection, resulting in a very potent, orally active hallucinogen. The drug α-ET was made available commercially by the Upjohn company under the name Monase, and was an effective antidepressant. It was withdrawn due to some side-effects, and shortly thereafter began to be sold on the street market as a substitute for MDMA. It was placed in Schedule I as a hallucinogen by the DEA.

Three related compounds deserve specific mention. The 4-methyl homologue of α-MT, α,4-dimethyltryptamine (α,4-DMT), is orally active in humans at 20mg and produces some feelings of unreality, with neurological toxicity including skin flushing and eye dilation. The 4-hydroxy analogue (4-HO-α-MT) has been observed by some to evoke visual changes, accompanied by dizziness and mild depression. In the 15–20mg range (orally) there is occasional tachycardia, headache, and diarrhea. And the complete relocation of the 3-(2-aminopropyl)-chain of α- MT to the 5-position creates an isomer called 5-IT which, with 20mg orally produces a state of increased heart rate, anorexia, diuresis and slight hyperthermia, all lasting about 12 hours. None of these materials could be called hallucinogens.

Quite a different story follows the inclusion of a methyl group at the indolic 2-position, in both of the two above groups.

With the aryl, N-substituted analogues of DMT, three compounds are of specific interest. These are the three 2-methyl homologues of DMT, 5-MeO-DMT and DET. The first of these, 2,N,N-trimethyltryptamine (2,N,N-TMT) is orally active in the 50 to 100mg range. This is a potency very similar to the prototype DMT, but with this compound there is oral activity. It is as if just the presence of some aliphatic entity in the space between the indole ring and the 3-side-chain, whether it is an alpha-methyl or a 2-methyl, effectively protects the molecule from the monoamine oxidase. A similar protection from oxidative deamination, although at a considerable drop in potency, is seen with the 2-methylation of 5-methoxy-N,N-dimethyltryptamine. This base, 5-MeO-2,N,N-TMT, is orally active in the dose range of 75–150mg, down by a factor of 10 from the 5-MeO-DMT parent. The corresponding homologue of DET is 2-Me-DET. It is orally active with doses in the 80 to 120mg range.

One example of a 2-methylated-alpha-methyltryptamine has been studied in humans. This is 2,alpha-dimethyltryptamine, or 2,α-DMT. It leads to a gentle, pleasurable intoxication with oral doses in the 300–500 mg range.

3.2.2 PSILOCYBIN ANALOGUES

An unusual collection of tryptamines has its origins in the fungal world. These compounds are unique in that they have a 4-indolic hydroxyl group (or an ester group that can be metabolized to a hydroxyl group) and yet they are orally active. The natural prototypes are psilocybin itself, and the two demethylated homologues baeocystin and norbaeocystin. (See Figure 3.12.)

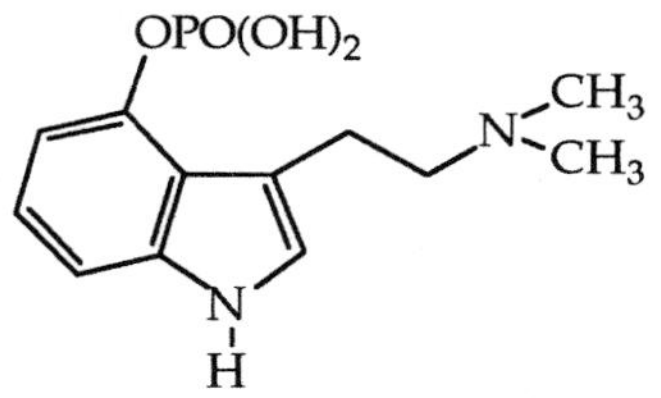

$OPO(OH)_2$

$N(CH_3)_2$

#8 Psilocybin

Figure 3.12

Chemical structure of psilocybin

These compounds have been found in various combinations, along with the dephosphorylated analogue psilocin, in perhaps a hundred species of mushrooms, largely of the *Psilocybin* genus. (See Table 3.22.)

It has now been established that the phosphate group is metabolically removed, revealing the free phenolic compound as the intrinsically active drug in humans. That it should be centrally active at all may involve a structurally

Table 3.22

Psilocybin analogues

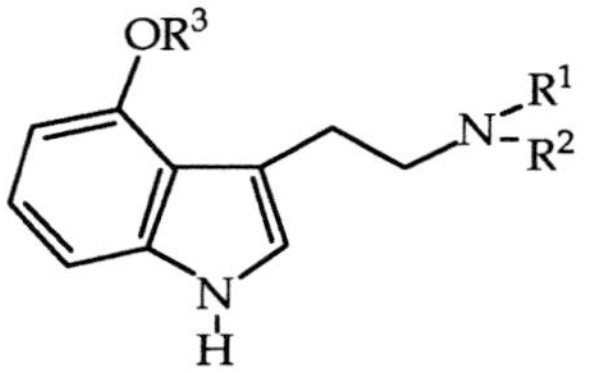

Psilocybin analogues

R^1	R^2	R^3	common name	code	potency (mg)	x-DMT
H-	H-	PO_3H_2	norbaeocystin		?	
H-	CH_3-	PO_3H_2	baeocystin		6–10	10
CH_3-	CH_3-	H-	psilocin	CX-59	7–10	10
CH_3-	CH_3-	PO_3H_2	psilocybin	CY-39	10–15	6
CH_3-	Et-	H-	4-hydroxymethylethyl-T[a]	4-HO-MET	10–20	6
CH_3-	Pr-	H-	4-hydroxymethylpropyl-T	4-HO-MPT	?	
CH_3-	iPr-	H-	4-hydroxymethylisopropyl-T	4-HO-MIPT	12–25	5
CH_3-	iPr-	CH_3-	4-methoxymethylisopropyl-T	4-MeO-MIPT	20–30	3
Et-	Et-	H-	4-hydroxydiethyl-T	CZ-74	10–20	6
Et-	Et-	PO_3H_2	4-phosphoryloxydiethyl-T	CEY-19	15–25	4
Pr-	Pr-	H-	4-hydroxydipropyl-T	4-HO-DPT	?	
iPr-	iPr-	H-	4-hydroxydiisopropyl-T	4-HO-DIPT	15–20	4
iPr-	iPr-	Ac-	4-acetoxydiisopropyl-T	4-AcO-DIPT	6–10	10
Bu-	Bu-	H-	4-hydroxydibutyl-T	4-HO-DBT	>20	?
- $(CH_2)_4$ -		H-	4-hydroxytetramethylene-T	4-HO-pyr-T	>20	?

[a] "T" stands for tryptamine

allowable close association and intermolecular neutralization between the acidic hydroxy group and the basic amine group. The generalization that all ester substituted 4-hydroxytryptamines are saponified before being active may not be valid. The doubled potency of the 4-AcO-DIPT over that of 4-HO-DIPT (in the above table) and the speed of onset (10 to 20 minutes) suggests that it might be absorbed directly from the stomach as the unsaponified ester.

Two compounds defy classification in the above tables, but they were designed for that exact reason – to challenge the structure-activity relationships (SAR) that derive from this data base. They incorporate the 4-oxygen atom of psilocin and the 5-oxygen atom of bufotenine. The compounds are the 4,5-methylenedioxy analogues of DMT and of DIPT. Both of them (4,5-MDO-DMT and 4,5-MDO-DIPT) are of unknown potency in humans, although the latter has been assayed up to 25mg with some activity. This challenge has not yet been answered.

3.3 SIGNIFICANT OTHERS

3.3.1 LSD ANALOGUES

The last large class of hallucinogenic compounds that has been extensively studied as to structure activity relationships is represented by the well-known prototype, LSD. (See Figure 3.13.)

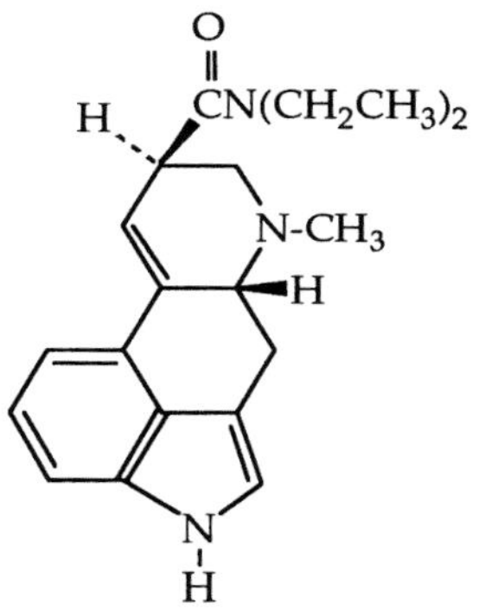

#9 LSD

Figure 3.13

Chemical structure of LSD

This is a synthetic base, derived from the lysergic acid nucleus that is best known from the ergot alkaloids. This family of compounds was originally discovered in the rye fungus and was responsible for the lethal outbreaks of Saint Anthony's Fire in the Middle Ages. (See Table 3.23.)

These compounds are now known to be present in morning glory seeds and helps explain their use in older Western cultures.

A great number of homologues and analogues of LSD have been studied. The most easily available group consists of variations of substituents on the amide group, often accompanied with substituents on the indolic pyrrole ring. No material in this group has shown a potency that exceeds that of LSD itself. It must be remembered that a number of these studies are statistically weak, as the number of subjects was very limited. All of these studies were conducted on compounds with the absolute configuration shown. Prolonged exposure of LSD to basic conditions promotes the isomerization of the carbon atom at the 8-position (that bears the carboxamide function) resulting in the generation of the inactive d-iso-LSD. Trials with the two unnatural diastereoisomers (l-LSD and l-iso-LSD) have been shown to be free of any LSD-like effects at rather

Table 3.23

Amide analogues and pyrrole derivatives of LSD

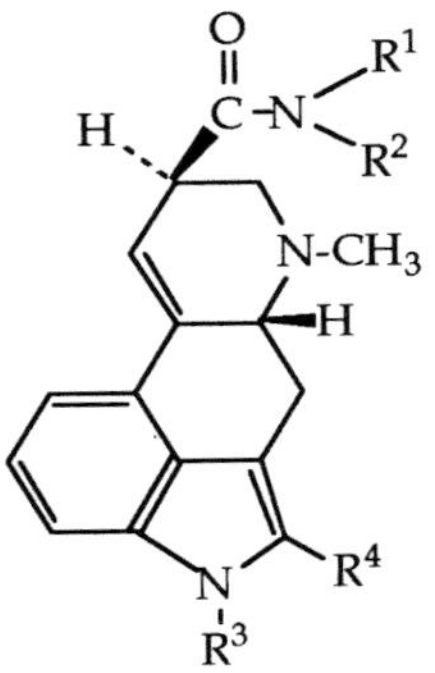

amide analogues and pyrrole derivatives of LSD

R¹	R²	R³	R⁴	common name	code	potency (mg)	potency x-LSD
H-	H-	H-	H-	LA-amide, ergine	LA-111	0.5–1	0.1[a]
Me-	H-	H-	H-	LA-methylamide		~0.5	0.2[a]
Me-	Me-	H-	H-	LA-dimethylamide	DAM-57	0.5–1.2	0.1
Et-	H-	H-	H-	LA-ethylamide	LAE-32	0-5–1.6	0.1
Et-	H-	Ac-	H.	1-acetyl-LA-ethylamide	ALA-10	~1.2	0.1
Et-	H-	Me-	H-	1-methyl-LA-ethylamide	MLA-74	~2	0.05
Et-	Me-	H-	H-	LA-methylethylamide	LME-54	[b]	
Et-	Et-	H-	H-	LA-diethylamide	LSD-25	0.05–0.2	1
Et-	Et-	H-	Br-	2-bromo-LSD	BOL-148	>1	<0.1
Et-	Et-	Ac-	H-	1-acetyl-LSD	ALD-52	0.1–0.2	1
Et-	Et-	Me-	H-	1-methyl-LSD	MLD-41	0.2–0.3	0.3
Et-	Et-	Me-	Br-	l-methyl-2-bromo-LSD	MBL-61	>10	<0.01
Et-	Et-	Me-	I-	l-methyl-2-iodo-LSD	MIL	?[c]	
Pr-	H-	H-	H-	LA-propylamide		>0.5	<0.2
Pr-	Me-	H-	H-	LA-methylpropylamide	LMP	>0.1	<1[d]
Pr-	Et-	H-	H-	LA-ethylpropylamide	LEP-57	[b]	
Pr-	Pr-	H-	H-	LA-dipropylamide		>1	<0.1
Al-	Al-	H-	H-	LA-diallylamide	DAL	>1	<0.1
[f]	H-	H-	H-	ergonovine		10	0.01
[g]	H-	H-	H-	methylergonovine		2	0.05
[g]	H-	Me-	H-	methysergid[e]	UML-491	4–8	0.02
$-(CH_2)_4-$		H-	H-	LA-pyrrolidineamide	LPD-824	~0.8	0.1
$-(CH_2)_4-$		Me-	H-	1-methyl-LA-pyrrolidineamide	MPD-75	>1.6	<0.05
$-(CH_2)_2O(CH_2)_2-$		H-	H-	LA-morpholineamide	LSM-775	0.3–0.6	0.2

Table 3.23 (continued)

"LA" is the abbreviation for lysergic acid

"LSD" is the abbreviation for lysergic acid diethylamide

[a] sedative action or autonomic changes in humans; not hallucinogenic

[b] active but less so than LSD; no numbers available; used in cross-tolerance studies with LSD

[c] often used in human ^{11}C PET scanning; activity unknown

[d] used as a legal decoy against LSD prosecutions

[e] this is the prescription antimigraine drug Sansert; when used at 10× the usual dose, there is an LSD-like intoxication

[f] 1-hydroxy-2-propyl

[g] 1-hydroxy-2-butyl

large dosages. This reinforces the extreme stereoselectivity of the LSD structure, and the isomeric purity of the compounds studied. Exposure of an aqueous solution of LSD to light, especially sunlight with its UV component, causes a rapid loss in potency. Here this is due to the addition of a molecule of water to the 9,10-double bond in ring D. This product is called lumi-LSD and it is also inactive in humans.

A second area of structural modification of the LSD molecule is the homologation of the alkyl group located on the nitrogen atom at the 6-position. These are best called 6-substituted nor-lysergic acid diethylamides, and are synthetically considerably more challenging than the amide homologues. (See Table 3.24.)

Here, apparently, there is considerable indifference to the nature of the alkyl group that is attached to the molecule, suggesting that this would be an ideal location for a radioactive label for PET scanning, with a β-haloethyl group carrying an iodine or a fluorine atom. Earlier work employed the halogen at the pyrrole 2-position which, although easily made, was an inactive molecule, at least as a hallucinogen.

There are several additional hallucinogens which are properly part of this review but which have been unexplored as to structural variations. They are shown in the sections below.

Table 3.24

5-Alkyl homologues of 5-nor-LSD

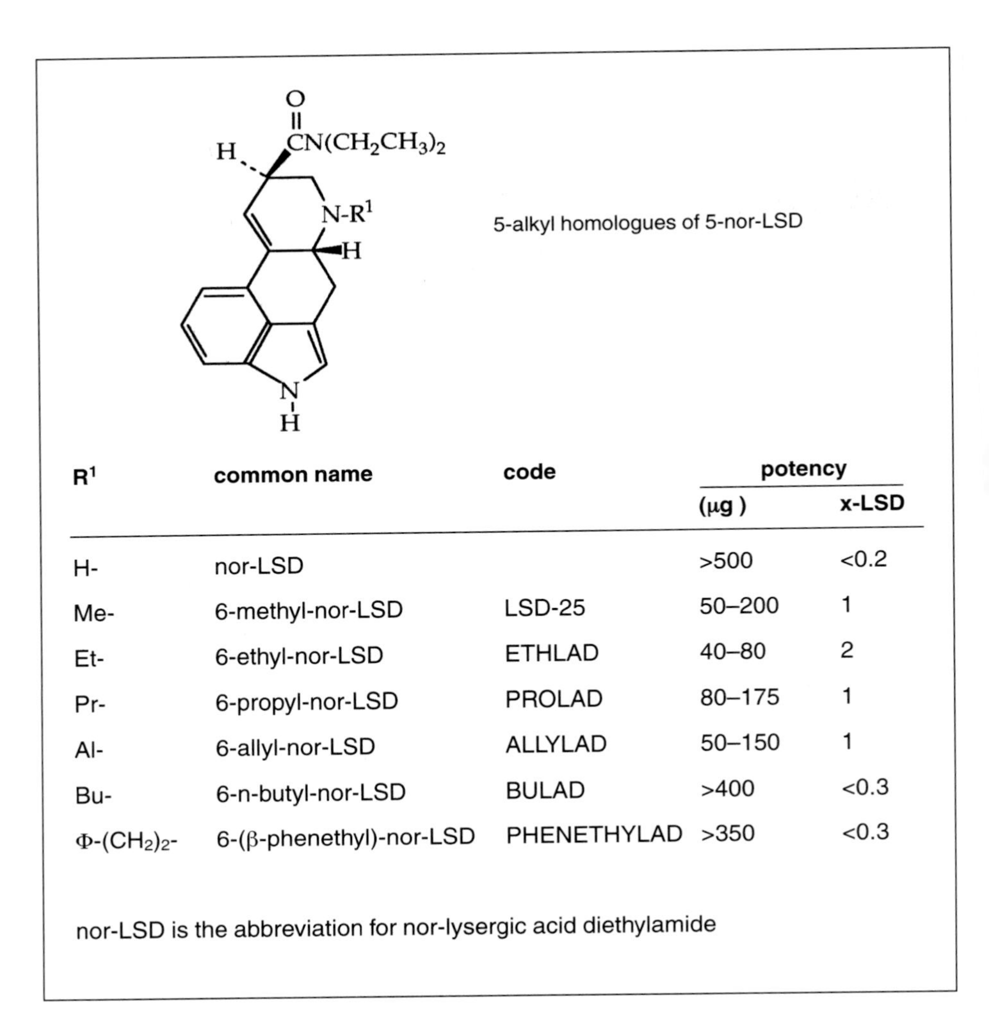

R^1	common name	code	potency (μg)	potency x-LSD
H-	nor-LSD		>500	<0.2
Me-	6-methyl-nor-LSD	LSD-25	50–200	1
Et-	6-ethyl-nor-LSD	ETHLAD	40–80	2
Pr-	6-propyl-nor-LSD	PROLAD	80–175	1
Al-	6-allyl-nor-LSD	ALLYLAD	50–150	1
Bu-	6-n-butyl-nor-LSD	BULAD	>400	<0.3
$\Phi\text{-}(CH_2)_2\text{-}$	6-(β-phenethyl)-nor-LSD	PHENETHYLAD	>350	<0.3

nor-LSD is the abbreviation for nor-lysergic acid diethylamide

3.3.2 IBOGAINE

Ibogaine is a complex alkaloid found as a major component in the African shrub Tabernanthe iboga. It has been drawn in the accompanying diagram in a manner that emphasizes its tryptamine skeleton, and the three-dimensional birdcage ring complex. (See Figure 3.14.)

This cage system contributes one of the elements of chirality to ibogaine that is not obvious to a non-chemist. Looking at it from one of the vertices, the N-E-I order on rotation can be clockwise or counterclockwise. The second chiral center is the carbon that holds the ethyl group. Thus there are two diastereoisomeric forms possible, each with two optical isomers. The positional isomer with the aromatic methoxyl group at the indolic 6-position rather than at the 5-position, is also found in the plant and is called tabernanthine.

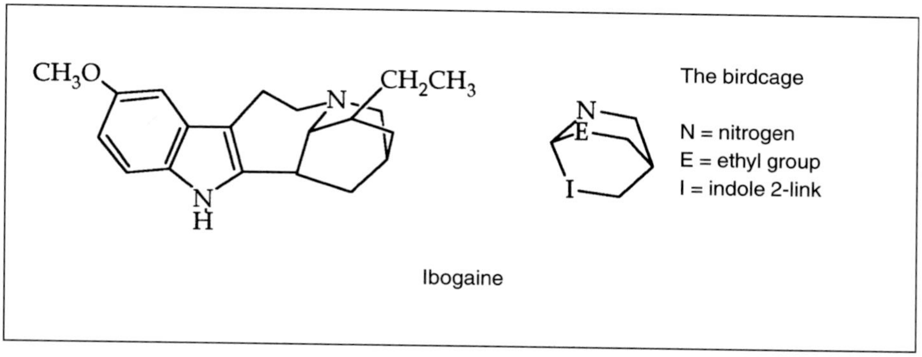

The birdcage

N = nitrogen
E = ethyl group
I = indole 2-link

Ibogaine

Figure 3.14

Chemical structure of ibogaine

Ibogaine is an active hallucinogen in the 400 milligram area and has been clinically studied for the treatment of heroin addiction. In this latter role, the dosages employed may range as high as 1500mg. A primary human metabolism is via O-demethylation to give the free phenol 12-hydroxyibogamine. This metabolite, misnamed nor-ibogaine in the literature, appears to be pharmacologically active in its own right.

The native use of this plant is in Gabon, where it is used in the Bwiti sacramental initiation ceremonies. The material used is called eboka, and is taken from the root of the *Tabernanthe iboga* plant, with the bark scrapings being the richest source. The ritual may last for two or three days with a continuous feeding of the sacrament to the initiate, at the maximum tolerated rate.

3.3.3 AYAHUASCA

This is an ancient decoction made from one or more South American plants. Originally the drink, Ayahuasca (Vine of the Soul) or Yaje, was prepared by the prolonged boiling of the fleshy parts of a Western Amazon liana known as *Banisteriopsis caapi.* The major alkaloids present are harmaline and harmine, both known to be with psychoactive properties. Both of these materials are classified as beta-carbolines. (See Figure 3.15.)

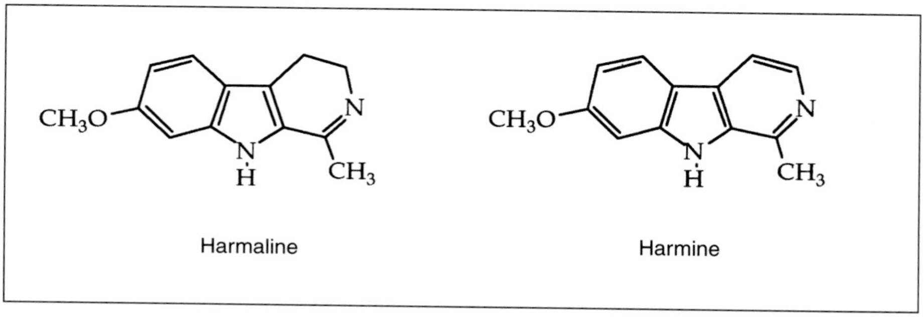

Harmaline

Harmine

Figure 3.15

Chemical structures of the beta-carboline alkaloids of ayahuasca

Harmaline is 7-methoxy-1-methyl-3,4-dihydro-β-carboline and harmine is the totally aromatic 7-methoxy-1-methyl-β-carboline. Their action in humans is one of intoxication with a visual component, but with a considerable burden of nausea and related toxic symptoms. But both compounds are effective monoamine oxidase inhibitors (MAOI) which explains their role in the more complex forms of ayahuasca. Many native tribes have, over the years, developed a pattern of adding additional plant materials to the ayahuasca brew. One of the most fascinating has been the addition of the material chacruna, the leaves of another plant *Psychotria viridis.* This addition greatly enhances the visual and sacramental impact of ayahuasca, without doubt because this latter plant has, as its major alkaloidal constituent, N,N-dimethyltryptamine (DMT). This is the ubiquitous tryptamine base that is not active orally because it is destroyed by the deamination enzymes in the gut. But in the mixture of these two plants, the beta-carboline alkaloids are effective inhibitors of the enzyme system that destroys the DMT. Neither component of the ayahuasca mixture works well alone; in combination they become an effective hallucinogenic agent.

Over the last decade the term "ayahuasca" has come to represent, in the popular jargon of the drug-oriented scene, a mixture of any two things that are, in combination, psychoactive. The choice of each of the two components has became increasingly loose. The caapi component is often replaced by the seeds of the Syrian rue plant, *Peganum harmala.* They contain the harmala alkaloid inventory similar to *Banisteriopsis caapi,* with the addition of several quinazoline alkaloids that may contribute to psychopharmacological differences. Also, both harmine and harmaline are available commercially as fine chemicals, and they have been reported as the components in the inhibitory side of the drink being called ayahuasca. The chacruna side of this combination has been supplied by any of the many botanical individuals, from grasses to bamboos to acacia trees, any of which can provide the DMT component. And there are totally different tryptamines that can serve this role. The base 5-methoxytryptamine (5-MeO-DMT) can be gotten from many sources such as the secreted toxins of the desert toad *Bufo alvarius,* to the catalogs offering fine chemicals from any of several chemical supply houses. I have analyzed one street sample that showed, by GCMS, only two components. It was a mixture of pure harmine and pure 5-MeO-DMT. As neither of these two chemicals is included in the Scheduled Drug listings, this combination is completely legal. The term "ayahuasca" has now come to describe any two individual things that, in combination, evoke a psychotropic state.

3.3.4 TETRAHYDROCANNABINOL

The final two compounds illustrated here are unusual in that they do not contain the nitrogen atom, an unusual property for a centrally active drug.

Tetrahydrocannabinol is the major active component of *Cannabis sativa*, or marijuana, and is psychoactive in the area of 10mg. It is the major active component within the tissues of the plant itself. The resinous material around the blossom area of the female plant is especially rich in THC, and the collection of this resin produces a potent form of marijuana known as hash, a term derived from the ancient term, hashish. Since THC cannot be isolated from the marijuana plant without considerable sophistication, the only practical source of it is through chemical synthesis. (See Figure 3.16.)

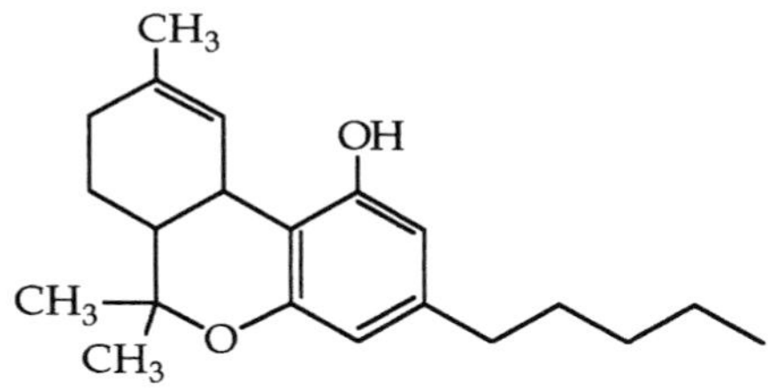

Figure 3.16

Chemical structure of tetrahydrocannabinol

This product is the prescription drug Marinol (l-Δ1-THC) which has been approved by the authorities for use as an anti-emetic. There are many dozens of related compounds in the plant itself, and all of them are potential pharmacological contributors to the action of the total plant. The single component THC is probably not the sole active component. Many structural variations of THC have been studied, some in the rabbit, some in the dog, and few in humans. Variations on the 5-pentyl group in the resorcinol half of the molecule have led to drugs of both legal and pharmacological interest. The 5-hexyl homologue is Parahexyl or Synhexyl, and it was brought into an illegal status because there was no evidence for its safety. The dimethyl-heptyl homologue is the most potent of these analogues, and was explored as a possible weapon in the military chemical warfare research studies.

There are a number of metabolic products known, but they have been used largely to confirm the illicit use of marijuana by urine tests. THC is an unusually lipophilic compound, and so tends to deposit in the fat tissues of the body. This removes it from availability for biological action, but it does provide a reservoir of "active" chemical that can only be excreted over a long time period. This means that there is a wait of several weeks for the clearance of sequestered THC from the body, long after all psychopharmacological effects have disappeared. In urine testing for drug impairment, there can well be a positive result from a person who is not under the influence of the drug at the time of the urine sampling. The psychological effect of THC spans a few hours. The presence of urinary THC spans a few weeks.

Hemp is the bulk fiber material of the plant *Cannabis sativa*, and has many commercial uses. It is a fiber, it is a cloth. It is the primary material for our cigarette paper and our $20 bills; and for the paper upon which the US Constitution was written.

3.3.5 SALVINORIN-A

The Mazatec Indians in northern Oaxaca have used for centuries a mint plant as a healing medicine and a sacred sacrament. The plant has the botanical taxon *Salvia divinorum* (Hojas de la Pastora) and is apparently raised as a cultigen as it has never been observed in the wild. The major active component is a diterpene, a neoclerodane, and is usually absorbed from the tissues of the mouth. It is not orally active, but the dried leaves can be smoked. (See Figure 3.17.)

Figure 3.17

Chemical structure of salvinorin A

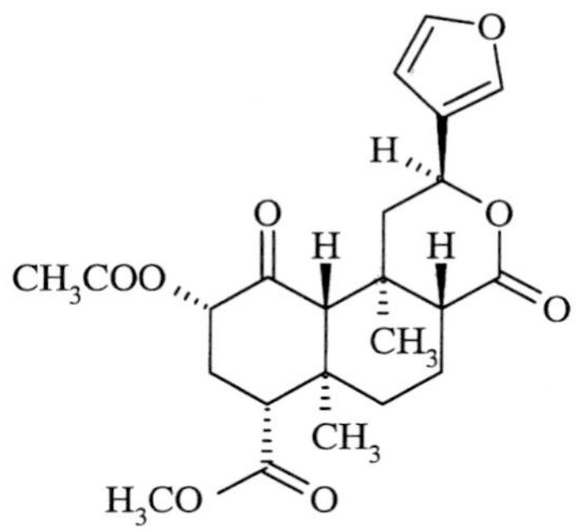

The pure, isolated active compound Salvinorin A is parentally active at a dose of less than a milligram. As of the present time, the mechanism of action and the metabolic fate in humans is totally unexplored. But with the commercial availability of the plants, the extracts, and the pure Salvinorin A itself, clinical research with this most potent of all natural hallucinogenic drugs will be done.

This has been little more than a touching of the surface of the world of the hallucinogens. There are many plants that have not yet received sufficient study to be included in any compilation such as this, plants that may be intoxicants or delusional agents, but which would be hallucinogenic in some people's definition. There are hallucinogenic components described in many synthetic analgesics or anesthetics. The pharmacological classification remains forever unsatisfactory as this particular property is one that demands the human animal for its identification.

3.4 FEDERAL US DRUG LAW

3.4.1 HISTORY

There have been two broad Federal drug laws, the Harrison Narcotics Act, enacted in 1914, which was succeeded by the Controlled Substances Act of 1970. The first drug that received Federal attention was opium. Along with the importation of tens of thousands of Chinese people as inexpensive labor for the construction of the trans-continental railroads, came the importation of smoking opium. Opium, and its major active alkaloid morphine, were medically approved and widely use in the United States, for pleasure as well as pain relief. It was available in over-the-counter medicines. But it was with the connection of the smoking custom, the opium dens, with the Chinese unemployment problem, that focused the legal attention more to a racial issue than to a drug issue. In the 1870s a number of cities passed ordinances, States passed laws forbidding the smoking of opium and outlawing opium dens. When these failed, the Federal Government passed a law prohibiting the importation of opium into the United States by any citizen of China. Importation from Canada was still allowed.

Over the next two decades, there was the formation of an Opium Commission (1903) to study the steadily increasing levels of opium and of coca leaves, and they recommended more legal action. A law was passed in 1909 prohibiting any importation of opium for other than medical purposes. A Hague Convention in 1912 concluded that controls should also be placed on domestic sales. This set the stage for the Harrison Narcotics Act.

In 1914, while the initial actions were taking place in Europe that exploded into World War I, the US Congress was arguing the wording of the Federal drug law. On 17 December 1914, President Wilson signed U.S.C. 4701, the Federal Narcotics Internal Revenue Regulations, commonly called the Harrison Narcotics Act. The measure was a tax and registration measure, not a prohibition one. It was carefully worded so as not to interfere with the practice of medicine. The goal was to register the dispensers of the narcotic drugs (mainly opium, morphine, heroin and cocaine) for the fee of $1 a year so as to allow orderly marketing. The law actually stated, "Nothing . . . in this section shall apply . . . to the dispensing and distribution of any of the aforesaid drugs to a patient by a physician, dentist, or veterinary surgeon registered under this Act in the course of his professional practice only."

There was a strange transition that occurred over the next six years that was intimately tied in with the proposal and passage of the Volstead Act and the ratification of the alcohol prohibition 18th Amendment, both in 1919. During this period there was a national change in attitude towards drugs. The Supreme Court (in 1916, US *v.* Jin Fuey Moy) heard a case where the enforcement authorities claimed that addiction was not a disease and therefore a physician is violating the law by prescribing a drug to support that addiction. The court decision went against the Revenue police with the comment that such an interpretation of the Act would be unconstitutional. Just three years later, in an almost identical case (US *v.* Web) the same court found that a physician doing the same thing should, according to the law, be criminally prosecuted. The Volstead Act gave the Internal Revenue Service a Prohibition Unit, with a subunit that was called the Narcotics Division. Opium and cocaine prohibition fell under the purview of the IRS. In 1929, President Coolidge signed the Porter Bill which established the first Narcotics Farms for the confinement and treatment of Federal prisoners who were addicts.

In June 1930, Congress established the Bureau of Narcotics, still within the Department of Treasury, and Herbert Hoover appointed Henry J. Anslinger to head the Bureau. The reefer madness era began with claims that marijuana caused madness and criminal behavior, that all kids would start using it. President Roosevelt signed the Marihuana Tax Act of 1937. So by an act of Congress, marijuana had become a narcotic.

For the next 25 years, a number of minor laws were passed, all designed to curb illicit production, increase penalties and to broaden the strength and the scope of the Bureau of Narcotics' power.

In 1962, a landmark bill was passed into law. This was the Racketeering and Corrupt Organizations Act (RICO, 18 U.S.C. 1962). Its purpose was to eliminate the infiltration of organized crime into legitimate organizations. There was no indication that this law had been specifically intended to address criminal activities such as drug trafficking. In 1970, Congress modified the RICO law to insert criminal forfeiture provisions. Although there is a rich history of civil forfeiture involved in proceedings against property which has been involved in some criminal action, the act of criminal forfeiture involves seizure of property of a person convicted of a felony. It was a common penalty in historic England but it was specifically prohibited in 1790 by the first Congress of the United States. As a result, criminal forfeitures were unknown in the United States for 180 years. This amendment to the RICO statute, and the Continuing Criminal Enterprise section of the Controlled Substance Act (see below) are the first inclusions of this penalty, ever, in American history. The need of a conviction before seizure was dropped in 1978 (see below).

In the mid-1960s some new developments occurred in the social scene. There was an ever-increasing protest to the Vietnam War, there was a musical and philosophical rebellion characterized by the "Summer of Love," and there were the hippies. And with them came a cascade of new drugs, stimulants and hallucinogens, certainly very little in the opium and cocaine narcotics world. Marijuana was everywhere, yes, but just about everything else was unknown to the law. The law-makers responded to this new crisis with the introduction of the Drug Abuse Control Amendments of 1965. These were approved by Congress and on 1 February 1966, there was established within the Food and Drug Administration, a Bureau of Drug Abuse Control (BDAC) under the direction of John H. Finlator. Suddenly the FDA was empowered to become a law enforcement police power. They put out a monthly newsletter called the *FDA Papers*. The flood of hallucinogens, as they were called, were being placed in the BDAC illegal drug listings. The rivalry became quite intense.

Territorial battles were routine. Some semblance of order came from a Presidential Order from Lyndon Johnson on 8 April 1968, dictating a plan to merge the BDAC group (in the FDA) with the BN group (still in the Department of the Treasury). This new group was named the Bureau of Narcotics and Dangerous Drugs (BNDD) and they were to answer to the Department of Justice.

In 1970, the passage of the Comprehensive Drug Abuse Prevention and Control Act, known as the Controlled Substances Act of 1970, effectively removed the Harrison Narcotics Act from the books, after a life of some 55 years. This new law, which is still current and in effect today, dictates that all illegal drugs are to be placed in one of five Schedules. These are to classify drugs with a high abuse potential to drugs with a low abuse potential. The definitions of the five Schedules follow from page 116.

Schedule I:

(a) The drug or other substance has a high potential for abuse,

(b) The drug or other substance has no currently accepted medical use in treatment in the United States,

(c) There is the lack of accepted safety for the use of the drug or other substance under medical supervision.

Schedule II:

(a) The drug or other substance has a high potential for abuse,

(b) The drug or other substance has a currently accepted medical use in treatment in the United States or a currently accepted medical use with severe restrictions,

(c) Abuse of the drug or other substance may lead to severe psychological or physical dependence.

Schedule III:

(a) The drug or other substance has a potential for abuse less than the drugs or other substances in Schedules I or II,

(b) The drug or other substance has a currently accepted medical use in treatment in the United States,

(c) Abuse of the drug or other substance may lead to moderate or low physical dependence or high psychological dependence.

Schedule IV:

(a) The drug or other substance has a low potential for abuse relative to the drugs or other substances in Schedule III,

(b) The drug or other substance has a currently accepted medical use in treatment in the United States,

(c) Abuse of the drug or other substance may lead to limited physical dependence or psychological dependence relative to the drugs or other substances in Schedule III.

Schedule V:

(a) The drug or other substance has a low potential for abuse relative to the drugs or other substances in Schedule IV,

(b) The drug or other substance has a currently accepted medical use in treatment in the United States.

(c) Abuse of the drug or other substance may lead to limited physical dependence or psychological dependence relative to the drugs or other substances in schedule IV.

Note that there is no place to put a drug that has no currently accepted medical use but which has a low potential for abuse. There is a listing of all currently scheduled hallucinogenic drugs later in this section, along with a tabulation of all listed chemicals that are legally associated with the manufacture of scheduled drugs.

There are two aspects of the Controlled Substances Act that deserve special mention, as they have become quite important in the last few years. The Continuing Criminal Enterprise (CCE) section was mentioned above, in the RICO discussion. A person is defined as engaging in a continuing criminal enterprise if he or she is the organizer or supervisor of a group of five or more persons who have obtained substantial income through a series of violations of this Act. He or she is not only subject to more severe penalties, but shall also forfeit both profits and properties associated with his or her felonious acts. A second aspect deals with the no-knock entry issue. Following the legal issuance of a search warrant, an authorized agent may enter, forcibly and without warning, if he or she feels that evidence might otherwise be disposed of, or if the warrant has allowed such action to be taken.

In March 1972, Congress passed the Drug Abuse Office and Treatment Act, which was dedicated to bringing about a reduction of the incidents of drug abuse within the shortest period of time. It called for the formation, within the Executive Office, of a Special Action Office for Drug Abuse Prevention (SAODAP). It also called for the establishment of a National Council for Drug Abuse Prevention which would provide for the creation of a National Drug Abuse Training Center. There would also be the creation of a National Institute on Drug Abuse (NIDA). The following year (1973), the Drug Enforcement Administration (DEA) was created by the merging of three groups: the Office of Drug Abuse Law Enforcement (ODALE, which was created in the Department of Justice in 1972), the Office of National Narcotics Intelligence (ONNI, also created in the Department of Justice in 1972) and the Bureau of Narcotics and Dangerous Drugs (BNDD, created from the merger of the Bureau of Narcotics, BN, and the Bureau of Drug Abuse Control, BDAC, in 1968). The drug enforcement and intelligence functions of the US Customs Service were also brought into the merger. The DEA answers to the Department of Justice.

Congress passed the Psychotropic Substance Act of 1978 as an amendment to the Controlled Substance Act, which dealt mainly with increased penalties for phencyclidine (PCP) and precursors such as piperidine. The very last section of the amendment is entitled "Forfeiture of Proceeds of Illegal Drug Transactions." This addition states, quite simply, that all proceeds from drug transactions may be seized as forfeiture. This has given the government immediate possession of boats, airplanes, real estate property and bank accounts. This asset seizure and forfeiture (before conviction) has been felt by many law enforcement groups to have proven itself to be an effective weapon in the area of drug use prevention.

The Posse Comitatus Act was passed into law just after the Civil War. It was a strategy to keep Federal soldiers from interacting with the civilians in the South, and the specific prohibition was that no military forces can be involved in the enforcement of civil law. The "Department of Defense Authorization Act of 1982" includes a provision that revises the Posse Comitatus statute. Prior to this law, any military involvement in civil law was prohibited unless authorized by the Constitution or by some specific action of Congress. With this law there was a clarification of the role of the military in civilian enforcement activities, and the assistance and support services which may be rendered by the military to law enforcement were defined. There was quick implementation in the formation of the President's Task Force South Florida in January, 1982. This operation was geared to the interdiction of narcotics being smuggled into Florida from the Caribbean and from Latin America. The military aid provided included complex logistic and vessel support, aviation and radar surveillance, and the loan of equipment and facilities. This "allowing" of military participation in drug law matters was further extended in 1986 and in 1989. On 8 April 1986, President Reagan signed a National Security Decision Directive stating that drug trafficking constituted a threat to the national security of the United States. This authorized the military to participate in all international law enforcement activity that was drug related, except for making seizures and arrests. These latter two restrictions were removed by an opinion published by the Department of Justice, on 3 November 1989.

The release of information concerning income tax returns and related financial records has been, at least up until 1982, severely restricted. In September of that year, the "Tax Equity and Fiscal Responsibility Act of 1982" opened everything up. Complex financial transactions are often associated with sophisticated criminal activity, and this information was not available to Federal law enforcement authorities. The new law included several provisions sought by the Department of Justice "to facilitate the appropriate disclosure of tax information to Federal law enforcement agencies for criminal investigative purposes while maintaining safeguards needed to protect the privacy of innocent citizens." Federal officials may now gain access to tax information which is a most valuable source of financial data necessary to prosecute narcotics trafficking and organized crime.

The "Comprehensive Crime Control Act of 1984" is commonly called the Emergency Scheduling Act. The conventional way for the DEA to place a new drug into the CSA has always been to publish an announcement of this intention in the Federal Register, and open a 60-day window, during which time all interested parties could request an input at public hearings. If there were no offended parties, the drug became scheduled in 60 days. If there were to be hearings and discussion, the delay would be longer. This new law is designed to put an immediate legal control on a new drug, before the hearings are requested and scheduled. The window authorized is one year, extendible by another six months, and at the end of this time, the drug enters its appropriate scheduled position. But during this entire period, it has the status of a scheduled drug and is, of course, illegal. It achieved that "illegal" status immediately with the first announcement in the Federal Register.

An amendment was made to the Controlled Substance Act which was originally called the Scheduled Drug Analogue Bill, but which finally evolved as the "Designer Drug Enforcement Act of 1986." It was initially inspired by the appearance on the street of several analogues of the drug Fentanyl; in its final form it was considerably broader. One of the strengths of the CSA was that it was explicit. It stood apart from many drug laws of other countries in that the drugs to be considered as being covered by the law were precisely named. True, there was an occasional "all possible isomers" but when there can only be six, there is an exact and unambiguous definition of just what structures were covered by the law. This explicitness was lost in 1986 with the introduction of the Analogue Drug Bill. The process of scheduling drugs changed from a list format, which was cumbersome with respect to meeting new and evolving trends, to a structure and activity format which provides a catch-all whether or not a compound related to a scheduled one has any abuse liability. Here is an exact transcription of the text of the law:

PUBLIC LAW 99-570
CONTROLLED SUBSTANCE ANALOGUE ENFORCEMENT ACT OF 1986

(32) (A) Except as provided in subparagraph (B), the term "controlled substance analogue means a substance –
(i) the chemical structure of which is substantially similar to the chemical structure of a controlled substance in Schedule I or II;
(ii) which has a stimulant, depressant, or hallucinogenic effect on the central nervous system that is substantially similar to or greater than the stimulant, depressant, or hallucinogenic effect on the central nervous system of a controlled substance in Schedule I or II; or
(iii) with respect to a particular person, which such person represents or intends to have a stimulant, depressant, or hallucinogenic effect on the central nervous system that is substantially similar to or greater than the stimulant depressant, or hallucinogen effect on the central nervous system of a controlled substance in Schedule I or II.

(B) Such term does not include –
(i) a controlled substance;

(ii) any substance for which there is an approved new drug application;
(iii) with respect to a particular person any substance, if an exemption is in effect for investigational use, for that person, under section 505 of the Federal Food, Drug, and Cosmetic Act (21 U.S.C. 355) to the extent conduct with respect to such substance is pursuant to such exemption; or
(iv) any substance to the extent not intended for human consumption before such an exemption takes effect with respect to that substance.

TREATMENT OF CONTROLLED SUBSTANCE ANALOGUES
SEC: 203. A controlled substance analogue shall, to the extent intended for human consumption, be treated for the purpose of this title and title III as a controlled substance in Schedule I.

Some comments are appropriate concerning this analogue drug law – points that may not be immediately apparent from the first reading. Let us assume that the drug in question is the correctly identified, but completely legal drug, ABC.

Many people, lawyers and defendants alike, come to the conclusion that if a drug fits the description of being an analogue, then that drug becomes a Schedule I drug. This is absolutely not so. If, in a situation where a case has gone to trial and the jury accepts that drug ABC is indeed an Controlled Substance Analogue of a Schedule I or II , then in that court case, ABC is treated as a Schedule I drug. But if, in an entirely separate situation, there is a criminal charge dealing with ABC and the Controlled Substance Analogue Enforcement Act is the basis of the prosecution's case, ABC is a virgin again. It is not a Controlled Substance Analogue because of the other trial. It is not a Schedule I drug. The establishment of it as a Controlled Substance Analogue must start all over from scratch. This law does not make drugs into Schedule I drugs; however, it does permit the control of "designer" compounds on a case-by-case basis.

Within the areas of chemical structure, "analogue" is a chemical term meaning, quite loosely, something that looks pretty much the same as something else. Unlike the rather exact chemical terms such as isomer (same weight), isotope (same atom different mass) or homologue (one carbon different), the chemical term analogue is quite loose and means similar. But the legal term "Controlled Substance Analogue" is rigidly defined by the criteria listed above and as such "Analogue" and "Controlled substance Analogue" are not interchangeable.

In general, a new drug law, usually as an amendment of the Controlled Substances Act, has been enacted every even-numbered year. Usually they add new drugs to the schedules and increase the penalties. In 1988, there was enacted the Chemical Diversion and Trafficking Act which brought into the legal record a listing of precursor chemicals and of essential chemicals. These were reclassified as List I and List II a few years later, and the current status of both lists is detailed below. In the same Act, there was the Anti-Drug Abuse Act amendment to the CSA that broadly increased the civil penalties associated with drug-related convictions, including the withdrawal of Federal benefits, the cancellation of FHA mortgages and student loans, and the suspension of drivers' licenses. This law also led to the creation of the Office of National Drug Control Program (ONDCP) which is now best known as the office of the Drug Czar. The mission of ONDCP was to coordinate the anti-drug efforts of the various agencies and departments of the Federal government, to consult with States and local agencies and assist their anti-drug efforts, to conduct a national media campaign, and to annually promulgate the National Drug Control Strategy.

The intended goal of the Comprehensive Methamphetamine Control Act of 1996 was to increase the penalties for the manufacture and trafficking in methamphetamine and other Scheduled drugs, and to increase regulatory controls. In this latter area, there was the establishment of yet another list of industrial chemicals, called a Special Surveillance, intended to alert chemical suppliers of possible misuse of the products that they sell, and to suggest that they may be held responsible if these products are misused. This list is given below (p. 122).

3.4.2 LISTED CHEMICALS

This is the current inventory of listed chemical (as of October 2000). As several of them are precursors to illegal drugs other than the hallucinogens, the target drugs are listed under the comments. Also included there are incidental bits of information that may be of value to the forensic chemist.

Prior to this listing process that was called for by the Chemical Diversion and Trafficking Act of 1988, a number of known precursors were recognized and were entered into the CSA Schedules. Although these are not drugs *per se*, they have remained in the schedules under a drug status, rather than having been relocated to the List I collection. They are tallied here for reference purposes.

SCHEDULED PRECURSORS WITH ILLEGAL DRUG STATUS

Precursor	*Commentary*
Lysergic acid	This is listed as a Schedule III depressant. It is not pharmacologically active but it is the immediate hydrolysis product of most of the ergot alkaloids, and is an immediate starting material for the synthesis of LSD, a Schedule I hallucinogen.
Lysergic acid amide	This is listed as a Schedule III depressant. It has a sedative amide action in humans, and is a documented component of morning glory seeds. On hydrolysis it gives rise to lysergic acid, and is an immediate starting material for the synthesis of LSD, a Schedule I hallucinogen.
Phenylacetone	This is listed as a Schedule II immediate precursor. It is not pharmacologically active, but it is an immediate precursor for the synthesis of amphetamine and methamphetamine, both Schedule II stimulants.
1-Phenylcyclohexylamine	This is listed as a Schedule II immediate precursor. It is not pharmacologically active, but it is an immediate precursor for the synthesis of phencyclidine (PCP), a Schedule II depressant.
1-Piperidinocyclohexane carbonitrile (PCC)	This is listed as a Schedule II immediate precursor. It is not pharmacologically active, but it is an immediate precursor for the synthesis of phencyclidine (PCP), a Schedule II depressant.

In Sec. 1300.02 (18) of the CSA, is the legal definition of List I chemicals. "The term List I chemical means a chemical specifically designated by the Administrator in Sec. 1310.02(a) of this chapter that, in addition to legitimate uses, is used in manufacturing a controlled substance in violation of the Act and is important to the manufacture of a controlled substance." These are entered below, valid as of October, 2000, with appropriate commentary.

LISTED CHEMICALS: LIST I – CHEMICAL PRECURSORS

Precursor	*Commentary*
(A) Anthranilic acid, its esters, and its salts	This is a precursor to N-acetylantranilic acid (also a listed chemical) which is, in turn, a precursor to either mecloqualone (a Schedule I depressant) or methaqualone (also a Schedule I depressant.)
(B) Benzyl cyanide	This compound in several steps is a precursor to phenylacetone, a listed Schedule II immediate precursor. See above.
(C) Ephedrine, its salts, optical isomers, and salts of optical isomers	This compound is an immediate precursor to methamphetamine, a Schedule II stimulant, using hydroiodic acid as the reducing agent. Phosphorous can be used to regenerate the hydroiodic acid, and elemental iodine can be used to initially generate the hydroiodic acid.
(D) Ergonovine and its salts	This compound is an intermediate to lysergic acid, an immediate precursor to Lysergic acid, a Schedule III depressant and precursor to LSD.
(E) Ergotamine and its salts	This compound is an intermediate to lysergic acid, an immediate precursor to Lysergic acid, a Schedule III depressant and precursor to LSD.
(F) N-Acetyl anthranilic acid, its esters, and its salts	This compound is mentioned under anthranilic acid (A above) as an immediate precursor to either mecloqualone (a Schedule I depressant) or methaqualone (also a Schedule I depressant).
(G) Norpseudo-ephredine, its salts, optical isomers,	This compound is an immediate precursor to amphetamine, a Schedule II stimulant, using hydroiodic acid as the reducing agent. Phosphorous can be used to regenerate the hydriodic acid

and salts of optical isomers	and elemental iodine can be used to initially generate the hydroiodic acid.
(H) Phenylacetic acid, its esters and its salts	This compound is a precursor to phenylacetone, a listed Schedule II immediate precursor. See above.
(I) Phenylpropanolamine, its salts, optical isomers, and salts of optical isomers	This compound is an immediate precursor to amphetamine, a Schedule II stimulant, and to 4-methylaminorex, a Schedule I stimulant.
(J) Piperidine and its salts	This compound is in several steps a precursor to 1-piperidino cyclohexane carbonitrile, the Schedule II immediate precursor mentioned above, and itself an immediate precursor for the synthesis of phencyclidine (PCP), a Schedule II depressant.
(K) Pseudoephedrine, its salts, optical isomers, and salts of optical isomers	This compound is an immediate precursor to methamphetamine, a Schedule II stimulant, using hydroiodic acid as the reducing agent. Phosphorous can be used to regenerate the hydroiodic acid, and elemental iodine can be used to initially generate the hydroiodic acid.
(L) 3,4-Methylenedioxyphenyl-2-propanone	This compound is an immediate precursor to MDA, MDMA, MDE and N-hydroxy-MDA, all Scheduled I hallucinogens.
(M) Methylamine	This compound is an immediate precursor to methamphetamine (a Schedule II stimulant) and to MDMA (a Schedule I hallucinogen).
(N) Ethylamine	This compound is an immediate precursor to ethylamphetamine (a Schedule I stimulant) or MDE (a Schedule I hallucinogen).
(O) Propionic anhydride	This compound is an immediate precursor to fentanyl (a Schedule II opiate) and ketobemidone (a Schedule I opiate).

(P) Isosafrole	This compound is in several steps a precursor to 3,4-methylene-dioxyphenyl-2-propanone (L, above) as an immediate precursor to MDA, MDMA, MDE or N-hydroxy-MDA, all Schedule I hallucinogens.
(Q) Safrole	This compound is in several steps a precursor to 3,4-methylene-dioxyphenyl-2-propanone (L, above) as an immediate precursor to MDA, MDMA, MDE or N-hydroxy-MDA, all Schedule I hallucinogens.
(R) Piperonal	This compound is in several steps a precursor to 3,4-methylene-dioxyphenyl-2-propanone (L, above) as an immediate precursor to MDA, MDMA, MDE or N-hydroxy-MDA, all Schedule I hallucinogens.
(S) N-Methyl-ephedrine	This compound is an immediate precursor to N,N-dimethyl-amphetamine, a Schedule I stimulant.
(T) N-Methyl-pseudoephedrine	This compound is an immediate precursor to N,N-dimethyl-amphetamine, a Schedule I stimulant.
(U) Hydroiodic acid	This compound is a frequently employed reducing reagent used in the synthesis of methamphetamine, a Schedule II stimulant.
(V) Benzalde-hyde	This compound is in several steps a precursor to phenylacetone, a Schedule II immediate precursor (see above) to methamphetamine (a Schedule II stimulant).
(W) Nitroethane	This compound is in several steps a precursor to phenylacetone, a Schedule II immediate precursor (see above) to methamphetamine (a Schedule II stimulant).
(X)	Any salt, optical isomer, or salt of an optical isomer of the chemicals listed in subparagraphs (M) through (U) of this paragraph.

In Sec. 1300.02 (19) of the CSA, is the legal definition of List II chemicals. "The term List II chemical means a chemical, other than List I chemical, specifically designated by the Administrator in Sec. 1310.02(b) of this chapter that, in addition to legitimate uses, is used in manufacturing a controlled substance in violation of the Act." These are entered below, valid as of October 2000, with appropriate commentary. It must be noted that these compounds on List II are widely used in industry as solvents or reagents widely used in research. Their gathering here is due to the fact that they have been used in the manufacture of illegal drugs as well.

LISTED CHEMICALS: LIST II – ESSENTIAL CHEMICALS

Essential chemical	*Commentary*
(A) Acetic anhydride	This compound is used in the conversion of morphine (a Schedule II opiate) to heroin (a Schedule I opium derivative); of anthranilic acid to N-acetyl anthranilic acid (both List I compounds) and can be used in the conversion of phenylacetic acid (I, List I, see above) to phenyl-2-propanone (a Schedule II immediate precursor, see above). It is a common laboratory chemical.
(B) Acetone	This compound is used in the conversion of cocaine base to cocaine hydrochloride, and in the purification of morphine base. It is a common laboratory solvent.
(C) Benzyl chloride	This chemical is used in the synthesis of phenyl-2-propanone, a Schedule II immediate precursor, see above. It is a common laboratory chemical.
(D) Ethyl ether	This chemical is used in the conversion of cocaine to cocaine hydrochloride, a Schedule II substance of vegetable origin. It is a common laboratory solvent.
(E)	Originally hydriodic acid. Moved from List II to List I in 1990.
(F) Potassium permanganate	This chemical is used to purify cocaine, a Schedule II substance of vegetable origin, and to oxidize ephedrine (List I, C) to methcathinone, a Schedule I stimulant. This is a common laboratory chemical.
(G) 2-Butanone	This chemical is used in the conversion of cocaine base to cocaine hydrochloride, a Schedule II substances of vegetable origin. It is a common laboratory solvent.

(H) Toluene	This chemical is used in the conversion of cocaine base to cocaine hydrochloride, a Schedule II substances of vegetable origin. It is a common laboratory solvent.
(I) Iodine	This chemical is used as a source of hydroiodic acid, a List I chemical. It is a common laboratory chemical.
(J) Hydrochloric gas.	This chemical, more often referred to as hydrogen chloride gas, is used as a reagent in the conversion of organic bases to their water-soluble hydrochloride salts. It is a common laboratory chemical.

The present status of the Special Surveillance List authorized by the Comprehensive Methamphetamine Control Act of 1998 is given here. This Act provides for a civil penalty of not more than $250,000 for the distribution by a business of a chemical to anyone who uses, or tries to use it in synthesizing either a controlled drug or a Listed chemical. The penalties would apply if the seller showed a "reckless disregard" of the possible illegal uses that their product could be put to. A number of chemical supply houses now will not sell any of the following, even to industrial and academic institutions, simply because of the increased paper work required.

SPECIAL SURVEILLANCE LIST

Ammonia gas
Ammonium formate
Bromobenzene
1,1-Carbonyldiimidazole
Cyclohexanone
1,1-Dichloro-1-fluoroethane (e.g. Freon 141B)
Diethylamine and its salts
2,5-Dimethoxyphenethylamine and its salts
Formamide
Hypophosphorous acid
Lithium metal
Lithium aluminum hydride
Magnesium metal (turnings)
Mercuric chloride
N-Methylformamide
Organomagnesium halides (Grignard reagents) e.g. ethylmagnesium bromide and phenylmagnesium bromide

Phenylethanolamine and its salts
Phosphorus pentachloride
Potassium dichromate
Pyridine and its salts
Red phosphorus
Sodium dichromate
Sodium metal
Thionyl chloride
ortho-toluidine
Trichloromonofluoromethane (e.g. Freon-11, Carrene-2)
Trichlorotrifluoroethane (e.g. Freon 113)

There are four equipment items on this list; hydrogenators, tableting machines, encapsulating machines, and 22 liter heating mantels. A previous anti-methamphetamine law had already made the possession of a three-neck round bottom flask illegal, if there is the intent to manufacture methamphetamine in it.

3.4.3 SCHEDULE I HALLUCINOGENS

Here is listed the full Schedule I inventory of Hallucinogenic Drugs with the exact printed name or names given. The primary printed name is followed by the DEA Controlled Substance Code Number. After a few of the 31 entries there is added detail and commentary, that will help flesh out the definition of unusual words and give some of the associated history that is involved.

> Sec. 1308.11(d) Hallucinogenic substances. Unless specifically excepted or unless listed in another Schedule, any material, compound, mixture or preparation, which contains any quantity of the following hallucinogenic substances, or which contains any of its salts, isomers, and salts of isomers whenever the existence of such salts, isomers, and salts of isomers is possible with the specific chemical designation (for purposes of this paragraph only, the term "isomer" includes the optical, position (sic) and geometric isomers):
>
> (1) Alpha-ethyltryptamine 7249 Some trade or other names: etryptamine; Monase, alpha-ethyl-1H-indole-3-ethanamine; alpha-ET; and AET.

Alpha-ethyltryptamine is a stimulant produced by the Upjohn Pharmaceutical Company. After it was discontinued by them, it was added to the Schedule I drug list on 12 September 1994 [59 FR 46757].

(2) 4-Bromo-2,5-dimethoxy-amphetamine 7391 Some trade or other names: 4-bromo-2,5-dimethoxy-[alpha]-methylphenethylamine; 4-bromo-2,5-DMA.

The positional isomer addition in the opening paragraph of this section dictates that the 15 additional isomers of DOB are all to be considered as Schedule I drugs. DOB was added to the Schedule I list on 21 September 1973 [38 FR 26447].

(3) 4-Bromo-2,5-dimethoxyphenethylamine 7392 Some trade or other names: 2-[4-bromo-2,5-dimethoxyphenyl]-1-aminoethane; alpha desmethyl DOB; 2C-B; Nexus.

The positional isomer addition in the opening paragraph of this section dictates that the 15 additional isomers of 2C-B are all to be considered as Schedule I drugs. 2C-B was added to the Schedule I list on 6 January 1994 [59 FR 671].

(4) 2,5-Dimethoxyamphetamine 7396 Some trade or other names: 2,5-dimethoxy-[alpha]-methylphenethylamine; 2,5-DMA.

The positional isomer addition in the opening paragraph of this section dictates that the five additional isomers of 2,5-DMA are all to be considered as Schedule I drugs. 2,5-DMA was added to the Schedule I list on 21 September 1973 [38 FR 26447]. There is a non-medical use for this compound in the photographic industry, which requires and received approval for the production of multi-ton quantities. Also, this chemical can serve as the immediate synthetic precursor of DOB (see above) or the unscheduled analogue, DOI.

(5) 2,5-Dimethoxy-4-ethylamphetamine 7399 Some trade or other names: DOET.

The positional isomer addition in the opening paragraph of this section dictates that the 15 additional isomers of DOET are all to be considered as Schedule I drugs. DOET was added to the Schedule I list on 16 February 1973 [58 FR 4316].

(6) 4-Methoxyamphetamine 7411 Some trade or other names: 4-methoxy-[alpha]-methylphenethylamine; paramethoxyamphetamine, PMA.

The positional isomer addition in the opening paragraph of this section dictates that both additional isomers of PMA are all to be considered as Schedule I drugs. PMA was added to the Schedule I list on 21 September 1973 [38 FR 26447]

(7) 5-Methoxy-3,4-methylenedioxyamphetamine 7401

The positional isomer addition in the opening paragraph of this section dictates that the five additional isomers of MMDA are all to be considered as Schedule I drugs. MMDA was #2 of the 17 hallucinogenic drugs placed in the CSA when it was written.

(8) 4-Methyl-2,5-dimethoxyamphetamine 7395 Some trade and other names: 4-methyl-2,5-dimethoxy-[alpha]-methylphenethylamine; "DOM"; "STP."

The positional isomer addition in the opening paragraph of this section dictates that the 15 additional isomers of DOM are all to be considered as Schedule I drugs. DOM was #7 of the 17 hallucinogenic drugs placed in the CSA when it was written.

(9) 3,4-Methylenedioxyamphetamine 7400

The positional isomer addition in the opening paragraph of this section dictates that there is one additional isomer of MDA which is to be considered as a Schedule I drug. MDA was #1 of the 17 hallucinogenic drugs placed in the CSA when it was written.

(10) 3,4-Methylenedioxymethamphetamine (MDMA) 7405

The positional isomer addition in the opening paragraph of this section dictates that there is one additional isomer of MDMA which is to be considered as a Schedule I drug. MDMA was temporarily placed in Schedule I on 1 July 1985 under the authority of the Emergency Scheduling Act [50 FR 23118]. After one year (on 17 June 1986), this placement was extended for six months [51 FR 21911] and the permanent placement in Schedule I occurred on 13 November 1986 [51 FR 36552]. Shortly thereafter an appeal was filed concerning this ruling which argued, amongst other points, that the method of deciding currently accepted medical use, a requirement for the assignment to Schedule I, was inappropriate. The United States Court of Appeals for the First Circuit agreed with the petitioner on this point and ordered the DEA to vacate this drug from Schedule I (18 September 1987). The drug was removed from Schedule I effective 27 January 1988 [53 FR 2225] and relocated back into Schedule I on 22 February 1988, effective 23 March 1988 [53 FR 5156]. The Controlled Substance Handbook of the Government Information Services makes no mention of any of the earlier dates or Federal Register citations, and only records the last entry above for the placement of MDMA into Schedule I.

It is possible that there could not have been any illegal activity involving MDMA until 23 March 1988, if indeed it was not until this date that it became officially a Schedule I drug.

(11) 3,4-Methylenedioxy-N-ethylamphetamine 7404 (also known as N-ethyl-alpha-methyl-3,4(methylenedioxy)phenethylamine, N-ethyl-MDA, MDE, MDEA.

The positional isomer addition in the opening paragraph of this section dictates that there is one additional isomer of MDE which is to be considered as a Schedule I drug. MDE was added to the Schedule I list on 13 April 1989 [54 FR 14797].

(12) N-Hydroxy-3,4-methylenedioxyamphetamine 7402 (also known as N-hydroxy-alpha-methyl-3,4(methylenedioxy)phenethylamine, and N-hydroxy-MDA.

The positional isomer addition in the opening paragraph of this section dictates that there is one additional isomer of MDOH which is to be considered as a Schedule I drug. MDOH was added to the Schedule I list on 13 October 1988 [53 FR 40061].

(13) 3,4,5-Trimethoxy amphetamine 7390

The positional isomer addition in the opening paragraph of this section dictates that there are five additional isomers of TMA which are to be considered as a Schedule I drug. TMA was #3 of the 17 hallucinogenic drugs placed in the CSA when it was written.

(14) Bufotenine 7433 Some trade and other names: 3-([beta]-dimethylaminoethyl)-5-hydroxyindole; 3-(2-dimethylaminoethyl-5-indolol; N,N-dimethylserotonin; 5-hydroxy-N,N-dimethyltryptamine; mappine.

The positional isomer addition in the opening paragraph of this section dictates that there are three additional isomer of bufotenine which are to be considered as a Schedule I drug. One of these is the schedule I drug psilocin which is #26, below. Bufotenine was #4 of the 17 hallucinogenic drugs placed in the CSA when it was written.

(15) Diethyltryptamine 7434 Some trade and other names: N,N-diethyltryptamine, DET.

The positional isomer addition in the opening paragraph of this section does not apply to DET as there are no substituents. DET was #5 of the 17 hallucinogenic drugs placed in the CSA when it was written.

> (16) Dimethyltryptamine 7435 Some trade and other names: DMT

The positional isomer addition in the opening paragraph of this section does not apply to DMT as there are no substituents. DMT was #6 of the 17 hallucinogenic drugs placed in the CSA when it was written.

> (17) Ibogaine 7260 Some trade and other names: 7-ethyl-6,6b,7,8,9,10,12,13-octahydro-2-methoxy-6,9-methano-5H-pyrido [1',2':1,2] azepino [5,4-b] indole; Tabernanthe iboga.

The positional isomer addition in the opening paragraph of this section dictates that there are three additional isomers of ibogaine which are to be considered as a Schedule I drug. The botanical binomial may be improperly entered here into the law. Ibogaine was #8 of the 17 hallucinogenic drugs placed in the CSA when it was written, but in that initial listing, the trivial name ibogaine alone appeared. On the full printing of the revised list in 1971, the name *Tabernanthe iboga* appeared as a synonym. It is, however, the name of a plant and not the name of the compound. To make a new material illegal requires a formal process (postings, hearing, etc.) and there is no record of this having been done. The compound is soundly illegal. The plant is quite possibly not illegal.

> (18) Lysergic acid diethylamide 7315

The positional isomer addition in the opening paragraph of this section does not apply to LSD as there are no substituents. LSD was #9 of the 17 hallucinogenic drugs placed in the CSA when it was written.

> (19) Marihuana 7360

Marihuana was entry #10 of the 17 hallucinogenic drugs placed in the CSA when it was written. It is not given any trade or other names in the Code of Federal Regulation — it is just as presented here, as one word. In the definition section, however, the following appears: "The term "marihuana" means all parts of the plant *Cannabis sativa* L., whether growing or not; the seeds thereof; the resin extracted from any part of such plant; and every compound, manufacture, salt, derivative, mixture, or preparation of such plant, its seeds or resin.

Such term does not include the mature stalks of such plant, fiber produced from such stalks, oil or cake made from the seeds of such plant, and other compound, manufacture, salt, derivative, mixture, or preparation of such mature stalks (except the resin extracted therefrom), fiber, oil or cake, or the sterilized seed of such plant which is incapable of germination." Entry #27 below deals with the synthetic equivalents of the tetrahydrocannabinols which are normally found in the resin of *Cannabis sativa.*

(20) Mescaline 7381

Mescaline is 3,4,5-trimethoxyphenethylamine. The positional isomer addition in the opening paragraph of this section dictates that there are five additional isomers of mescaline which are to be considered as a Schedule I drug. Mescaline was #11 of the 17 hallucinogenic drugs placed in the CSA when it was written. It is also the major active component of the cactus Peyote, which is entry #22 below.

(21) Parahexyl 7374 Some trade or other names: 3-hexyl-1-hydroxy-7,8,9,10-tetrahydro-6,6,9-trimethyl-6H-dibenzo [b,d] pyran; Synhexyl.

The positional isomer addition in the opening paragraph of this section dictates that there are eleven additional isomers of parahexyl and, with a cis-trans possibility about the existing double bond there may be some 23 isomers which are to be considered as Schedule I drugs. Parahexyl was added to the Schedule I list on 22 December 1982 [47 FR 52432] as a direct consequence of the United States becoming (in 1980) a party to the International Drug Control Treaty "Convention of Psychotropic Substances, 1971." See the initial proposal [47 FR 33986] for details for this required scheduling.

(22) Peyote 7415 Meaning all parts of the plant presently classified botanically as *Lophophora williamsii* Lemaire, whether growing or not, the seeds thereof, any extract from any part of such plant, and every compound, manufacture, salts, derivative, mixture, or preparation of such plant, its seeds or extracts (Interprets 21 USC 812(c), Schedule Ic(c) (12))

This is one of the four plants that are listed in the Federal Drug Law with probable correctness. The other three are *Cannabis sativa* (marihuana, Schedule I above), *Papaver somniferum* (opium) and *Erythroxylon coca* (coca). The *Tabernanthe iboga* (#17 above) and *Catha edulis* (associated with Cathinone, Schedule I stimulants) are controversial. As peyote is a plant, the positional isomer detail in the opening paragraph has no meaning. It was #12 of the 17 hallucinogenic drugs placed in the CSA when it was written. Peyote is also the sacrament of an approved religion. Sec. 1307.31 exempts peyote from control when used in *bona fide* religious activities by the Native American Church, which is primarily made up of American Indians. However, persons who manufacture or distribute peyote to the Native American Church must register with DEA and otherwise comply with the regulations.

(23) N-Ethyl-3-piperidyl benzilate 7482

Both this compound and its N-methyl quaternary salt, as well as the N-methyl compound mentioned below in #24, have been studied as potential intoxicating agents of use in chemical warfare. The Lakeside code for this is JB-318. The positional isomer addition in the opening paragraph of this section dictates that two positional additional isomers of JB-318 can be classified as Schedule I drugs. The quaternary salt was explicitly excluded from consideration as a scheduled drug (see [35 FR 7069]). The presence of this compound, and the N-methyl directly below, is due to the fact that they were known and talked about during the Haight Ashbury times as compounds with wild potential. There is no evidence that they were ever drugs of abuse. JB-318 was #13 of the 17 hallucinogenic drugs placed in the CSA when it was written.

(24) N-Methyl-3-piperidyl benzilate 7484

As mentioned above, this compound and its N-methyl quarternary salt have been studied as potential intoxicating agents of use in chemical warfare. The Lakeside code for this compound is JB-336. See the entry above, #23, for some of the history and absence of abuse record. JB-336 was #14 of the 17 hallucinogenic drugs placed in the CSA when it was written.

(25) Psilocybin 7437

Psilocybin is the phosphate ester of N,N-dimethyl-4-hydroxytryptamine. The positional isomer addition in the opening paragraph of this section dictates that there are three additional isomers of psilocybin which are to be considered as a Schedule I drug. Psilocybin was #15 of the 17 hallucinogenic drugs placed in the CSA when it was written. It is a major active component of some hundred psychoactive mushrooms, many of the *Psilocybe* Genus, but none of these plants are recognized in the Federal law.

(26) Psilocyn

Psilocyn is N,N-dimethyl-4-hydroxytryptamine, and is usually spelled psilocin in the scientific literature. The positional isomer addition in the opening paragraph of this section dictates that there are three additional isomers of psilocin which are to be considered as a Schedule I drug. One of these is the Schedule I compound bufotenine mentioned previously (#14). Psilocin was #16 of the 17 hallucinogenic drugs placed in the CSA when it was written. It is present in many of the psychoactive mushrooms mentioned under #25. But again, none of these mushrooms are recognized in the Federal law.

(27) Tetrahydrocannabinols 7370 Synthetic equivalents of the substances contained in the plant, or in the resinous extracts of Cannabis, sp. and/or synthetic substances, derivatives, and their isomers with similar chemical structures and pharmacological activity such as the following:

Δ 1 cis or trans tetrahydrocannabinol, and their optical isomers
Δ 6 cis or trans tetrahydrocannabinol, and their optical isomers
Δ 3,4 cis or trans tetrahydrocannabinol, and their optical isomers

(Since nomenclature of these substances is not internationally standardized, compounds of these structures, regardless of numerical designation of atomic positions covered.) (sic.)

Tetrahydrocannabinol is one of the major components of cannabis resin. This entry covers all the major stereoisomers, but follows the synthetic approach. Tetrahydrocannabinol was #17 of the 17 hallucinogenic drugs placed in the CSA when it was written. The Δ 6 trans isomer, dronabinol or Marinol, in sesame oil in a soft gelatin capsule, is an FDA approved product, and was rescheduled on 13 May 1986, to Schedule II [51 FR 17576. It was rescheduled again on July 2, 1999, to Schedule III [64 FR 35928].

Some other names for dronabinol: (6a, 10a-trans)-6a,7,8,10a-tetrahydro-6,6,9-trimethyl-3-pentyl-6H-dibenzo [b,d]pyran-1-ol, or (-)-delta-9-(trans)-tetrahydrocannabinol. The other isomers remain, presumably, in Schedule I.

(28) Ethylamine analogue of phencyclidine 7455 Some trade or other names: N-ethyl-1-phenylcyclohexylamine, (1-phenyl-cyclohexyl)ethylamine, N-(1-phenylcyclohexyl) ethylamine, cyclohexamine, PCE.

This and the next three entries are all substitutional analogues of PCP. None of them have substitution, so there are no position isomers involved. All four of these analogues have a structure that can be related directly to PCP, a Schedule II depressant. The parent PCP skeleton has a 1-phenyl-1-piperidinyl substitution pattern.

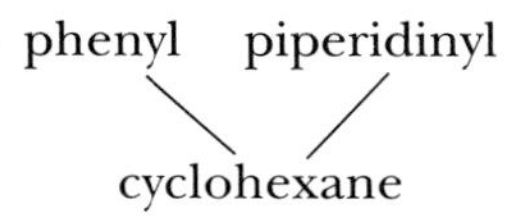

There are no ring substituents, thus there are no positional isomers. PCE has a 1-phenyl-1-aminoethyl substitution pattern. It can be symbolized as

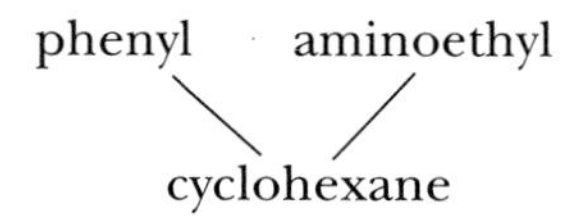

It was entered into Schedule I on 25 October 1978 [43 FR 43295].

(29) Pyrrolidine analogue of phencyclidine 7458 Some trade or other names: 1-(1-phenylcyclohexyl)-pyrrolidine, PCPy, PHP.

This is the second of the four most similar PCP analogues. The PCPy skeleton has a 1-phenyl-1-pyrrolidyl substitution pattern. It can be symbolized as:

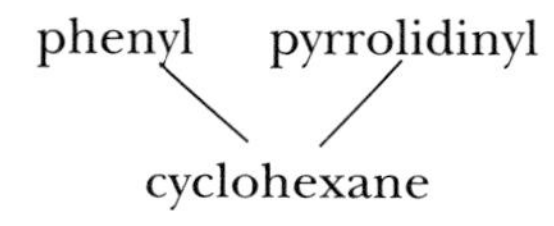

It was entered into Schedule I on 25 October 1978 [43 FR 43295].

(30) Thiophene analogue of phencyclidine 7470 Some trade or other names: 1-[1-(2-thienyl)-cyclohexyl]-piperidine, 2-thienylanalogue of phencyclidine, TPCP, TCP.

This the third of the four most similar PCP analogues. The TCP skeleton has a 1-(2-thienyl)-1-piperidyl cyclohexyl substitution pattern, and is known by the name of Tenocyclidine. It can be symbolized as:

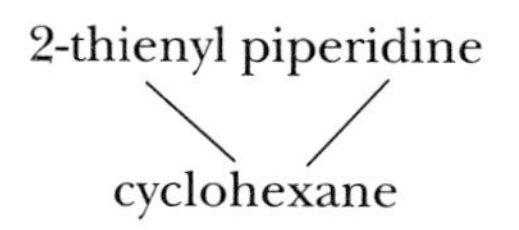

It was entered into Schedule I on 11 August 1975 [40 FR 28611].

(31) 1-[1-(2-Thienyl)cyclohexyl]pyrrolidine 7473 Some other names: TPCP, TCPy

This is the fourth of the four most similar PCP analogues. The TCPy skeleton has a 1-(2-thienyl) 1-pyrrodinyl cyclohexyl substitution pattern. It can be symbolized as:

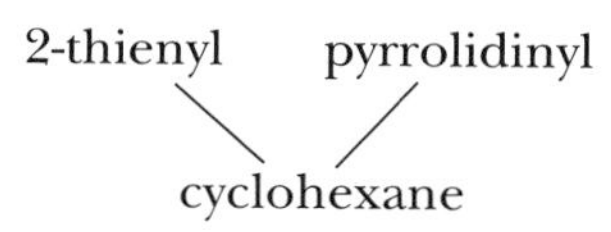

It was entered into Schedule I on 6 July 1989 [54 FR 28414].

REFERENCES

Rather than referencing a detailed bibliography, the reader wishing more detailed information is encouraged to seek out the citations included in the following reviews, which were called upon for this present chapter.

Controlled Substances Handbook (2000), Washington DC: Government Information Services, Thompson Publishing Group.

Shulgin, A. T. (1978) Psychotomimetic drugs: Structure activity relationships, in *Handbook of Psychopharmacology*, Vol. 11, L. L. Iversen, S. D. Iversen and S. H. Snyder (eds), New York: Plenum Press, pp. 243–333.

Shulgin, A. T. (1981a) *Hallucinogens, Burger's Medicinal Chemistry*, 4th edn, M.E. Wolff (ed.), New York: Wiley & Co., pp. 1109–1137.

Shulgin, A. T. (1981b) Chemistry of psychotomimetics, in *Handbook of Experimental Pharmacology*, Vol. 55/ III, F. Hoffmeister and G. Stille (eds), Berlin: Springer-Verlag, pp. 3–29.

Shulgin, A. T. (1992) *Controlled Substances; Chemical and Legal Guide to Federal Drug Laws*, 2nd edn, Berkeley, CA: Ronin Press.

Shulgin, A. T. and Shulgin, A. (1991) *PIHKAL, A Chemical Love Story*, Berkeley, CA: Transform Press.

Shulgin, A. T. and Shulgin, A. (1997) *TIHKAL, The Continuation*, Berkeley, CA: Transform Press.

CHAPTER 4

METHODS OF ILLICIT MANUFACTURE

Richard Laing
John Hugel

4.0 INTRODUCTION

Hallucinogenic plants and toxins have been have been exploited throughout the ages. This exploitation has been unparalleled in the twentieth century. From the pharmaceutical industry introducing new therapeutic agents as a treatment for mental illness to the popularity of drug abuse by individuals, the collective knowledge of hallucinogenic drug production has increased tremendously in the last 60 years. The dissemination of this knowledge has allowed individuals to cultivate and manufacture hallucinogenic compounds for illicit use. This chapter describes the common syntheses of many classes of hallucinogens to provide an overview of illicit manufacturing.

4.0.1 THE ERGOT ALKALOIDS

The poisonous properties of ergot, *Claviceps*, a parasitic fungus common to edible grains, have long plagued civilized humans. Outbreaks of ergot poisoning dubbed Holy Fire or Saint Anthony's Fire have been well documented in ancient cultures. The result of ingestion of the ergot alkaloids included "fire" or burning feeling of the extremities followed by gangrenous infection and blackening of the appendage as if it were consumed by fire. It was not until 1670 that ergot was proved the source of these epidemics which had raged for centuries uncontrolled. Some of ergot's medicinal properties were known prior to it being found as the source of Saint Anthony's Fire. Ergot-infected grain was used as a medicinal herb in aiding the childbirthing process. It was proven effective as an oxytoxic, inducing contractions within the womb, and shortly afterwards it was shown to cause vasoconstriction and in toxic dosages hallucinations and delirium. Its wide use as an oxytoxic in the 1800s led to an increase in the number of stillbirths and in 1824 the Medical Society of New York, after an inquiry into its usage, recommended it be used for postpartum hemorrhage control only. Ergot alkaloids, however, are now widely used to control migraine headaches through its action as a cerebral vasoconstrictor, which decreases the amplitude of the pulsation of the cranial arteries.

The three main biogenetically related groups of ergot alkaloids all have a common ergoline moiety (Figure 4.1) (Cordell 1981):

1. The clavine type (e.g. elymoclavine, argroclavine, and chanoclavine-I). Approximately 32 alkaloids have been characterized and have little use in the pharmaceutical industry. The clavines are characterized by not having a carbonyl functionality anchored at the C-8 carbon and the double bond may either be in the 8,9- position or the 9,10- position.
2. Water soluble lysergic acid derivatives (e.g. ergine (lysergamide), ergonovine) have a carboxylic functionality attached to the C-8 carbon with simple amide groups.
3. Water insoluble lysergic acid derivatives (e.g. ergotamine, ergocornine, and ergocryptine) consist of the largest alkaloids and are sometimes referred as peptide ergot alkaloids because of the resulting production of amino acids upon hydrolysis. For example, the hydrolysis of ergotamine gives lysergic acid, proline, pryruvic acid, and L-phenylalanine.

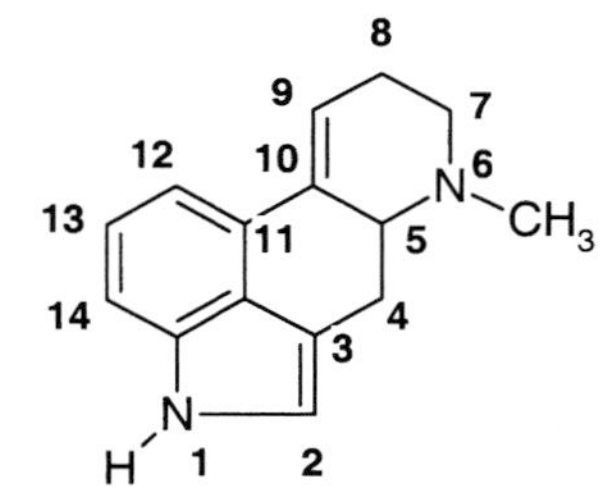

Figure 4.1

Basic structure of the ergoline moiety

A characteristic of the lysergic acid derivatives and peptide alkaloids is the existence of two isomeric forms based on the configuration of the C-8 carbon. The peptide alkaloids exist in nature as both C-8 diastereoisomers with the iso forms (αC-8) having names ending in -inine. (See Figure 4.2.)

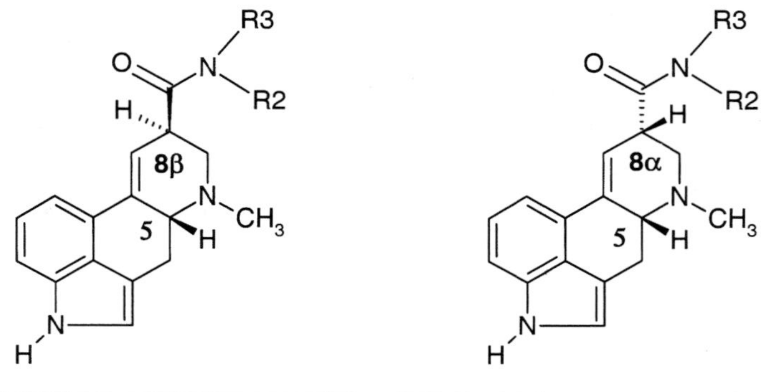

Figure 4.2

Normal and iso-N-substituted lysergic acid

In the lysergic acid derivative series the native form is physiologically active while the iso diastereoisomer either shows little or no activity. In the synthesis of LSD and other alkylamide derivatives of lysergic acid, some synthetic routes favor the production of one form over the other. However, the iso form can easily be converted into the active and desired form by a simple reaction. (See Figure 4.3.)

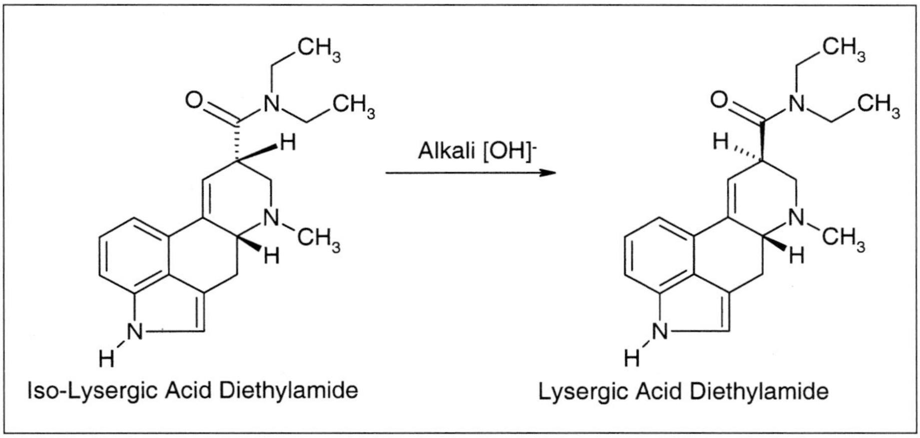

Iso-Lysergic Acid Diethylamide

Alkali $[OH]^-$

Lysergic Acid Diethylamide

Figure 4.3

Simple conversion of iso-LSD to LSD under alkali conditions

4.1 ILLICIT MANUFACTURE

4.1.1 LSD

The synthesis of LSD and related amides of lysergic acid requires the condensation of an organic amine with either the carboxylic acid group of lysergic acid or its derivatized intermediate forming an amide bond. Numerous synthetic routes can be employed in the formation of this amide bond and many techniques have evolved with advances in peptide chemistry. Much research into the discovery of new lysergamides was spurred by the pharmaceutical industry in its search for more effective agents. In order to create an amide the hydroxyl group of the acid functionality must be removed. Since this hydroxyl is a poor leaving group it is either derivatized or converted to a functionality that does make a good leaving group. Compounding this difficulty is the fact that lysergic acid is sensitive to light and oxygen and is highly sensitive to many reagents.

The common synthetic routes, Figure 4.4, can be classified based on the type of intermediate compound upon which the alkylamine acts as a nucleophile in forming the desired amide bond: azide, mixed anhydride, acid chloride and imidazole intermediates.

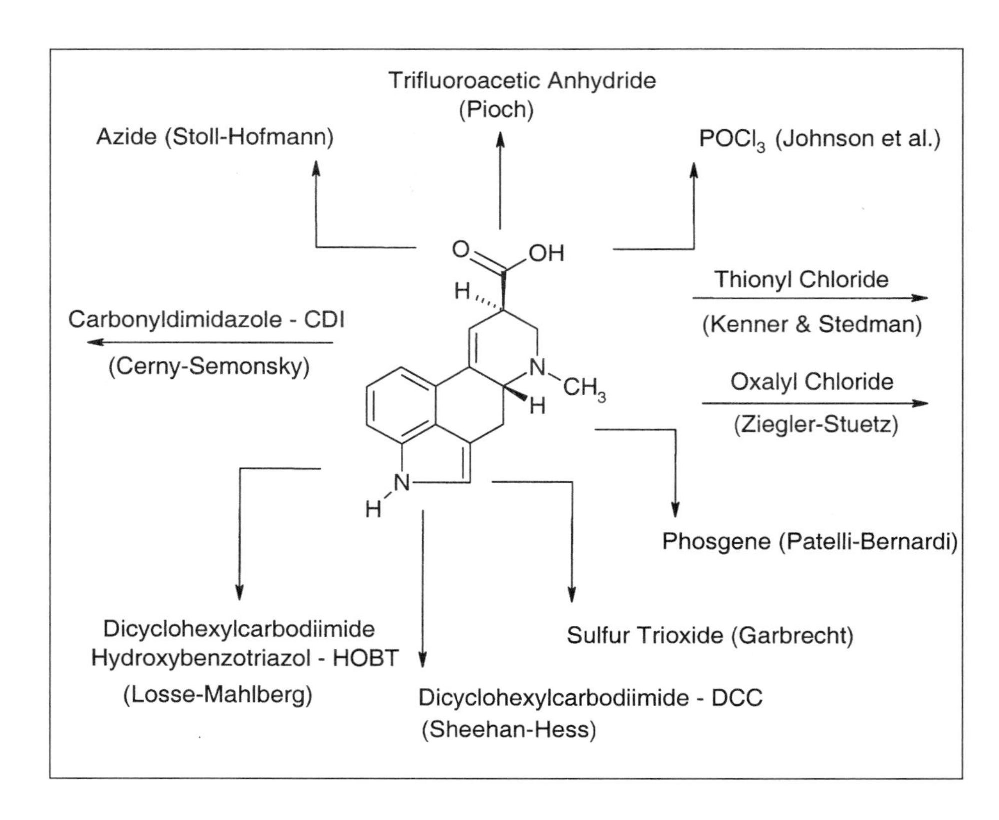

Figure 4.4

Basic synthetic routes for the manufacture of LSD

Azide method

Stoll and Hofmann published and patented extensively their work at Sandoz Research Laboratories, Basel, Switzerland, in regard to lysergic acid amide derivatives. They first described the condensation of organic amines with lysergic acid azide at low temperature (Stoll and Hofmann, 1937). They found that numerous types of amines reacted well provided that one labile hydrogen was present on the amine. The lysergic acid azide was found to be very stable and could be recrystallized optically pure. The azide resulted from the treatment of the lysergic acid hydrazide, derived by the cleavage of ergotamine using anhydrous hydrazine resulting in a 70% yield with sodium nitrite under acidic conditions (Stoll and Hofmann, 1944). The treatment of ergotamine with hydrazine produces predominantly the iso form. They also described the epimerization reaction between the non-active iso form to the active lysergic form by subjecting the iso forms to an alkali environment using sodium or potassium hydroxide (Stoll and Hofmann, 1937). In 1943 Stoll and Hofmann described the synthesis of dialkylamine derivatives of lysergic acid including diethylamide, dimethylamide, dipropylamide, dibutylamide, diamylamide, methyl ethylamide, ethyl allylamide, butyl amylamide (Figure 4.5). In 1944 Stoll and Hofmann applied for a US patent on the synthesis of LSD which was

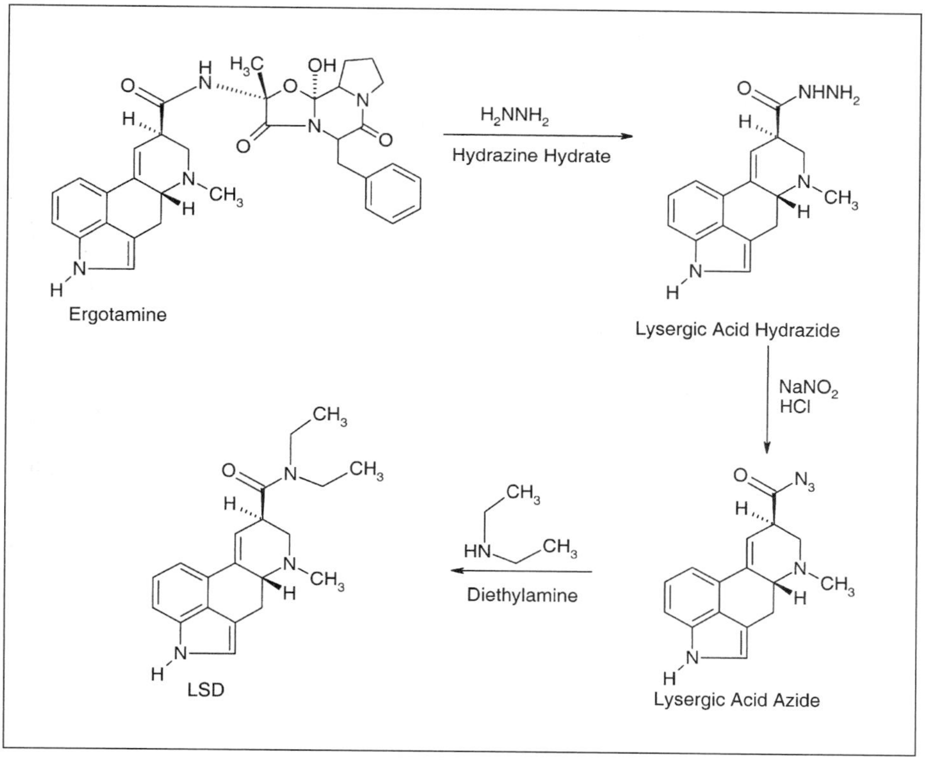

Figure 4.5

Azide route from ergotamine by Stoll and Hofmann (1936)

awarded to Sandoz Ltd in 1948 (Stoll and Hofmann, 1948). They would go on to synthesize and characterize 46 amide derivatives (Stoll and Hofmann, 1955). This reaction produces a mixture of the normal (about 80%) and iso forms with a yield of 69% from the lysergic acid hydrazide.

Mixed anhydride

An anhydride is formed when two carboxylic acid groups condense losing a molecule of water. The resulting anhydride is reactive and many types are used in synthetic chemistry. In the early 1950s there was great interest in developing new synthetic routes for peptides. Boissannas (1951, 1952) and Bourne *et al.* (1952) described the use of mixed anhydrides in the formation of amide bonds. A mixed anhydride occurs when the second carboxylic acid group is different from the first, and the more electronegative it is in nature the better it is in leaving during nucleophilic attack. It was observed that acetic acid and trifluoroacetic anhydride under certain conditions produced significant amounts of n-acetylation which the authors noted useful in peptide synthesis. Pioche (1956) described this reaction on lysergic acid at temperatures less than 0°C using trifluoroacetic anhydride. It was found that the trifluoroacetyl derivative of lysergic acid is unstable at room temperature and, as such, the conversion to the amide with the appropriate alkylamine was undertaken in situ or as soon as the mixed anhydride had formed in solution. The overall yield was reported to be 58% with the production of significant quantities of the iso form (Figure 4.6).

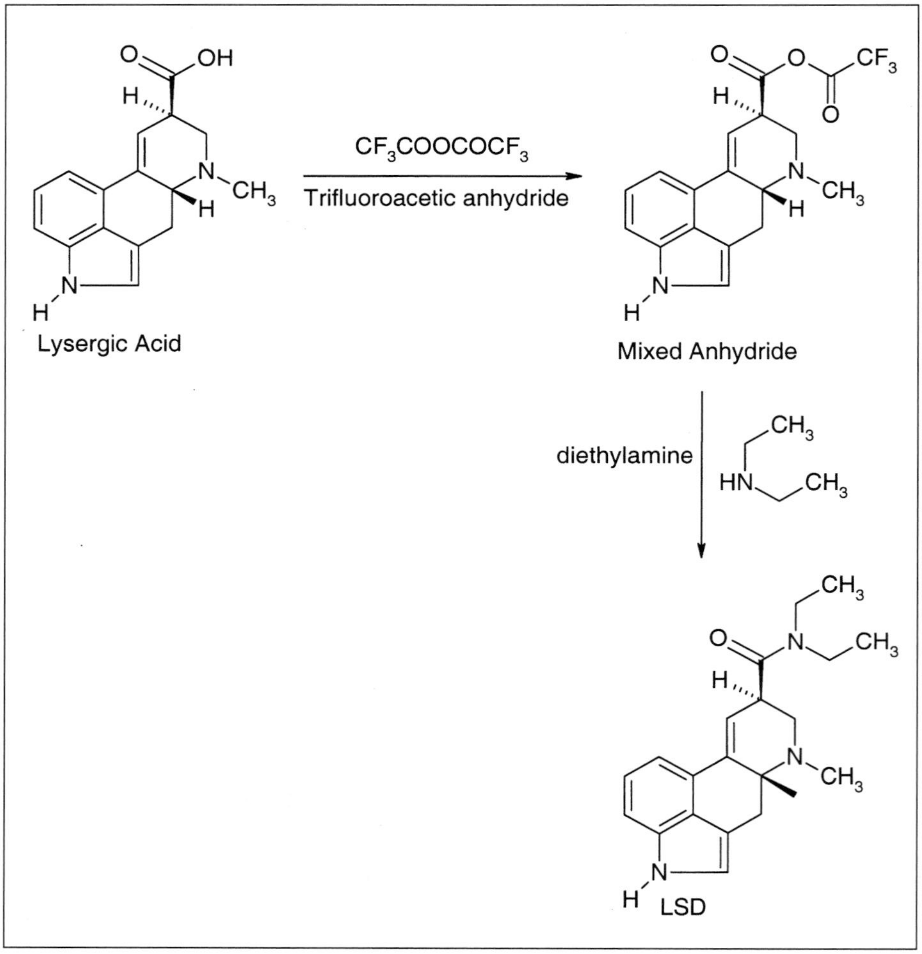

Figure 4.6

Mixed anhydride synthesis by Pioche (1956)

Lysergic Acid

$CF_3COOCOCF_3$

Trifluoroacetic anhydride

Mixed Anhydride

diethylamine

LSD

Sulfur trioxide-dimethylformamide complex

In the continuing search for peptide coupling agents Kenner and Stedman (1952) described the mixed anhydride of sulfuric acid: sulfur trioxide in dimethylformamide. Garbrecht (1956) successfully synthesized LSD by the sulfur trioxide-dimethylformamide (SO_3-DMF) route. In this mixed anhydride reaction the stoichiometry of the reactants was found to be critical in the success of the reaction.

Garbrecht (1959) described the advantages of the SO_3-DMF complex over the mixed anhydrides described by Pioche (1956) and the hydrazide method described by Stoll and Hofmann. The hydrazide cleavage of ergotamine method causes racemization at the 8-carbon position requiring isolation and recrystallization. One difficulty with the mixed anhydride reaction is the acidic nature of the reagents which decomposes lysergic acid, thus reducing yields. Furthermore, the mixed anhydride is a non-specific acetylating agent forming unwanted by-products and few anhydride reagents can effectively be used in the synthesis of LSD.

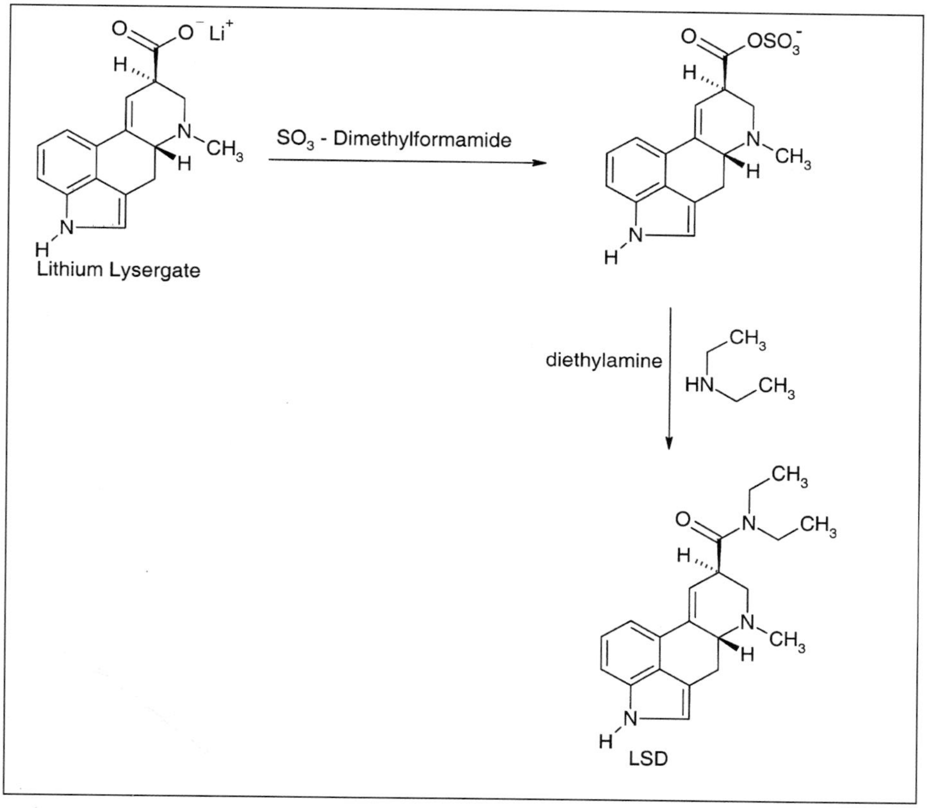

Figure 4.7

Sulfur trioxide-dimethylformamide synthesis by Garbrecht (1956)

The sulfur trioxide reaction relies on the preparation of a SO_3-DMF complex by placing freshly distilled dimethylformamide in a dried reaction flask and adding carefully sulfur trioxide drop wise over a 4–5 hour period with stirring while maintaining 0–5°C. The resulting complex was found to be stable for 2–3 months when kept cooled. Lysergic acid monohydrate was treated with lithium hydroxide hydrate in methanol to produce glass-like lithium lysergate. The lysergate when reacted with SO_3-DMF reagent at or near 0°C produced the mixed anhydride of sulfur trioxide which then was reacted with the alkylamine to produce the amide (Figure 4.7). Typically the maleate or tartrate salts were readily recrystallized from methanol-ether mixtures.

Acyl chloride

Patelli and Bernardi (1962) used a phosgene-dimethylformamide complex to manufacture several lysergic acid N-alkylamines including LSD. In this process anhydrous lysergic acid in dimethylformamide is cooled to –10°C and is reacted with the phosgene-dimethylformamide complex to which the alkylamine is added. The solution is extracted and the tartaric salt is made after an alkaline extraction. Johnson *et al.* (1973) described a facile and convenient reaction producing predominately the normal form (8β-) in relatively high yields of the n-substituted alkylamines of lysergic acid. Lysergic acid is refluxed in chloroform and is treated simultaneously with phosphorus oxychloride ($POCl_3$) in chloroform and the alkylamine by way of separate addition funnels and adding reagents drop wise (Figure 4.8). Shulgin and Shulgin (1997) reported a 66% yield of the recrystallized tartrate salt of LSD using this method.

Figure 4.8

LSD via acyl chloride intermediate of lysergic acid

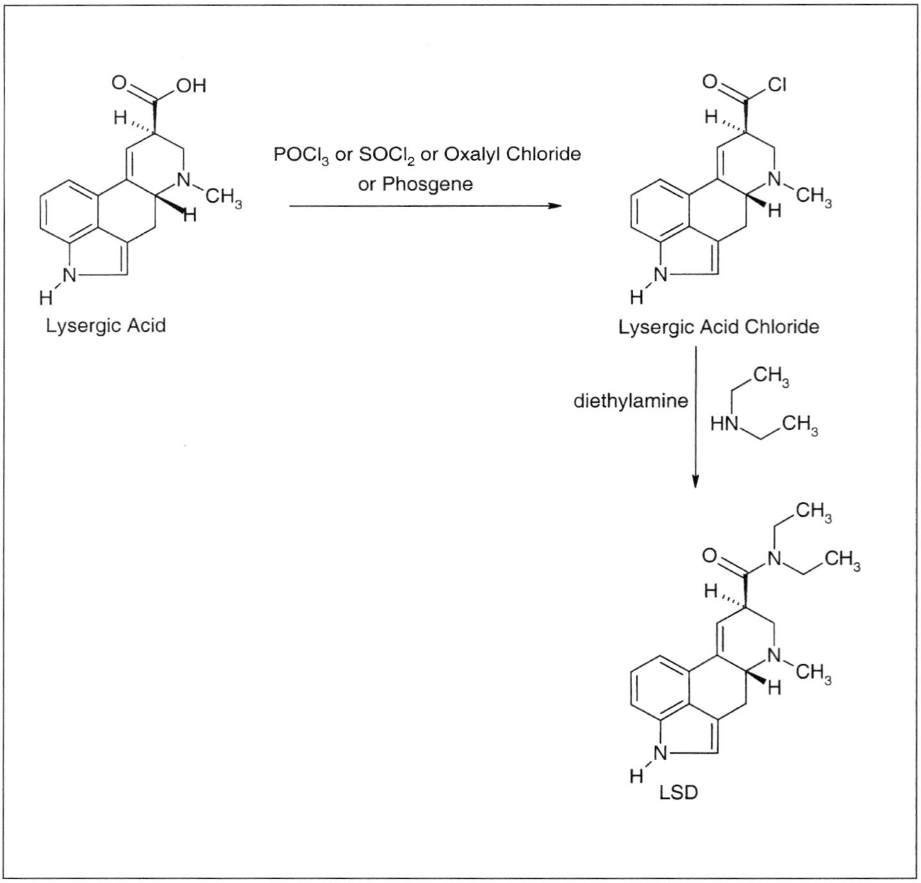

Lysergic Acid

$POCl_3$ or $SOCl_2$ or Oxalyl Chloride or Phosgene

Lysergic Acid Chloride

diethylamine

LSD

Ziegler and Stuetz (1983) used oxalyl chloride on lysergic acid (6-methyl-8β-ergolinecarboxylic acid) followed by treating the in-situ intermediate with aminopyrazine to give 6-methyl-N-pyrazinyl-8β-ergolinecarboxamide.

N,N-dicyclohexylcarbodimide

Sheehan and Hess (1955) described the condensation of a carboxylic acid group of an amino acid with the free amine group of another using N,N-dicyclohexylcarbodimide (DCC) at room temperature. This reaction afforded high yields and allowed the formation of the amide in aqueous solutions. When lysergic acid is treated with DCC it forms an intermediate by attaching itself to the carboxylic acid, thus providing a good leaving group. A second molecule of lysergic acid subsequently displaces the DCC intermediate forming a symmetric anhydride. This anhydride is unfavored due to steric hindrance as shown in Figure 4.9 and therefore this reaction has a poor overall yield.

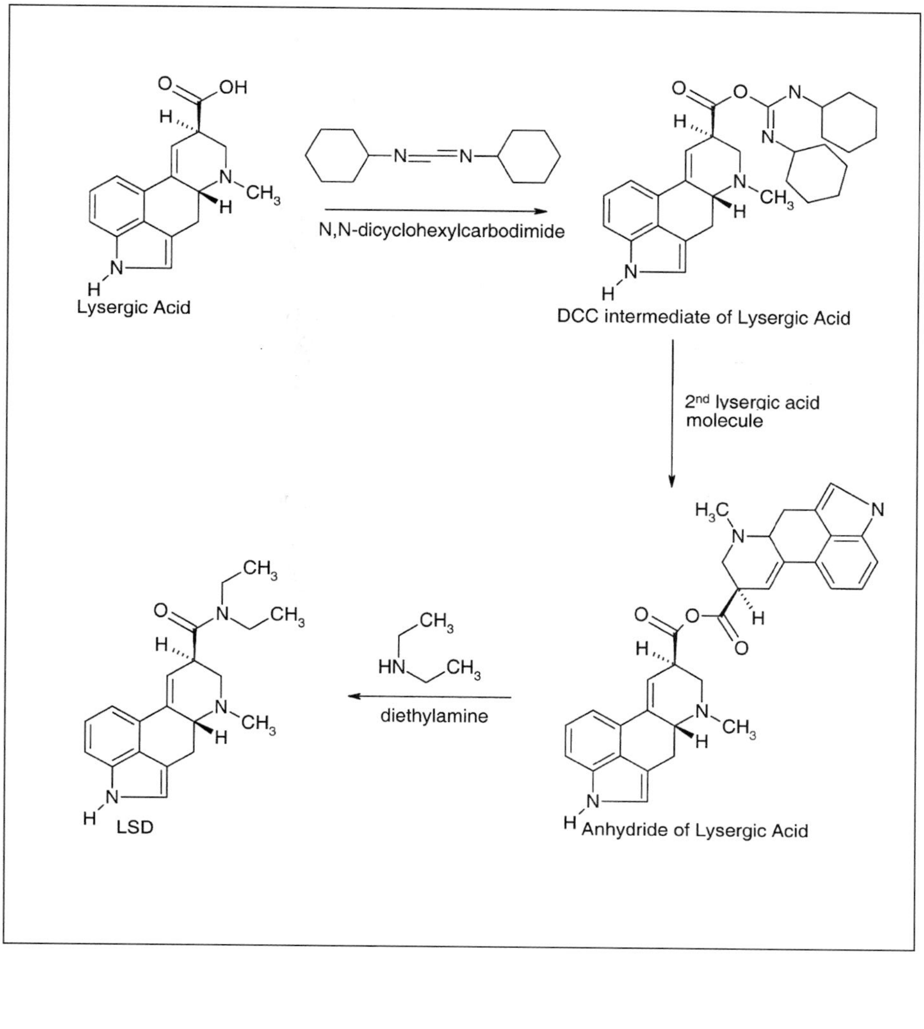

Figure 4.9

N,N-Dicyclohexylcarbodimide route by Sheehan and Hess (1955)

A modification of this reaction by Losse and Mahlberg (1978) surmounts the steric hindrance issue through the addition of 1-hydroxybenzotriaole (HOBT) to the reaction mixture. HOBT attaches to the carboxylic acid functionality and thus forms an excellent leaving group susceptible to the nucleophilic attack by the alkylamine. This reaction gives predominately, about 66%, the iso (α-C8) form with a total yield of about 80%.

N,N-Carbonyldiimidazole

Cerny and Semonsky studied the use of N,N-carbonyldiimidazole in the synthesis of lysergamide analogues (1954, 1963) (Figure 4.10). They found that the reaction provided fewer by-products than the Stoll and Hofmann azide synthesis and that the reaction proceeded at various temperatures (–10°C to 130°C) in dimethylformamide; however, optimal temperature was found to be 20°C. The yield of 81% was observed and the reaction predominantly favors the normal form.

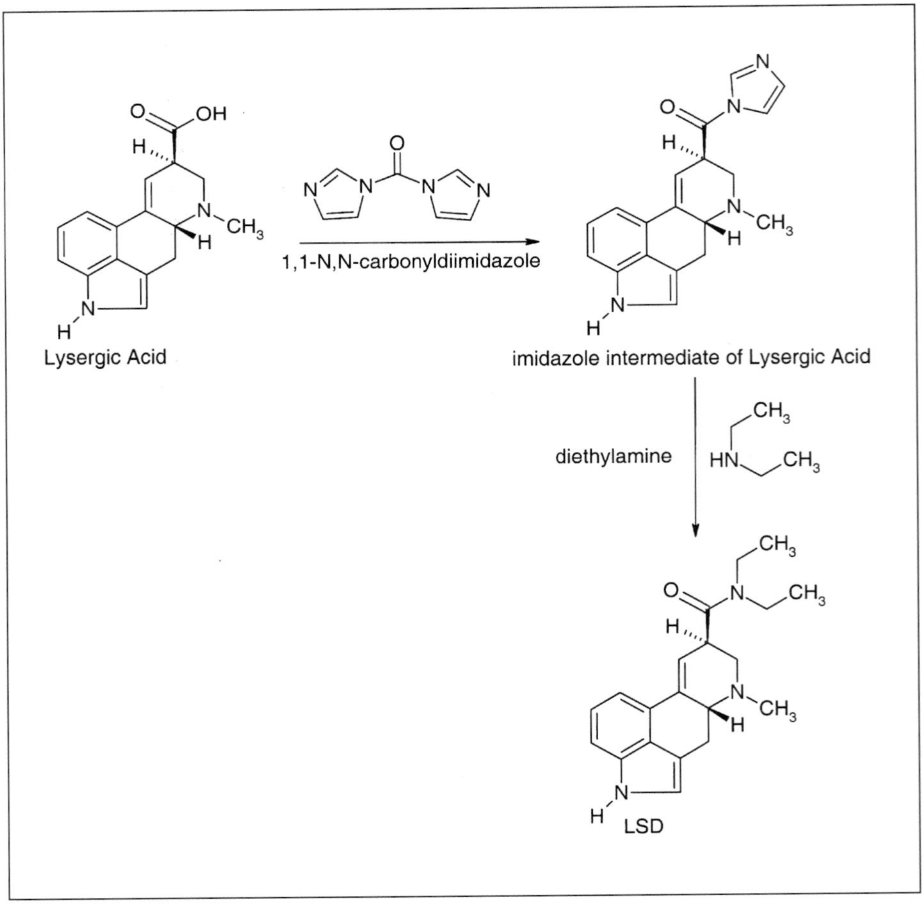

Figure 4.10

N.N-Carbonyldiimidazole synthesis by Cerny and Semonsky (1954)

Lysergic Acid

1,1-N,N-carbonyldiimidazole

imidazole intermediate of Lysergic Acid

diethylamine

LSD

Synthesis of lysergic acid

The synthesis of lysergic acid has been a scientific curiosity with much research in novel synthetic routes and techniques even though the cultivation of ergot fungi has proven to be a reliable and inexpensive source. The difficulty in the synthesis lies in maintaining the chiral configuration around C5 position of the fused heterocycle and the requirement of protecting the indole nitrogen. As such, most syntheses rely on the use of a suitable starting material. While it is not felt that a dissertation of all successful synthetic routes fits within the scope of this chapter, a few notable routes will be used to present an overview of the complexities of this synthesis (Figure 4.11). Kornfeld *et al.* (1956) successfully synthesized lysergic acid through a 15-stage reaction starting with 3β-carboxyethylindole which was converted to Kornfeld's ketone (1-benzoyl-5-keto-

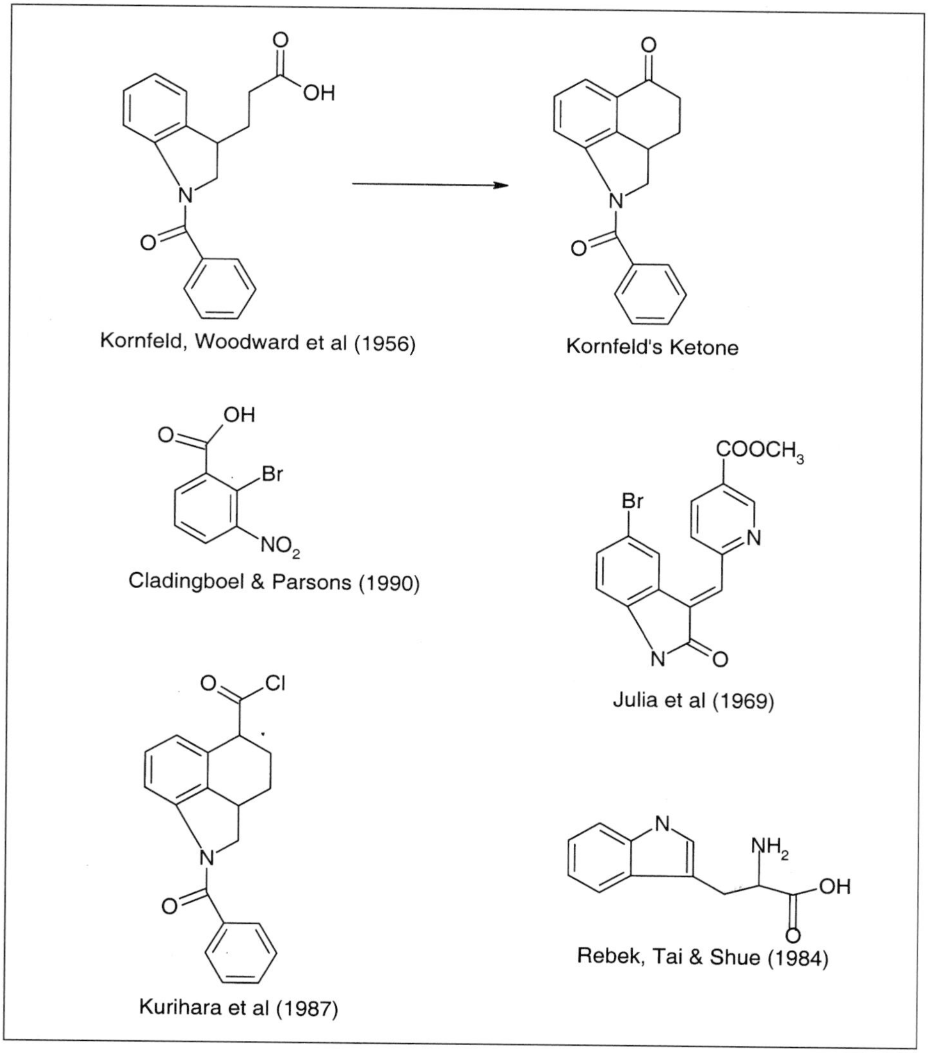

Kornfeld, Woodward et al (1956)

Kornfeld's Ketone

Cladingboel & Parsons (1990)

Julia et al (1969)

Kurihara et al (1987)

Rebek, Tai & Shue (1984)

Figure 4.11
Various different starting materials used in the synthesis of lysergic acid and related compounds

1,2,2a,3,4,5-hexahydrobenz[cd]indole). This important intermediate contains three of the four rings of lysergic acid. Julia *et al.* (1969) used a 12-stage sequence starting with the condensation of 5-bromoisatin with 6-methylnicotinate which formed the backbone with again three of the four rings of lysergic acid (Figure 4.11). Rebek *et al.* (1984) successfully synthesized lysergine and setoclavine starting with L-tryptophan in nine stages. The importance of this sequence is with the starting material of natural source, thereby making the reaction more economically viable. Kurihara *et al.* (1987) reported a facile synthetic route to (±)-lysergic acid through the intermediate 1-benzoyl-1,2,2a,3-tetrahydrobenz[cd]indole-5-carbonyl chloride. The advantage of their sequence is that it was carried out without isolation and purification of each individual intermediate. The final product is obtained after purification by column chromatography at the end of the synthesis. While the preceding reaction depended upon the cyclization of indole containing compounds, Cladingboel and Parsons (1990) devised a sequence in which the indole ring was constructed from a readily available precursor.

4.1.2 INDOLEALKYLAMINES

Tryptamines are ubiquitous in nature and many are hallucinogenic to humans. Throughout history ancient cultures have used hallucinogenic plant and animal preparations in religious rituals. The tryptamines (3-(2-aminoethyl)indole) have an indole base with possible substitutions around the six-membered ring (Figure 4.12), and a 2-ethylamine functionality attached to 3-position on the 5 membered nitrogen containing ring.

Figure 4.12

Basic structure of the indole moiety

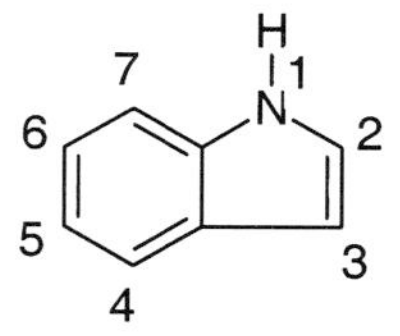

Indole

Many biologically important molecules are indoles such as the amino acid L-tryptophan or the neurotransmitter serotonin.

Bufotenine, psilocin and psilocybin

Serotonin is very similar in structure to bufotenine and psilocin, as shown in Figure 4.13.

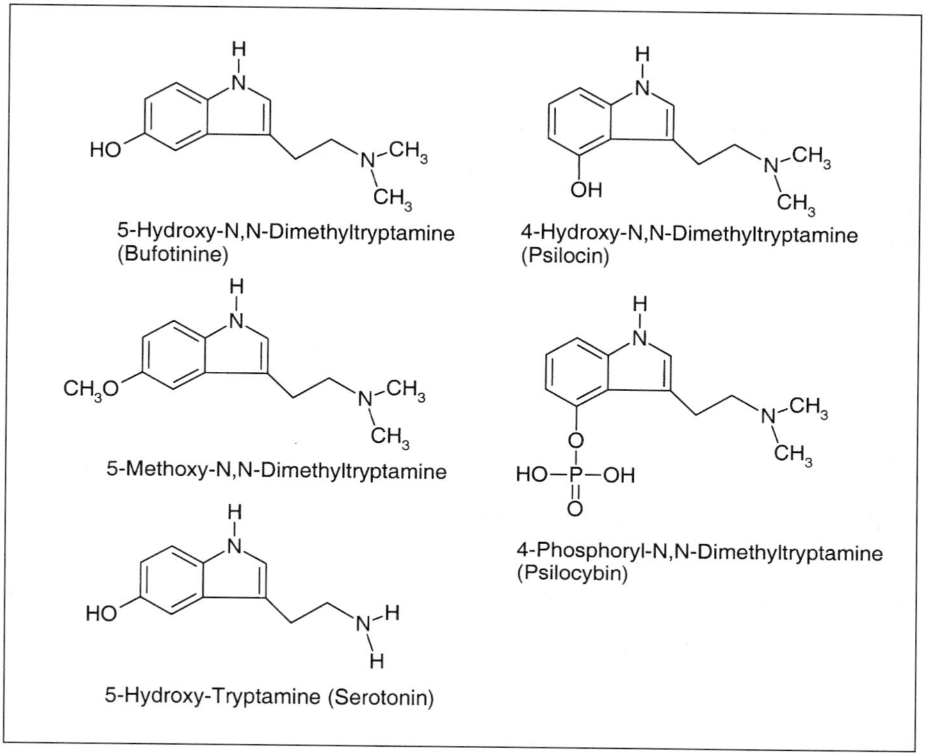

5-Hydroxy-N,N-Dimethyltryptamine (Bufotinine)

4-Hydroxy-N,N-Dimethyltryptamine (Psilocin)

5-Methoxy-N,N-Dimethyltryptamine

4-Phosphoryl-N,N-Dimethyltryptamine (Psilocybin)

5-Hydroxy-Tryptamine (Serotonin)

Figure 4.13

Chemical structures of serotonin, bufotinine, psilocin, psilocybin and 5-methoxy-N,N-dimethyltryptamine

The difficulty in this synthesis is in having to protect the hydroxy group in the 4 or 5 position. This is accomplished by using a starting material in which these groups are already protected as in the case with 4 or 5-benzyloxy-indole. Typically the last step of the reaction is a catalytic debenzylation producing the desired hydroxy group in either of the two positions. Hamlin and Fischer (1951) synthesized 5-hydroxy-tryptamine through an acetamide intermediate. The treatment of 5-benzyloxyindole with formaldehyde and dimethylamine produced 5-benzyloxygramine which when treated with sodium cyanide in ethanol 5-benzyloxyindole-3-acetamide formed. Upon reduction with lithium aluminum hydride the tryptamine was realized. N-Substituted tryptamines can be easily formed through the action of oxalyl chloride on indole (Speeter and Anthony, 1954). The glyoxyl chloride intermediate forms at the 3- position of the indole moiety and was recoverable in near quantitative yields. Bufotenine was prepared through the reaction of the intermediate 5-benzyloxy-3-indole-glyoxyl chloride with dimethylamine to form 5-benzyloxy-N,N-di-methyl-3-indoleglyoxylamide. Through the reduction of the glyoxylamide with lithium aluminum hydride the amine is realized and catalytic debenzylation gives bufotenine. The synthesis of psilocybin by Hofmann *et al.* (1958) followed the same oxalyl chloride reaction but instead they started with 4-benzyloxy-indole and as a last step the synthesis required the phosphorylation of the 4-hydroxy group.

5-Methoxy-N,N-dimethyltryptamine

Starting with 5-benzyloxy indole Benington *et al.* (1958) synthesized various 5 substituted tryptamines including 5-methoxytryptamine. In this reaction sequence the starting indole was debenzylated via a hydrogenation reaction using palladium on carbon catalyst. The resulting hydroxy group was converted to a methoxy group by reacting the hydroxy group with dimethylsulfate followed by sodium hydrosulfite. The resulting 5-methoxyindole can then be reacted with oxalyl chloride and dimethylamine followed by reduction of the gloxylamide with lithium aluminum hydride to produce 5-methoxydimethyl-tryptamine.

N,N-Dimethyltryptamine and homologues

The oxalyl chloride reaction has proven to be very efficient for dimethyltryptamine and related N-substituted tryptamines (Figure 4.14) (Shulgin and Shulgin 1997).

Figure 4.14

Synthesis of N-substituted tryptamines using oxalyl chloride by Speeter and Merril (1954)

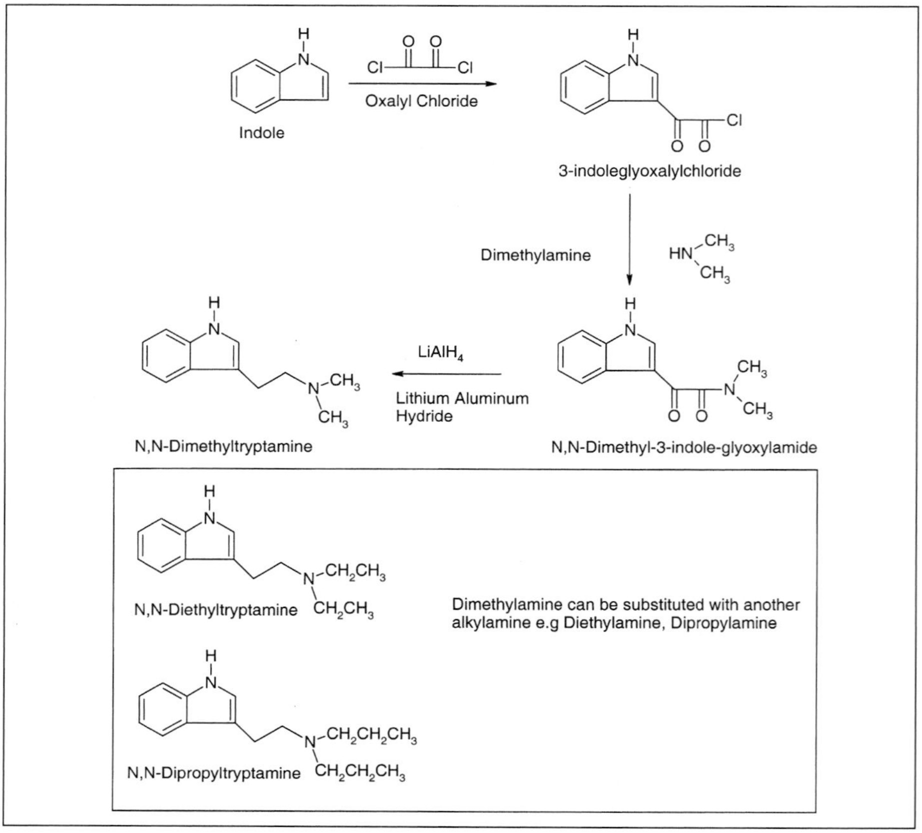

Cowie *et al.* (1982) identified pyrrolidine substituted tryptamine in street exhibits and determined the route of synthesis being the oxalyl chloride route due to the impurity profile. Another route to the tryptamines is through an intermediate aldehyde at the 3- position on the indole. Young (1958) made the 3-substituted aldehyde using phosphorus oxychloride and dimethylformamide that was recovered in high yields (85–95%) and purity. It was shown that the aldehyde could then be condensed with various nitroparaffins to produce indole nitro-olefins that would be reduced to the corresponding amine. An example is the condensation of the indole-3-aldehyde with nitromethane that when subsequently reduced results in tryptamine.

α-*Ethyltryptamine*

In the search for potential psychoactive drugs in the treatment of mental illness, Heinzelman *et al.* (1960) synthesized α-alkyltryptamines via the 3-carboxylaldehyde indole and the appropriate nitroalkane. α-Ethyltryptamine is produced when 1-nitropropane is condensed with the aldehyde and subsequently reduced using lithium aluminum hydride (Figure 4.15).

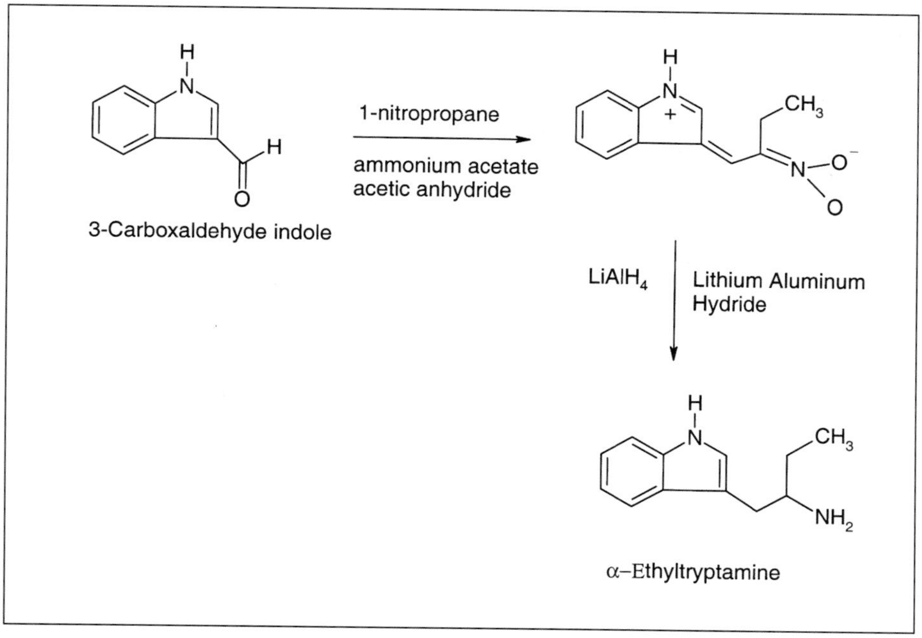

Figure 4.15

Synthesis of a-ethyltryptamine via a nitro-olefin by Heinzelman et al. *(1960)*

4.1.3 HALLUCINOGENIC PHENYLALKYLAMINES

Phenylalkylamines consist of a phenethylamine moiety with ring substitutions and / or extensions to the ethyl side chain (Figure 4.16).

Figure 4.16

Structures of phenethylamine, amphetamine, and 1-phenyl-2-butylamine

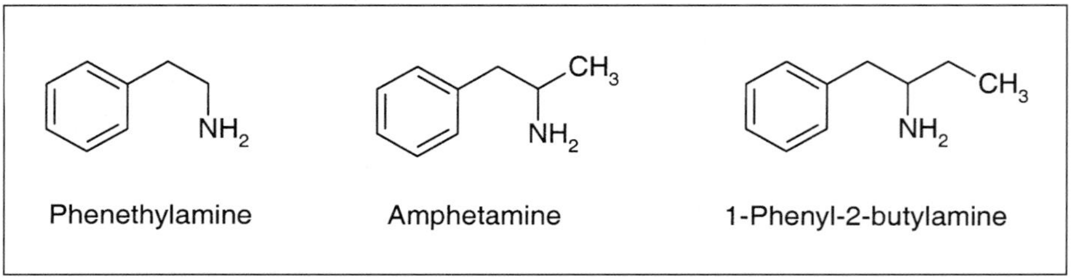

Phenethylamine   Amphetamine   1-Phenyl-2-butylamine

While phenethylamine itself is not active, amphetamine and its N-methylated analogue, methamphetamine, are well known for their stimulant, but not hallucinogenic, properties. Substituents on the benzene ring occasionally lead to hallucinogenic properties. Shulgin and Shulgin (1991) described over 150 different phenethylamines, amphetamines, and phenylbutylamines including, their effects, dosage levels and synthesis. This section will provide an overview of some of these syntheses, particularly those that have been used in clandestine laboratories.

The two most well-known hallucinogenic phenylalkylamines are mescaline and ecstasy (MDMA) whose structures are depicted in Figure 4.17.

Figure 4.17

Structures of mescaline and MDMA

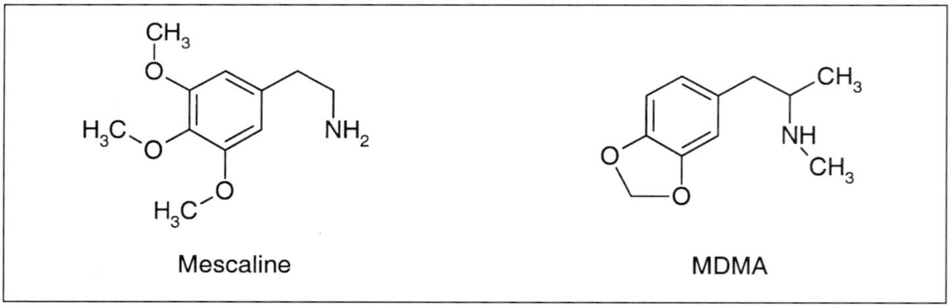

Mescaline   MDMA

Mescaline has been known and used in religious rituals for centuries as a hallucinogen. In discussions regarding hallucinogenic activity, it has become the standard against which all other phenylalkylamines are measured. Ecstasy, on the other hand, has a short history. In the late 1990s and early 2000s with the increasing popularity of rave parties, the use of Ecstasy increased in North America and Europe. Due to its popularity and tablet form many look-alike preparations have been found on the street containing other drugs including not only the MDA family but amphetamines, ketamine, PCP and the highly dangerous 4-methoxyamphetamine and 4-methoxy-N-methylamphetamine (PMA and PMMA respectively), to name a few.

Most of the synthetic routes depicted in this chapter can be employed to produce differently substituted phenylalkylamines provided one begins with the corresponding substituted benzaldehyde or phenylalkylketone. The challenge for the clandestine cook is to obtain these precursor chemicals. There are numerous published methods for the synthesis of the desired phenylalkylketone. In some clandestine laboratories, the phenylalkylketone is synthesized from the corresponding benzaldehyde or often there is a naturally occurring correspondingly substituted allylbenzene that can be synthesized into the appropriate benzaldehyde or phenylalkylketone. In addition, there are synthetic methods for making substituted phenylalkylamines directly from the correspondingly substituted allylbenzene without going through an intermediate aldehyde or ketone. While it is beyond the scope of this chapter to discuss all possible routes of manufacture for all phenylalkylamines, the most common synthesis have been grouped below:

1. phenylalkylamines from benzaldehydes;
2. phenylalkylamines from phenylalkylketones;
3. phenylalkylketones from benzaldehydes;
4. benzaldehydes and phenylalkylketones from allylbenzenes;
5. miscellaneous syntheses.

Dal Cason (1990) described the various synthetic routes to MDA and its homologues and analogues in the context of the likelihood of being prepared by a clandestine chemist.

Part I: Phenylalkylamines from benzaldehydes

Phenylalkylamines can be produced from the corresponding benzaldehydes through the formation of an intermediate nitrostyrene (phenyl-nitroethylene). The substituted benzaldehyde is reacted with nitromethane, nitroethane, or nitropropane to form an intermediate nitrostyrene, β-methyl-β-nitrostyrene or β-ethyl-β-nitrostyrene. Nitrostyrenes are readily recognized in clandestine settings by their characteristic bright yellow or orange color. Figure 4.18 illustrates the structures of the nitrostyrenes.

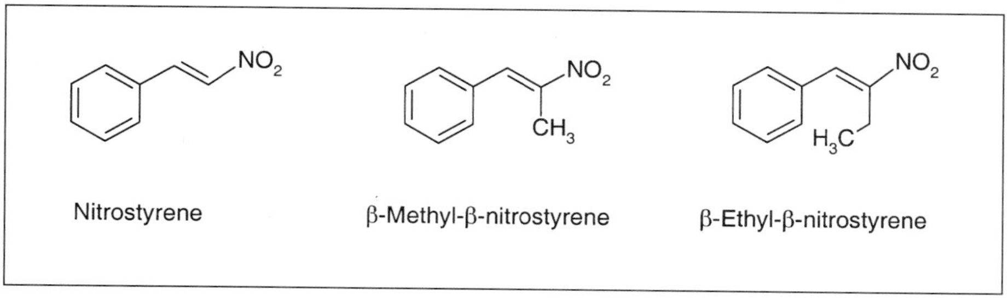

Nitrostyrene | β-Methyl-β-nitrostyrene | β-Ethyl-β-nitrostyrene

Figure 4.18
Nitrostyrene structures

Benington *et al.* (1958) discussed the synthesis of nitrostyrenes from benzaldehydes as one of three ways of making ring substituted phenethylamines. Hamlin and Weston (1949) followed this method in synthesizing 3-methoxy-4,5-methylenedioxyphenethylamine. Figure 4.19 is the first of a two-step synthesis of mescaline which illustrates the condensation of nitromethane with 3,4,5-trimethoxybenzaldehyde to form the corresponding nitrostyrene. In the second step, the nitrostyrene is reduced to form mescaline usually using lithium aluminum hydride as Figure 4.20 illustrates (Shulgin and Shulgin, 1991).

Figure 4.19
3,4,5-Trimethoxy-β-nitrostyrene by PIHKAL (1991)

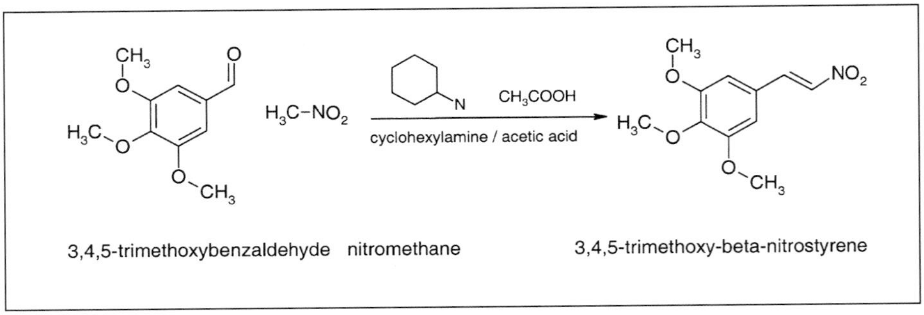

H_3C-NO_2 — CH_3COOH — cyclohexylamine / acetic acid

3,4,5-trimethoxybenzaldehyde nitromethane 3,4,5-trimethoxy-beta-nitrostyrene

Figure 4.20
Mescaline by PIHKAL (1991)

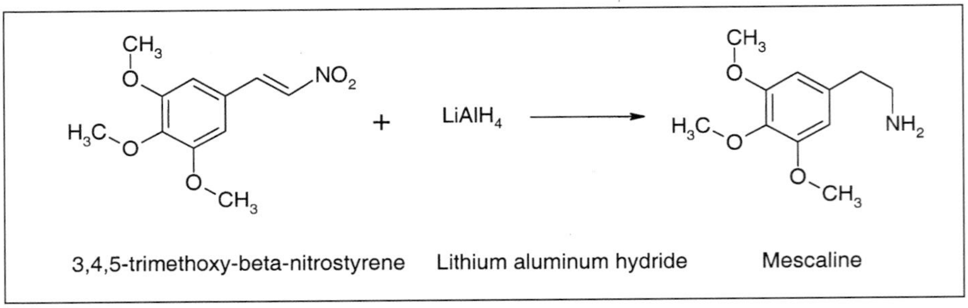

\+ $LiAlH_4$ →

3,4,5-trimethoxy-beta-nitrostyrene Lithium aluminum hydride Mescaline

A similar reaction encountered in clandestine settings, beginning with piperonal and nitroethane, yields the intermediate 3,4-methylenedioxy-β-methyl-β-nitrostyrene which is then reduced to 3,4-methylenedioxyamphetamine (MDA) with lithium aluminum hydride.

In the formation of the nitrostyrene, amines other than cyclohexylamine, depicted in Figure 4.19, have been used successfully. In particular, ammonium acetate was used with acetic acid in the work published by Tomita (1968) and Gairaud and Lappin (1953). In the latter paper, the formation of 24 nitrostyrenes including the formation of β-methyl and β-ethyl-β-nitrostyrenes was outlined. Hey (1947) used n-butylamine in order to obtain 3,4,5-trimethoxyamphetamine via the corresponding nitrostyrene while Heinzelman (1963) also used n-butylamine to synthesize o-methoxyphenyl-β-methyl-β-nitrostyrene from o-methoxybenzaldehyde. Hamlin and Weston (1949) used n-amylamine to make 3-methoxy-4,5-methylenedioxynitrostyrene from the corresponding aldehyde. Lerner (1958) used ethylenediamine without acetic acid to produce the nitrostyrenes and β-methyl-β-nitrostyrenes of five different substituted benzaldehydes including anisaldehyde and vanillin.

Using a different approach, Heacock and Hutzinger (1963) made the corresponding 1-phenyl-2-nitroethanol from a substituted benzaldehyde using a sodium hydroxide dissolved in ethanol or ethanol/dioxane solution. The reaction is halted by the addition of acetic acid in a short (<3 minute) reaction time. In a separate article, Heacock *et al.* (1961) described how substituted 1-phenyl-2-nitroethanols can be readily dehydrated using sodium acetate in acetic anhydride to form the corresponding nitrostyrene.

In the reduction of the nitrostyrenes, lithium aluminum hydride is not the only effective reagent. Hydrogen gas and a metal catalyst can also work quite effectively. In particular, Kawanishi (1957) used a nickel formate paraffin and Raney nickel, Tindall (1954) used Raney nickel, and Green (1962) used palladium on barium sulfate.

Part II: Phenylalkylamines from phenylalkylketones

The most common clandestine synthesis of phenylalkylamines is usually the reductive amination of the corresponding phenylacetaldehyde, phenyl-2-propanone, or phenyl-2-butanone. Several reducing agents can and have been used in clandestine settings.

Leuckart reaction

The Leuckart method uses formic acid as the reducing agent. Crossley and Moore (1944) discussed the application of the Leuckart reaction to unsubstituted phenyl-2-propanones and phenyl-2-butanones to form primary amines. In general, formamide (or ammonium formate) was reduced with the phenyl-2-propanone to form the N-formyl analogue of the phenylalkylamine which can be isolated. The N-formyl part of the molecule was removed by refluxing with strong acid or strong base. For example, Elks and Hay (1943) reacted 1-(3,4-methylenedioxyphenyl)-2-propanone (MD-P-2-P) with ammonium formate to synthesize MDA. As discussed by Novelli (1939), the use of N-methyl formamide produced the N-methyl analogue of the final product and N-ethyl-formamide produced the N-ethyl phenethylamine analogue. The synthesis of higher molecular weight analogues can, however, sometimes be problematic. For instance, the N-ethyl analogue of 3,4-methylenedioxyamphetamine cannot be made by this method. N,N-Dimethyl-phenylalkylamines also cannot be made by this method as the N-formyl intermediate would require the amino nitrogen to form four bonds. If, however, the N-formyl derivative is reduced, rather than cleaved, the N-formyl becomes N-methyl. Interestingly, Shulgin and Shulgin (1991) detailed a method of formylating MDA and then reducing the N-formyl-MDA as a synthesis of MDMA. Dal Cason (1990) included it as a way of obtaining N,N-dimethyl-MDA. Van Haeren *et al.* (2000) noted that the Leuckart reduction of MD-P-2-P with N-methylformamide is a popular reaction in Europe (Figure 4.21).

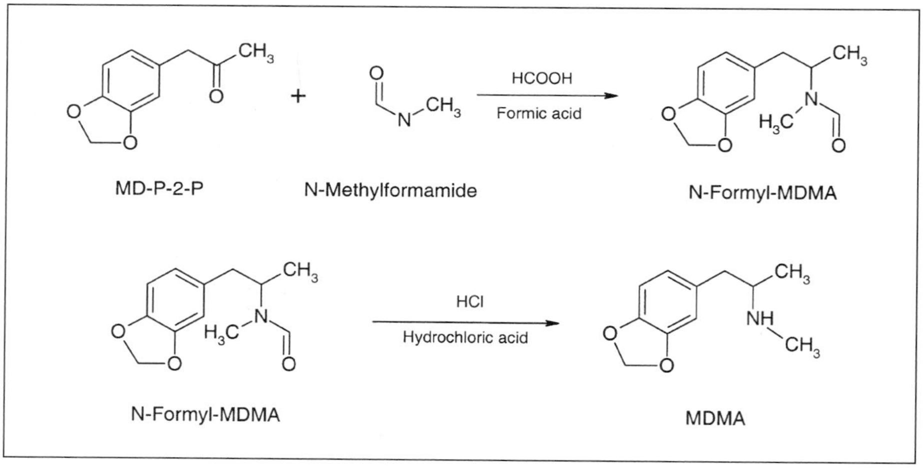

Figure 4.21

MDMA via the Leuckart reaction

If the final step is not taken to completion, residues of the N-formyl analogue of the final product can be found in the finished product. Presence of the N-formyl analogue is considered a marker for the Leuckart reaction.

In most cases, this reaction will work with phenyl-1-propanones, phenylacetaldehyde, phenylpropylaldehyde, and the phenylbutanones in an analogueous fashion.

Reduction using sodium borohydride or sodium cyanoborohydride

The reductive amination of the substituted phenyl-2-propanones can proceed using sodium borohydride or sodium cyanoborohydride as the reducing agent. Figure 4.22 illustrates the reaction with MD-P-2-P, ammonium acetate and sodium cyanoborohydride. The method was published by Braun *et al.* (1980) and Shulgin and Shulgin (1991).

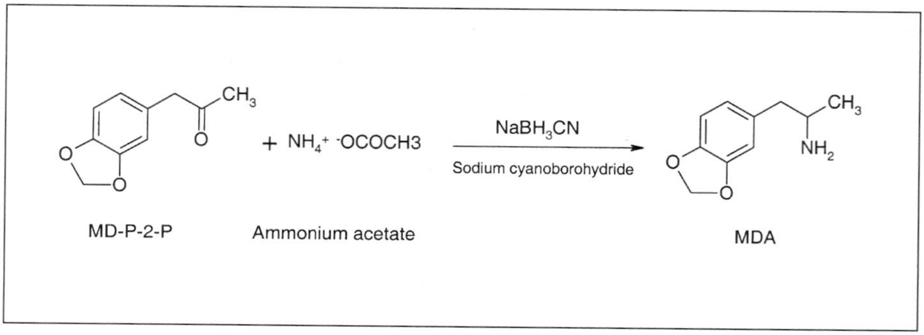

Figure 4.22

MDA by PIHKAL (1991)

The method works for N-substituted analogues as well. According to Van Haeren *et al.* (2000), a recently appearing method in the clandestine synthesis of MDMA used sodium borohydride in the reductive amination of MD-P-2-P and methylamine. This "Cold Method" involved keeping the reactants in a freezer to keep the reaction temperature down (Figure 4.23).

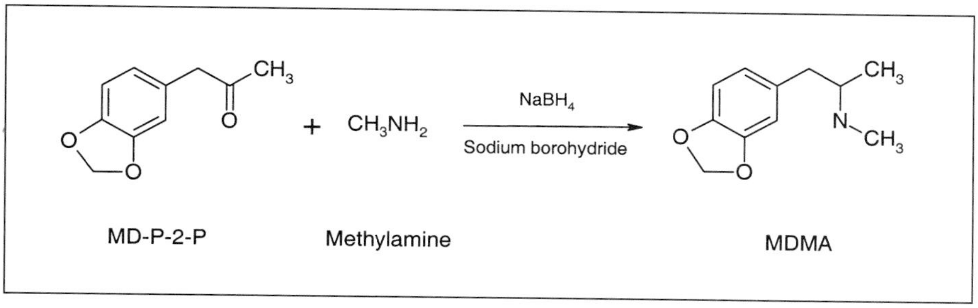

CH_3 + CH_3NH_2 $\xrightarrow{NaBH_4}$ Sodium borohydride CH_3 N CH_3

MD-P-2-P Methylamine MDMA

Figure 4.23

MDMA by Van Haeren et al. *(2000) cold method*

With both sodium borohydride and sodium cyanoborohydride, there is a competing side reaction which is the reduction of the carbonyl to the corresponding alcohol. For instance in the reaction depicted in 4.23, the side product would be 1-(3,4-methylenedioxyphenyl)-2-propanol (MD-P-2-Pol) (Figure 4.24).

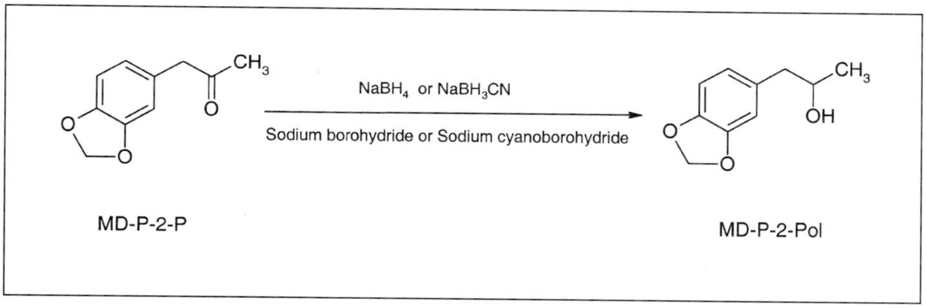

CH_3 $\xrightarrow{NaBH_4 \text{ or } NaBH_3CN}$ Sodium borohydride or Sodium cyanoborohydride CH_3 OH

MD-P-2-P MD-P-2-Pol

Figure 4.24

Formation of MD-P-2-pol

Increasing the concentration of the amine can reduce the effect of the competing side reaction. For instance, in the MDA synthesis (Figure 4.22), increasing the ammonium acetate will increase the yield of MDA versus the MD-P-2-Pol. Using sodium cyanoborohydride as opposed to sodium borohydride will also increase MDA yields relative to MD-P-2-Pol. Notably, any alcohol formed by this side reaction can be easily removed from the reaction product with a liquid-liquid extraction from acid aqueous solution.

Analogous reductive aminations using substituted phenylacetaldehydes and phenylbutanones as starting materials are, in general, workable.

Aluminum amalgam

The reductive amination of substituted phenyl-2-propanones can proceed using aluminum amalgam as the reducing agent. In general, about 0.1g of mercuric chloride is needed to amalgamate with about 4g of aluminum. The reaction is straightforward and is well known in clandestine circles. Figure 4.25 illustrates the synthesis of MDMA from MD-P-2-P and methylamine using the aluminum amalgam as the reducing agent as published by Shulgin and Shulgin (1991).

Figure 4.25
MDMA by PIHKAL (1991)

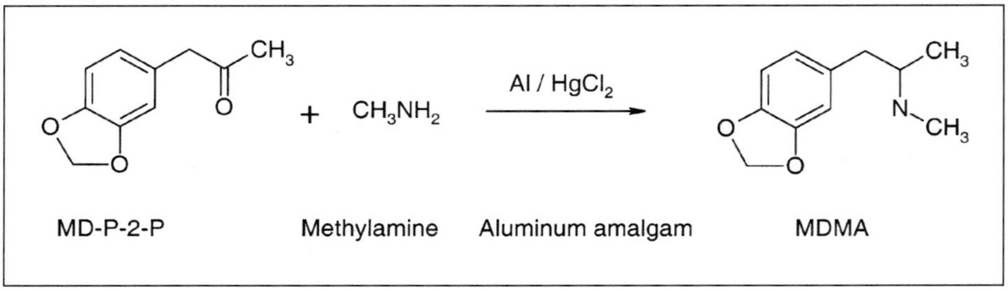

MD-P-2-P + CH_3NH_2 —Al / $HgCl_2$→ MDMA

MD-P-2-P | Methylamine | Aluminum amalgam | MDMA

Side products such as the carbonyl being reduced to the corresponding alcohol are often seen in these reactions unless the clandestine chemist removes it. See the discussion under reduction using sodium borohydride and sodium cyanoborohydride.

The analogous reactions in terms of using ammonium, and N-ethyl analogues are all feasible. Since the reaction appears to go through an intermediate imine step, di-substituted amines, like N,N-dimethylamine do not work with this reaction. The intermediate imine step as depicted in Figure 4.26 when applied to N,N-dimethylamine will require four bonds of nitrogen which prevents its formation.

Figure 4.26
Imine intermediate formation

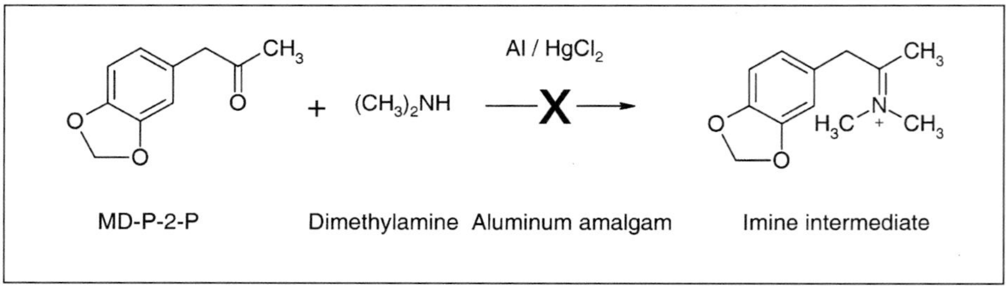

MD-P-2-P + $(CH_3)_2NH$ —Al / $HgCl_2$ X→ Imine intermediate

MD-P-2-P | Dimethylamine | Aluminum amalgam | Imine intermediate

Hydrogen gas and a metal catalyst

As reported by Van Haeren *et al.* (2000), a common European method for the synthesis of MDMA was the reductive amination using hydrogen and a platinum metal catalyst (Figure 4.27).

As with the sodium borohydride/cyanoborohydride and aluminum amalgam reductions, a common side product is the alcohol resulting from the reduction of the carbonyl. For example in Figure 4.27, an expected side product would be MD-P-2-Pol.

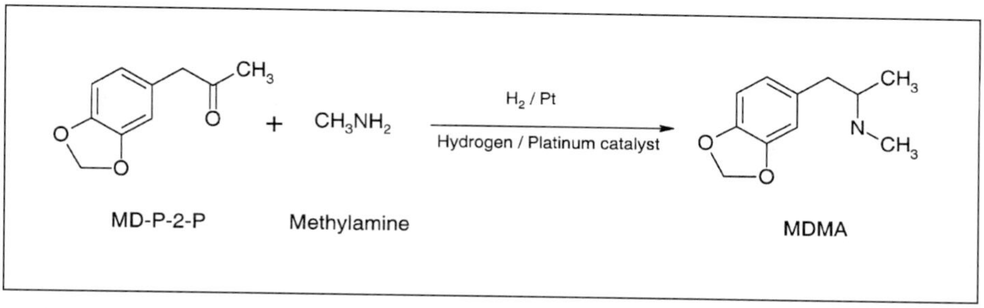

Figure 4.27

MDMA by Van Haeren et al. *(2000)*

In place of a platinum catalyst, Raney nickel can be used to reduce (unsubstituted) phenyl-2-propanone and methylamine according to Weaver and Yeung (1995).

Part III: Phenylalkylketones from benzaldehydes

Phenylalkylketones can be synthesized from the corresponding benzaldehyde by first making the nitrostyrene or β-methyl-β-nitrostyrene or β-ethyl-β-nitrostyrene. That nitrostyrene is then reacted with electrolytic iron and acetic or hydrochloric acid to make the phenylalkylketone. Figure 4.28 depicts the two step synthesis of MD-P-2-P from Piperonal using Shulgin and Shulgin (1991) method. Heinzelman (1963) used this method to obtain o-methoxyphenyl-2-propanone.

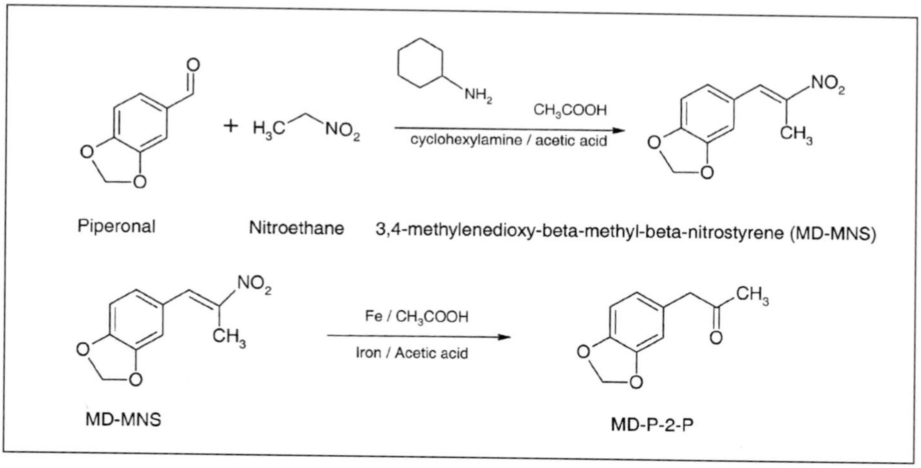

Figure 4.28

MD-P-2-P by PIHKAL (1991)

Part IV: Benzaldehydes and phenylalkylketones from allylbenzenes

There are several routes that start with the substituted allylbenzene to synthesize the corresponding phenylalkylamine. These routes are popular in clandestine settings because several substituted allylbenzenes are made in nature and are readily available. For example, *The Merck Index* (2001) notes that safrole is a 75% constituent of sassafras oil, while Hamlin and Weston (1949) mention that oil of nutmeg contains 7.5% myristicin. Shulgin (1963) discussed the presence of myristicin as well as methyl isoeugenol and elemycin in oil of nutmeg. An article by Schenk and Lamparsky (1981) analyzed the contents of nutmeg oil, finding that myrisiticin makes up 14%. Kalbhen (1971) discussed the contents of nutmeg oil from a possible hallucinogenic view. Figure 4.29 depicts the structures of several naturally occurring substituted allylbenzenes from Shulgin (1967). All could potentially react as described in this section to make the corresponding phenylalkylamine.

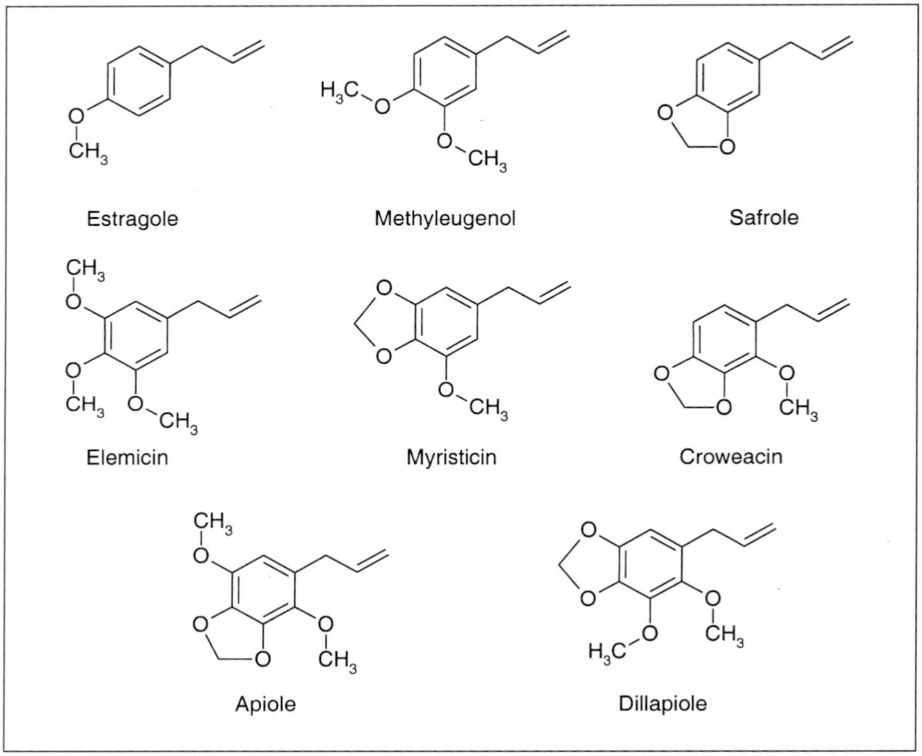

Figure 4.29
Substituted allylbenzene names from Shulgin (1967)

Estragole Methyleugenol Safrole
Elemicin Myristicin Croweacin
Apiole Dillapiole

In most cases, the substituted allylbenzene will go through an intermediate substituted phenylalkylketone or benzaldehyde that is reacted as described previously to form the phenylalkylamine.

Phenylalkylketone from corresponding allylbenzene

The most common example encountered in clandestine laboratories of the synthesis of phenylalkylketones from the corresponding allylbenzene is the synthesis of MD-P-2-P from its corresponding allylbenzene, safrole. In the first step, the allylbenzene (safrole) is isomerized to the 2-propenylbenzene (isosafrole) by boiling with potassium hydroxide according to Shulgin (1967) as depicted in Figure 4.30. The isosafrole is oxidized to MD-P-2-P using hydrogen

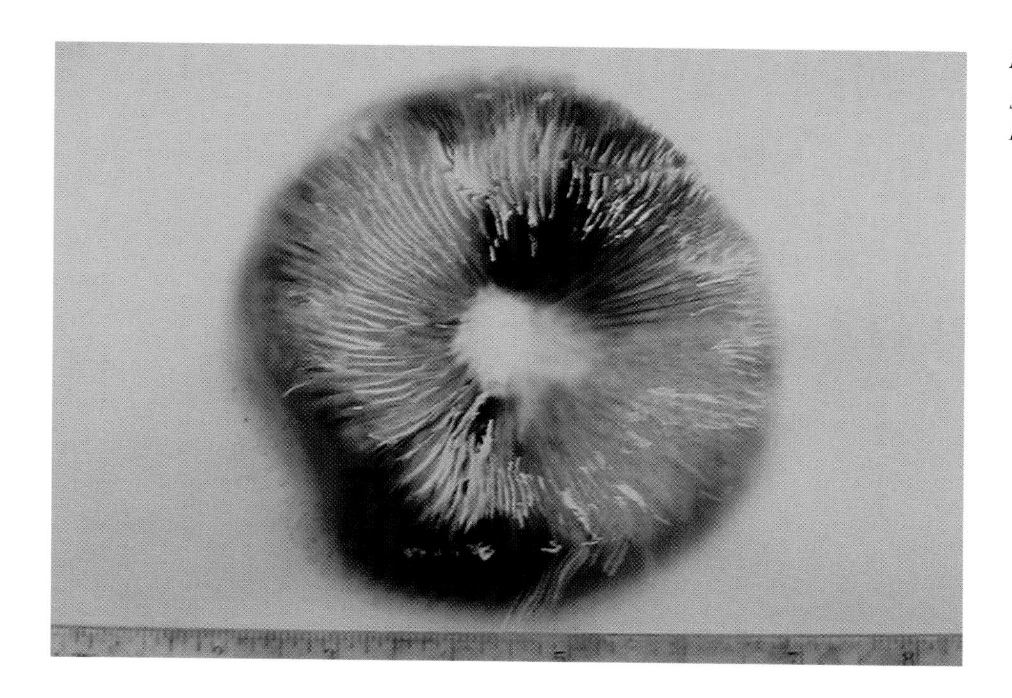

Plate 4.1

Spore print from Psilocybe cubensis

Plate 4.2

Mycelium cultivation on grain in Mason jar

Plate 4.3

P. cubensis *fruiting on a straw "log"*

Plate 4.4 (opposite, above)

A map of western Europe and Benelux region

Plate 4.5 (opposite, below)

A trafficking map showing recent major seizures of MDMA and precursors

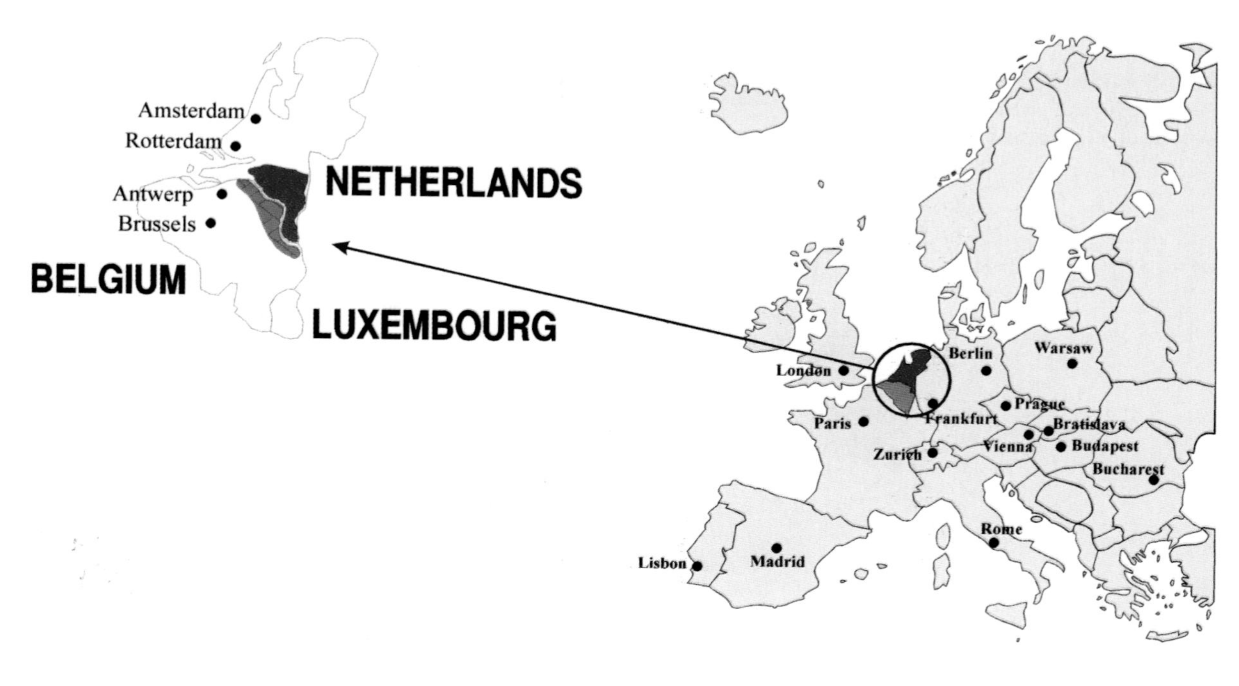
Amsterdam
Rotterdam
Antwerp
Brussels
NETHERLANDS
BELGIUM
LUXEMBOURG
London
Berlin
Warsaw
Paris
Frankfurt
Prague
Bratislava
Vienna
Budapest
Zurich
Bucharest
Rome
Lisbon
Madrid

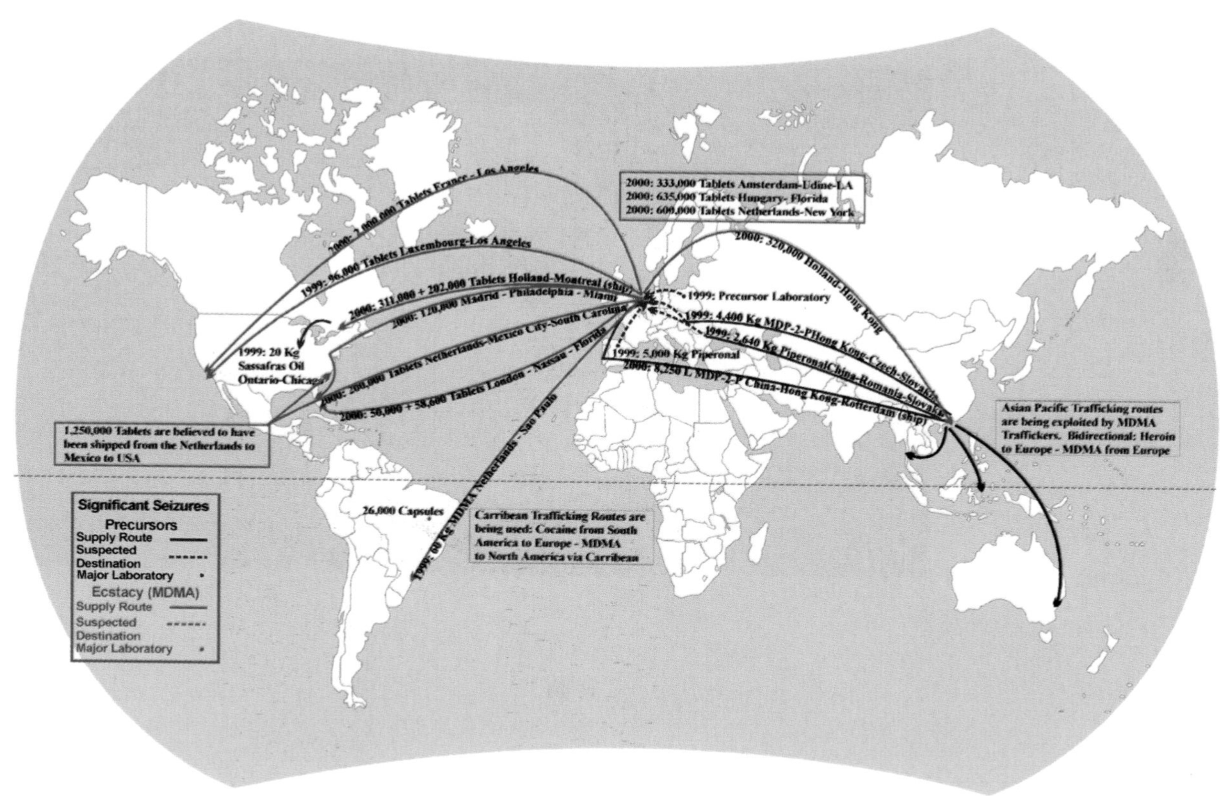
2000: 2,000,000 Tablets France - Los Angeles
1999: 96,000 Tablets Luxembourg-Los Angeles
2000: 311,000 + 202,000 Tablets Holland-Montreal (ship)
2000: 120,000 Madrid - Philadelphia - Miami
2000: 200,000 Tablets Netherlands-Mexico City-South Carolina
2000: 50,000 + 58,600 Tablets London - Nassau - Florida
1999: 60 Kg MDMA Netherlands - Sao Paulo
1999: 20 Kg
Sassafras Oil
Ontario-Chicago
2000: 333,000 Tablets Amsterdam-Udine-LA
2000: 635,000 Tablets Hungary- Florida
2000: 600,000 Tablets Netherlands-New York
2000: 320,000 Holland-Hong Kong
1999: Precursor Laboratory
1999: 4,400 Kg MDP-2-PHong Kong-Czech-Slovakia
1999: 2,640 Kg PiperonalChina-Romania-Slovakia
1999: 5,000 Kg Piperonal
2000: 8,250 L MDP-2-P China-Hong Kong-Rotterdam (ship)
1,250,000 Tablets are believed to have been shipped from the Netherlands to Mexico to USA
Asian Pacific Trafficking routes are being exploited by MDMA Traffickers. Bidirectional: Heroin to Europe - MDMA from Europe
26,000 Capsules
Carribean Trafficking Routes are being used: Cocaine from South America to Europe - MDMA to North America via Carribean
Significant Seizures
Precursors
Supply Route
Suspected Destination
Major Laboratory
Ecstacy (MDMA)
Supply Route
Suspected Destination
Major Laboratory

Plate 4.6

LSD laboratory upper floor showing equipment, work benches and lattice

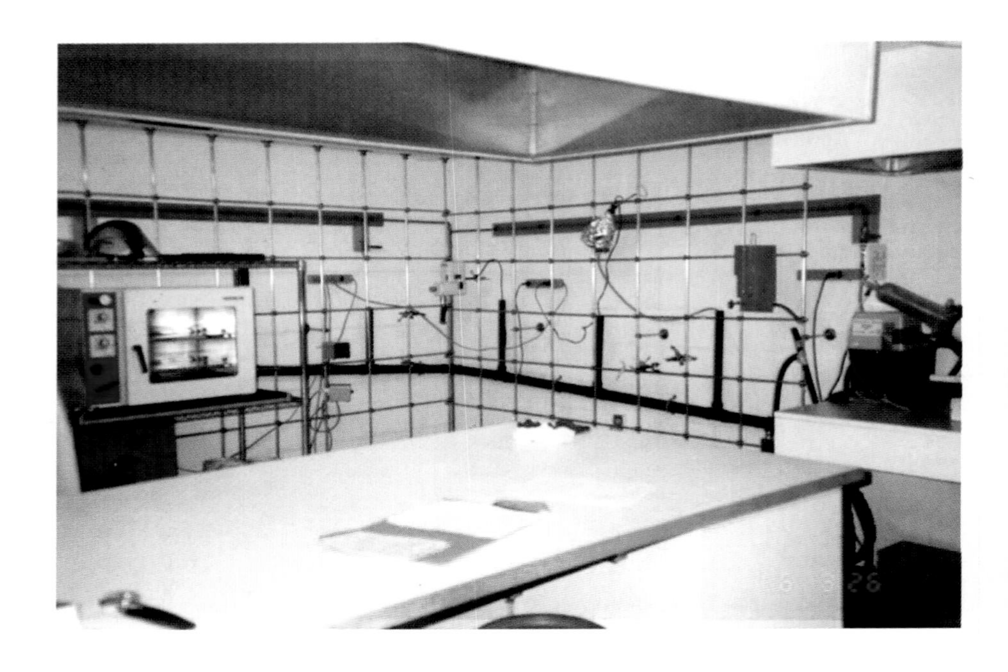

Plate 4.7

"Standards" of LSD and iso-LSD with rhomboid crystals of DMT

peroxide and formic acid and then sulfuric acid. Specifics as depicted in Figure 4.31 are taken from Shulgin and Shulgin (1991).

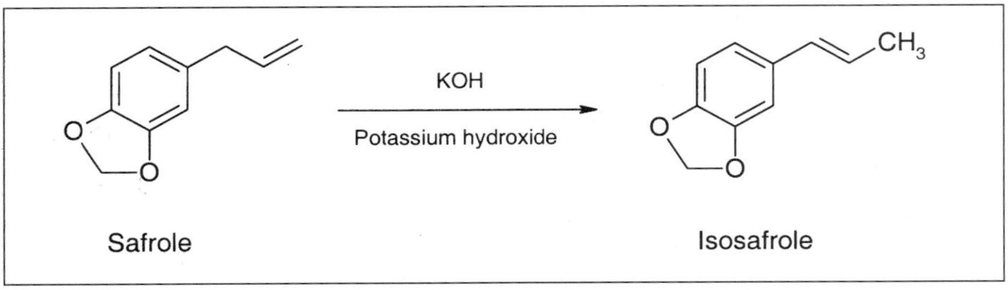

Figure 4.30
Isosafrole by Shulgin (1967)

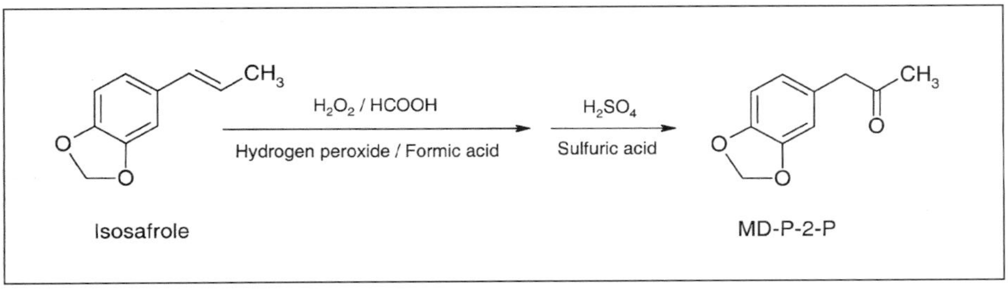

Figure 4.31
MD-P-2-P by PIHKAL (1991)

A newer one-step synthesis of phenylalkylketones from their corresponding allylbenzenes was published by Nakai and Enoralya in US Patent 4,638,094. Methyl nitrite was generated in situ and then introduced to a stirred solution of the allylbenzene containing the chloride or bromide salts of palladium. Cupric chloride can be used as a less expensive co-catalyst, but the authors have not had success with this variance. Figure 4.32 illustrates a synthesis of MD-P-2-P from safrole using example 24 from US Patent 4,638,094.

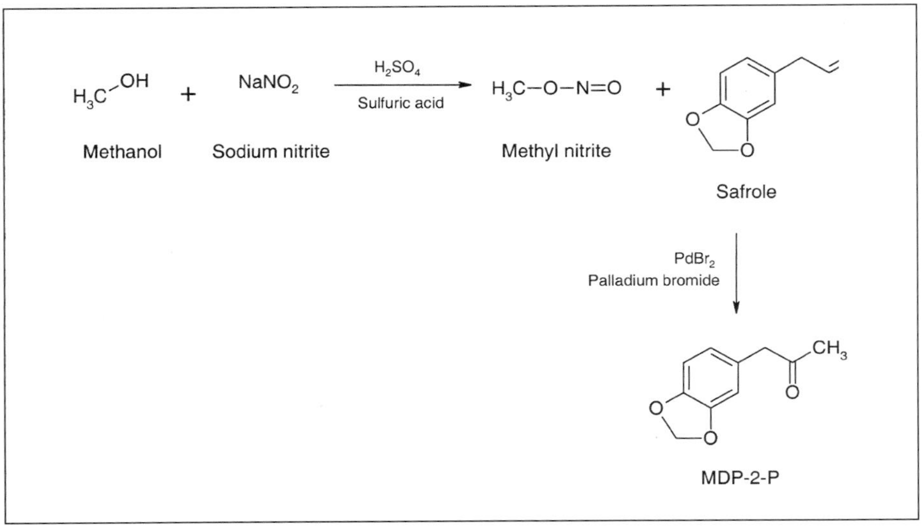

Figure 4.32
MD-P-2-P by Nakai and Enoralya (1987)

The use of the latter method of generating methyl nitrite and using palladium chloride as a catalyst has been encountered in several occasions in Canadian clandestine laboratories.

Benzaldehyde from corresponding allylbenzene

In an article by Milas (1937), the oxidation of three substituted propenylbenzenes with hydrogen peroxide and vanadium pentoxide or chromium trioxide is outlined. In particular, anethole is oxidized to anisaldehyde, isoeugenol to vanillin and isosafrole to piperonal (Figure 4.33). Hamlin and Weston (1949) use a similar method to obtain myristicinaldehyde.

Figure 4.33

Vanillin by Milas (1937)

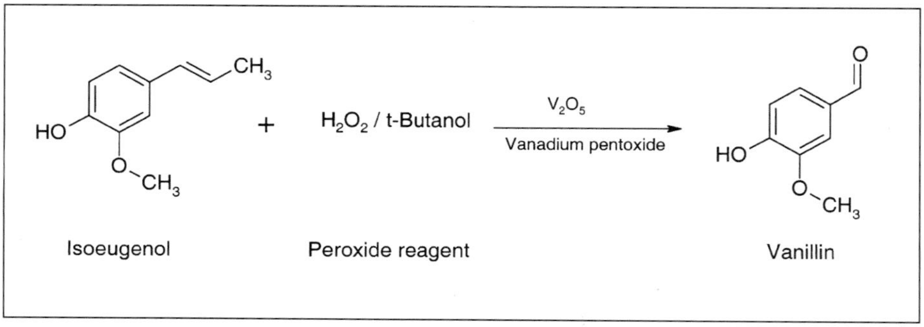

In a different approach Shulgin (1968) began with myristicin to isomyristicin to the corresponding nitrostyrene to myrisiticinaldehyde (Figure 4.34). Notably, the intermediate nitrostyrene could be reduced directly to form 3-methoxy-4,5-methylenedioxyamphetamine.

Figure 4.34

Myristicinaldehyde by Shulgin (1968)

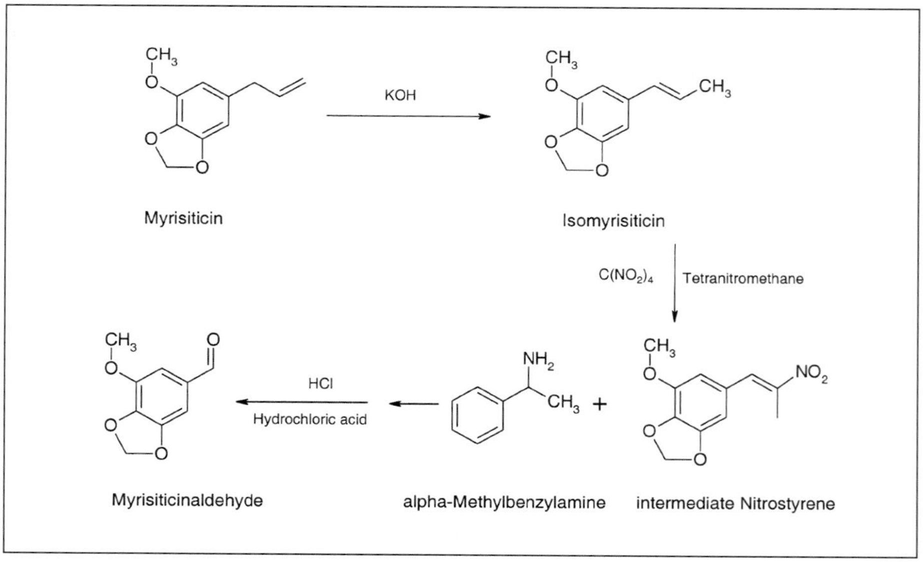

Part V: Miscellaneous syntheses

Phenylalkylamine from allylbenzene via phenylalkylbromine

Biniecki and Krajewski (1961) brominate safrole with hydrobromic acid which is reacted with methylamine to make MDMA (Figure 4.35).

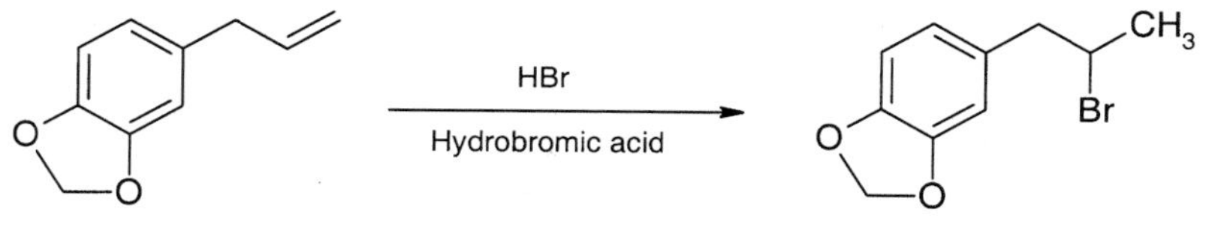

Safrole MD-P-2-bromopropane

Figure 4.35
MDMA by Biniecki and Krajewski (1961)

This reaction has turned up in clandestine laboratory investigations, with the clandestine cooks having mixed success.

Precursors via corresponding Grignard reagent

Beginning with the appropriately substituted bromobenzene, one can make the corresponding Grignard reagent which, in turn, can be converted into the corresponding benzaldehyde, phenylalkylketone, or the allylbenzene. For example, Tiffaneau (1907) outlined the formation of (unsubstituted) phenyl-2-propanone from phenylmagnesium bromide and chloroacetone. Feugas (1964) discussed the synthesis of safrole and piperonal among others from 3,4-methylenedioxybromobenzene. The corresponding Grignard reagent is formed and by a proper choice of reactant, several 3,4-methylenedioxy precursors can be made. The article included the synthesis of safrole, piperonal, and isosafrole among others from 3,4-methylenedioxybromobenzene. Figure 4.36 illustrates the synthesis of safrole.

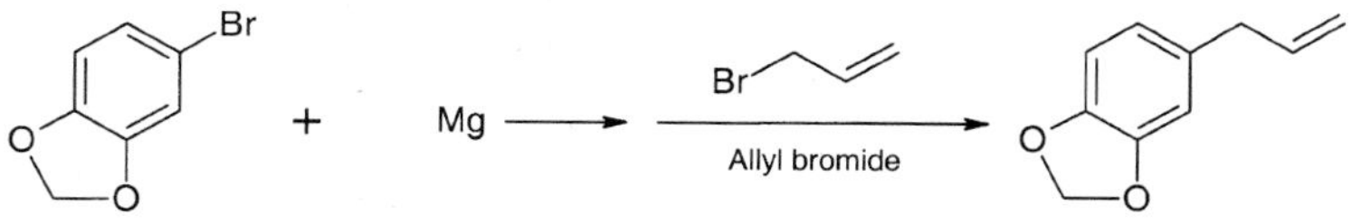

3,4-Methylenedioxybromobenzene Magnesium Safrole

Figure 4.36
Safrole by Feugeas (1984)

Nitrostyrenes from styrenes

Nitrostyrenes as well as β-methyl-β-nitrostyrenes and β-ethyl-β-nitrostyrenes can be synthesized by nitrating the correspondingly substituted styrene. Sy and By (1985) used nitryl iodide to nitrate substituted styrenes to form the corresponding nitrostyrenes (Figure 4.37).

Figure 4.37

Nitrostyrene by Sy and By (1985)

Anethole

N_2I

Nitryl Iodide

4-Methoxy-β-methyl-β-nitrostyrene

Precursors by modifying ring substitutents

Remarkably, in clandestine settings, some individuals will modify the ring substituents to obtain the desired precursor. This can involve methylating hydroxy groups, creating methylenedioxy bridges with di-hydroxy substituted benzenes and adding an aldehyde group to an already partially substituted benzene ring. Shulgin and Shulgin's compendium of reactions (1991) contained many such syntheses. The synthesis of 2,5-dimethoxy-4-methylthiobenzaldehyde is illustrated in Figure 4.38.

Figure 4.38

2,5-Dimethoxy-4-(methylthio)benzaldehyde by PIHKAL (1991)

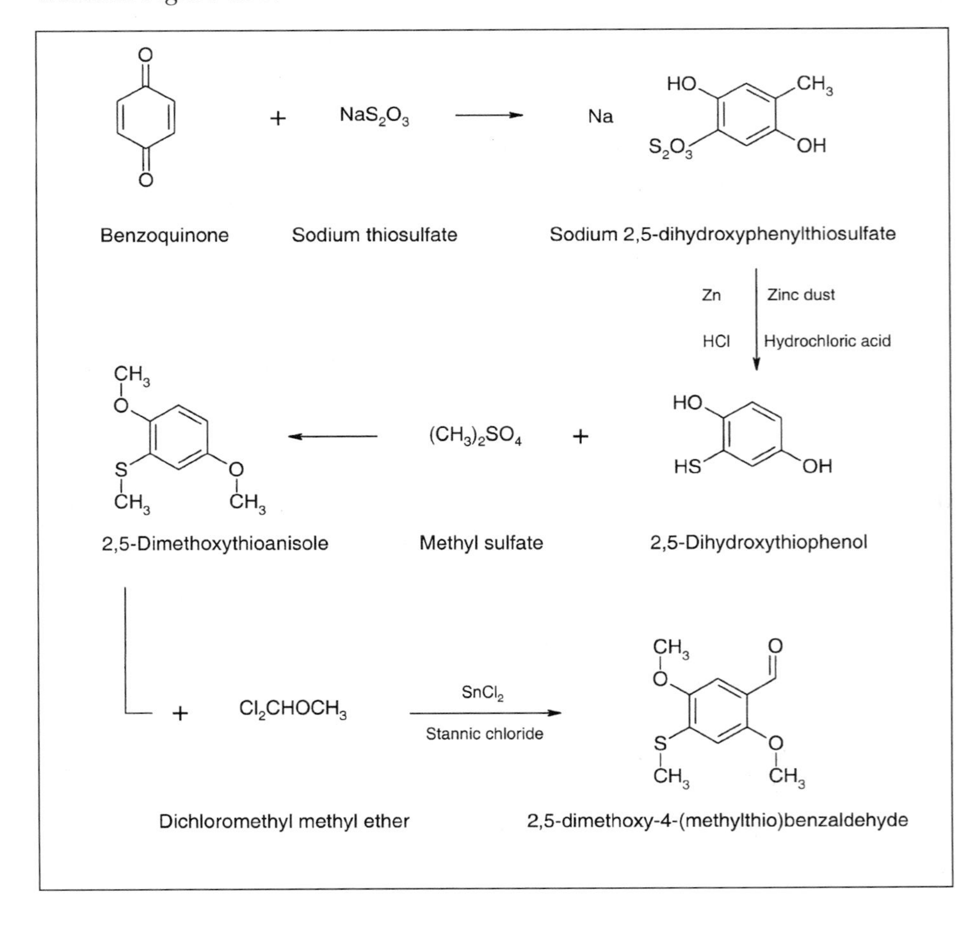

Campbell *et al.* (1951) discussed the synthesis of several methylenedioxy-methoxybenzaldehydes. An example of a synthesis described in the article is the synthesis of 2-methoxy-3,4-methylenedioxybenzaldehyde from 2,3-dihydroxyanisole. Earlier articles by Baker *et al.* (1932) and Trikojus and White (1948) discussed similar routes to myrsiticin. Baker *et al.* (1934) discussed the synthesis of dill apiole (2,3-dimethoxy-3,4-methylendioxyallylbenzene from 1,2,3,4-tetrahydroxybenzene. In another article, Baker *et al.* (1939) discussed the identification of the compound croweacin (2-methoxy-3,4-methylene-dioxyallylbenzene) and in doing so described its synthesis from 2-hydroxy-3,4-methylenedioxyallylbenzene. Clark *et al.* (1976) discussed a quick way of making the methylenedioxy group from 1,2-dihydroxybenzenes. Tomita and Aoyagi discuss a similar synthesis using methylene halides catalyzed by cupric oxide in dimethylformamide.

4.1.4 PCP, KETAMINE AND RELATED COMPOUNDS

Phencyclidine synthesis

Phencyclidine has been abused since the early 1970s. It remains common in certain geographical areas such as the province of Quebec in Canada. Notably, its well-deserved bad reputation has resulted in phencyclidine, or PCP, being often sold as other drugs such as mescaline or powdered THC.

An excellent review of synthetic routes to PCP is found in the article by Allen *et al.* (1993). The article reviews routes to phencyclidine and rates them from yield, difficulty, and hazard points of view. The discussion in this chapter emphasizes routes that are likely to be followed by a clandestine cook. The synthesis of phencyclidine usually involves the formation of an intermediate 1-piperidinocyclohexanecarbonitrile, commonly referred to as PCC. PCC is formed by the addition of potassium cyanide, cyclohexanone, and piperidine. There are two variations on the method to produce PCC. Maddox *et al.* (1965) specifies the formation of the cyclohexanone bisulfite addition product and then the addition of potassium cyanide and piperidine, as depicted in Figure 4.39. Kalir *et al.* (1969) call for the mixing of piperidine with hydrochloric acid, adjusting the pH to 3–4 and then the addition of cyclohexanone and potassium cyanide. The mixture is stirred for two hours and allowed to stand overnight. The off-white coloured PCC forms and is filtered and dried before being used. Yields using either method are in the 85 to 90% range.

PCC is dissolved in a solvent such as isooctane as described by Maddox *et al.* (1965) and then reacted with phenylmagnesium bromide. (Phenylmagnesium bromide is either purchased as a commercially prepared Grignard reagent or is made from bromobenzene and magnesium turnings.) The addition product so formed is hydrolyzed with 4N hydrobromic acid to form PCP hydrobromide.

Figure 4.39

1-Piperidinocyclohexane-carbonitrile (PCC) by Maddox et al. *(1965)*

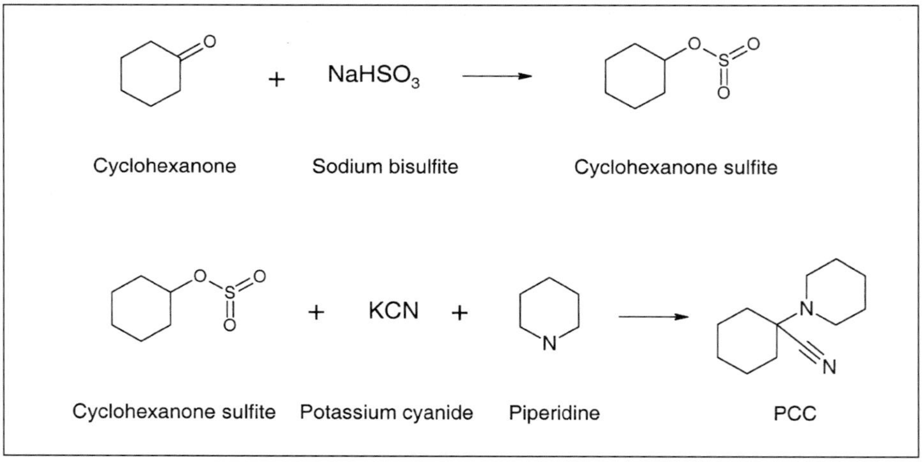

Cyclohexanone + $NaHSO_3$ → Cyclohexanone sulfite

Cyclohexanone | Sodium bisulfite | Cyclohexanone sulfite

Cyclohexanone sulfite + KCN + Piperidine → PCC

Cyclohexanone sulfite | Potassium cyanide | Piperidine | PCC

Figure 4.40 illustrates this synthesis of phencyclidine. Large quantities of by-product, typically magnesium salts, are generated along with the PCP. In some cases, the clandestine cook will purify the PCP by liquid–liquid extraction by dissolving the PCP hydrobromide in dilute acid, extracting to eliminate some of the biphenyl which forms, then making basic and extracting. Phencyclidine hydrochloride is then produced by adding hydrochloric acid. In other cases, the clandestine cook will not bother with an extraction clean up, but will dilute the PCP hydrobromide to a "safe" level and put it out on the street.

Figure 4.40

Phencyclidine (PCP) by Maddox et al. *(1965)*

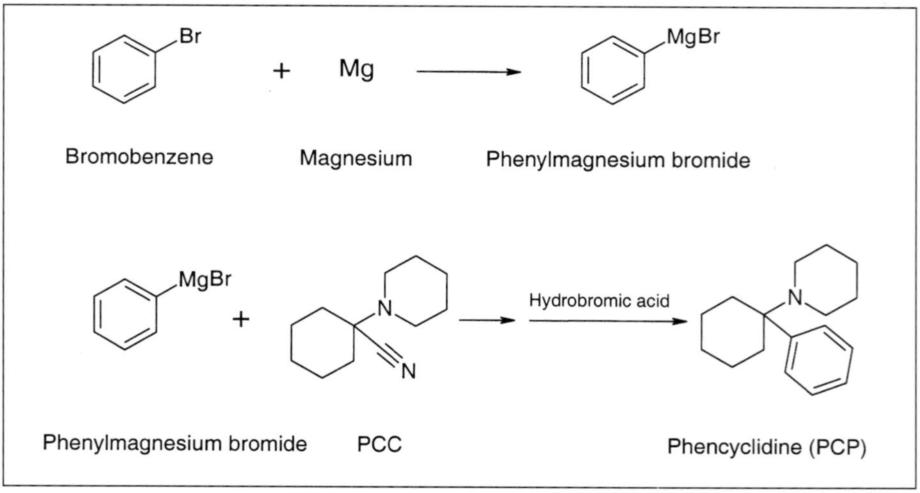

Bromobenzene + Mg → Phenylmagnesium bromide

Bromobenzene | Magnesium | Phenylmagnesium bromide

Phenylmagnesium bromide + PCC → Hydrobromic acid → Phencyclidine (PCP)

Phenylmagnesium bromide | PCC | Phencyclidine (PCP)

Generally, in Canada, the concentration of PCP in a powder street sample is in the 1–3% by weight range. Variations on this method call for the use of ammonium chloride (Kalir, 1969) to hydrolyze the addition product of PCC and phenylmagnesium bromide. In this case, an extraction is made to isolate the phencyclidine base and hydrochloric acid added to make the salt. In clandestine laboratories, the methods can be modified without regard for the chemistry behind the detailed procedure. An interesting discussion of a clandestine recipe in Allen *et al.* (1993) is an example. Other clandestine lab chemists have used xylenes in place of isooctane to dissolve the PCC prior to addition of phenylmagnesium bromide.

One notable ploy used by clandestine cooks is to make PCC in one location and either make or have phenylmagnesium bromide in a second location. They bring the two components together in a field or parking lot with lots of open space around. Should law enforcement show up, the chemicals are dumped and the suspects flee.

There are other procedures that do not involve the formation of PCC. In a US Patent by Godefoi *et al.* (1963), and a British Patent assigned to Parke Davis (1960), piperidine, cyclohexanone, and benzene are azeotropically distilled to form 1-(1-cyclohexenyl)-piperidine. Para-toluenesulfonic acid is then added and the addition product so formed is reacted with phenylmagnesium bromide. Figure 4.41 illustrates. Again, the addition product is decomposed with ammonium chloride, cleaned up by liquid–liquid extraction and the hydrochloride salt formed.

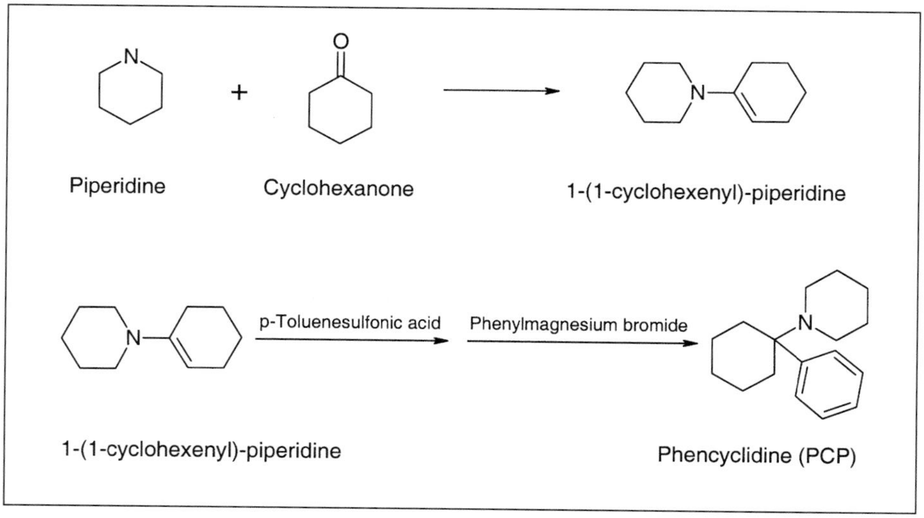

Piperidine + Cyclohexanone → 1-(1-cyclohexenyl)-piperidine

1-(1-cyclohexenyl)-piperidine —p-Toluenesulfonic acid→ —Phenylmagnesium bromide→ Phencyclidine (PCP)

Figure 4.41
Phencyclidine (PCP) by Godefoi et al. *(1963)*

Other less common methods for the synthesis of phencyclidine are discussed in the paper by Allen *et al.* (1993).

Synthesis by-products

In an article by Angelos *et al.* (1990), the synthetic impurities of clandestinely produced phencyclidine are described. The impurities are divided into starting materials, reaction intermediates, and reaction side products. The method of Maddox *et al.* (1965) and of Kalir *et al.* (1969) are described as the nitrile method while the method of Godefoi *et al.* (1965) is described as the enamine method. Structures of the various synthetic impurities are depicted in Figure 4.42. 1-Phenylcyclohexene has not been included as it is an analysis artifact when using gas chromatography. Only the presence of the reaction intermediates, namely 1-(1-cyclohexyl)piperidine and 1-piperidinocyclohexanecarbonitrile (PCC) are considered to be indicative of the particular method (nitrile or enamine) followed by the clandestine cook.

Figure 4.42

Impurities in clandestinely produced PCP by Angelos et al. *(1990)*

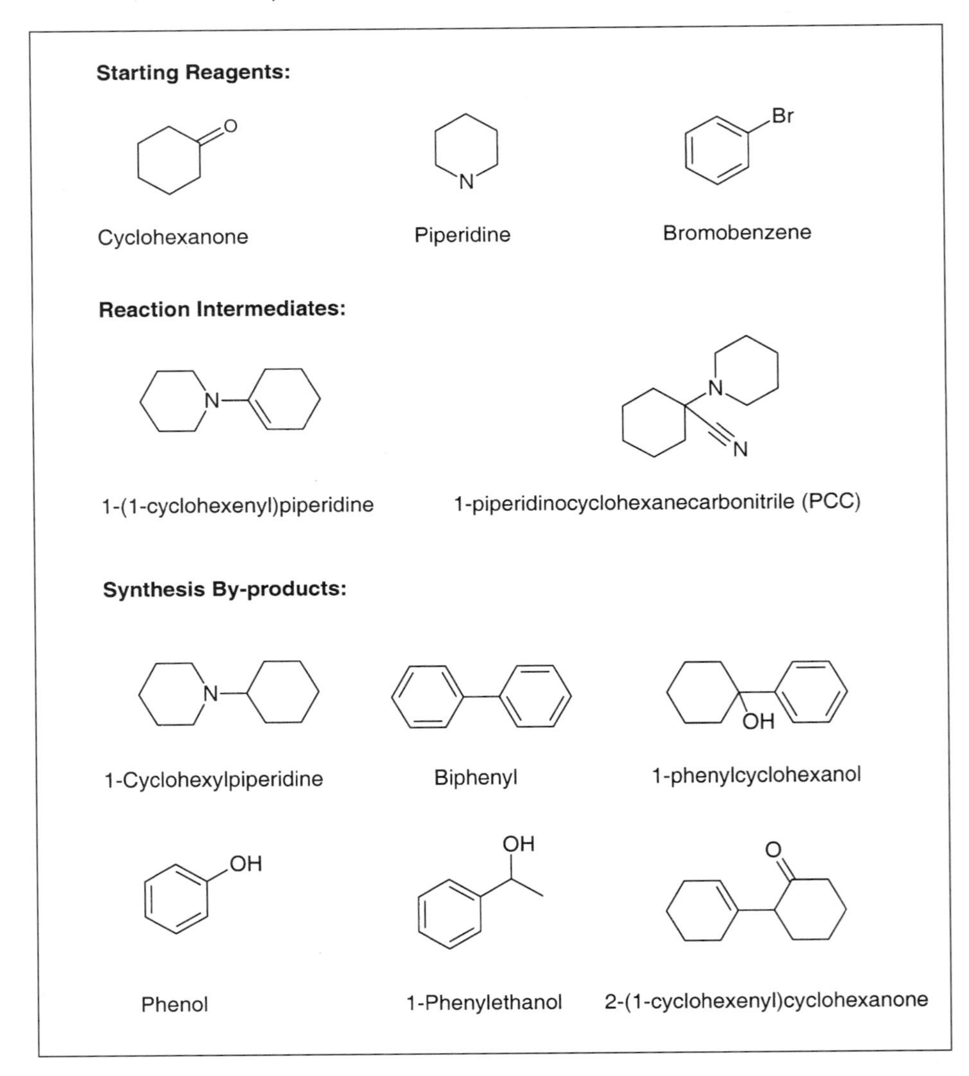

Analogues of phencyclidine

Several analogues of phencyclidine have been synthesized and sold at street level. Either the piperidine group is substituted or the phenyl group is substituted. Figure 4.43 depicts structures where the piperidine group has been substituted and Figure 4.44 illustrates structures where the phenyl group has been substituted.

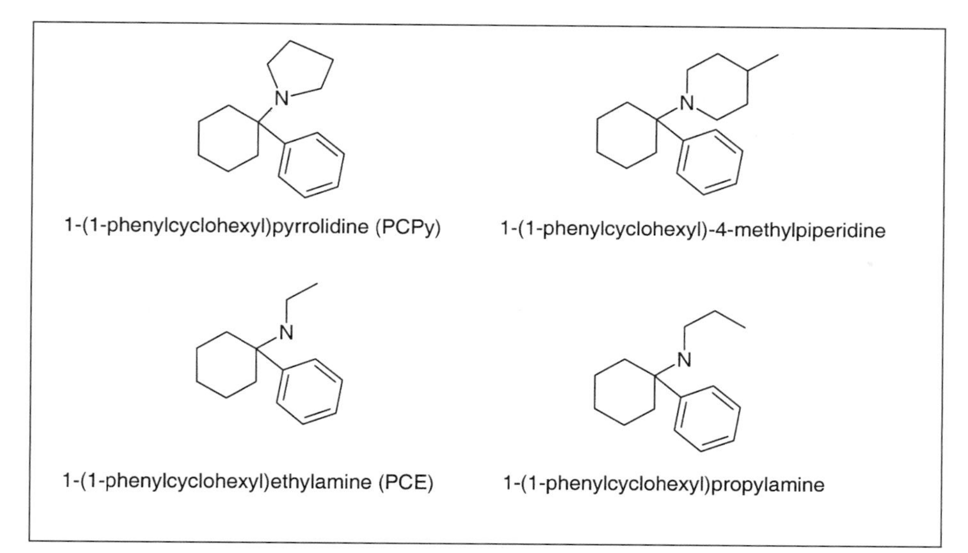

Figure 4.43

Street analogues of phencyclidine with piperidine substituted

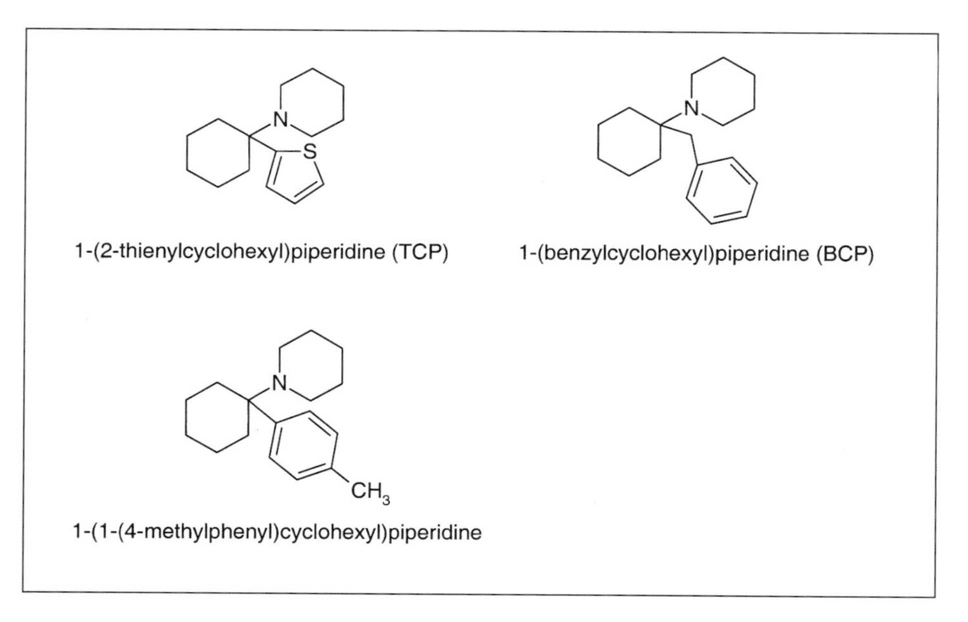

Figure 4.44

Street analogues of phencylclidine with phenyl group substituted

Shulgin and MacLean (1976), while discussing the synthesis of PCP analogues in general, mention that 1-(2-thienylcyclohexyl)piperidine (TCP), 1-(1-phenylcyclohexyl)ethylamine (PCE), and 1-(1-phenylcyclohexyl)pyrrolidine (PCPy) had been identified in street drug exhibits. Bailey and Legault (1979) during their description of the synthesis of derivatives of 1-phenylcyclohexylamine, state that, in addition to those already noted, that 1-(1-phenylcyclohexyl)propylamine had been identified in street exhibits. An article by Allen *et al.* (1980) discusses the possibility that 1-(1-phenylcyclohexyl)-4-methylpiperidine would appear on the street. Lodge *et al.* (1992) describe the appearance of the 1-(benzylcyclohexyl)piperidine (BCP) and 1-(1-(4-methylphenyl)cyclohexyl)piperidine.

The analogues where the piperidine group has been substituted by another secondary or cyclic amine (pyrrolidine or 4-methylpiperidine), are synthesized by producing the analogue of PCC. Figure 4.45, following the method of Kalir *et al.* (1969), illustrates the synthesis of 1-pyrrolidinocyclohexanecarbonitrile. That intermediate analogue is then reacted with phenylmagnesium bromide to form the final analogue product. The method of Maddox *et al.* (1965) and Godefoi *et al.* (1963) will work in a similar fashion.

Figure 4.45

1-Pyrrolidinocyclohexanecarbonitrile by Kalir et al. *(1969)*

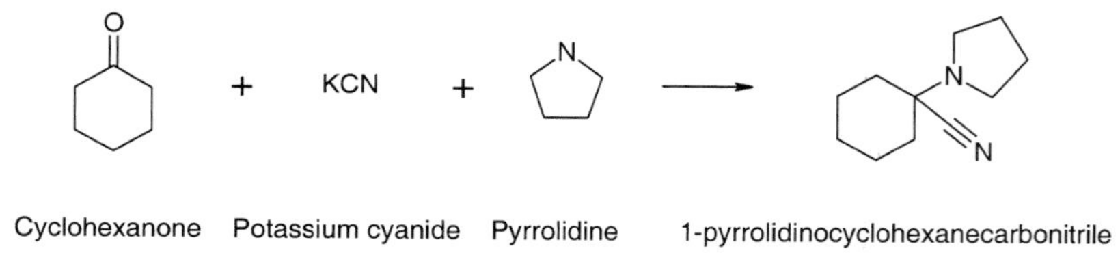

The synthesis of PCP analogues where the piperidine is replaced by a primary amine can be accomplished by the formation of a Schiff base that is reacted with phenyl magnesium bromide. The procedure of Maddox *et al.* (1965) for PCE is illustrated in Figure 4.46. As Shulgin and MacLean (1976) discuss in their article, phenyl lithium when reacted with a PCC (e.g. 1-ethylaminocyclohexanecarbonitrile) analogue with a primary amine adds to the nitrile group instead of displacing it. Bailey and Legault (1979) were, however, able to form the PCC analogues and then react those intermediate analogues with phenyl lithium to produce PCE and 1-(1-phenylcyclohexyl)propylamine (Figure 4.47).

Replacing the phenyl magnesium bromide with the appropriate Grignard reagent is the most common way to make PCP analogues in which the phenyl ring has been substituted. Kalir *et al.* (1969) includes TCP in his investigation by reacting PCC with 2-thienylmagnesium bromide. Bailey *et al.* (1976) follows the same method for making TCP, but uses the Godefoi *et al.* (1963) method to make the morpholine and pyrrolidine analogues of TCP. Lodge *et al.* (1992) start with benzyl bromide and 4-methyl-bromobenzene to synthesize BCP and 1-(1-(4-methylphenyl)cyclohexyl)piperidine. The bromo compounds are made into the corresponding Grignard reagent and reacted with PCC. Figure 4.48 illustrates the BCP example.

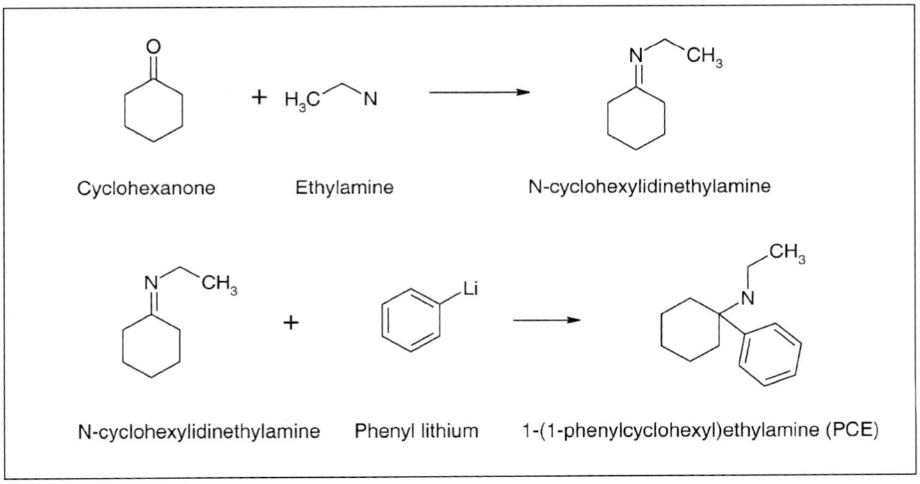

Cyclohexanone + Ethylamine → N-cyclohexylidinethylamine

N-cyclohexylidinethylamine + Phenyl lithium → 1-(1-phenylcyclohexyl)ethylamine (PCE)

Figure 4.46

1-(1-Phenylcyclohexyl) ethylamine (PCE) by Maddox et al. *(1965)*

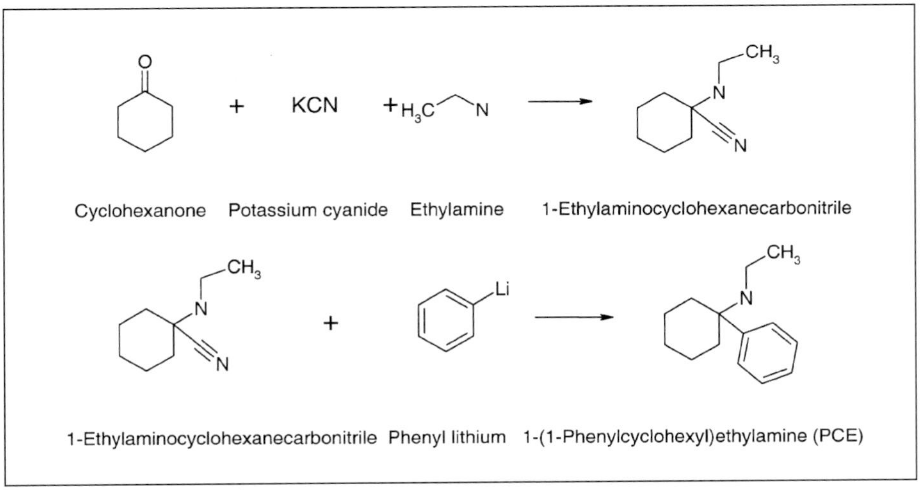

Cyclohexanone + Potassium cyanide (KCN) + Ethylamine → 1-Ethylaminocyclohexanecarbonitrile

1-Ethylaminocyclohexanecarbonitrile + Phenyl lithium → 1-(1-Phenylcyclohexyl)ethylamine (PCE)

Figure 4.47

1-(1-phenylcyclohexyl) ethylamine (PCE) by Bailey and Legault (1979)

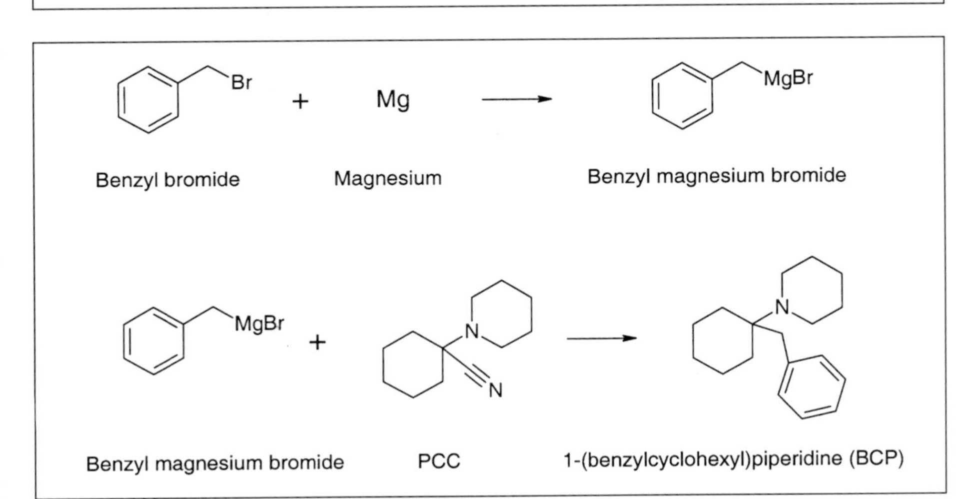

Figure 4.48

1-(Benzylcyclohexyl) piperidine (BCP) by Lodge et al. *(1992)*

Ketamine synthesis

Ketamine, although similar in structure to PCP, is synthesized via a completely different and more complex route (Lednicer and Mitscher, 1977). The reaction proceeds through a bromo-ketone intermediate formed through the reaction of cyclopentylmagnesium bromide with o-chlorobenzonitrile followed by a bromination step (Figure 4.49).

Figure 4.49

Synthesis of ketamine by Lednicer and Mitscher (1977)

Br

CH_3NH_2

Methylamine

Thermolysis of the imine hydrochloride causes a bond shift to produce a cyclohexyl group

Ketamine

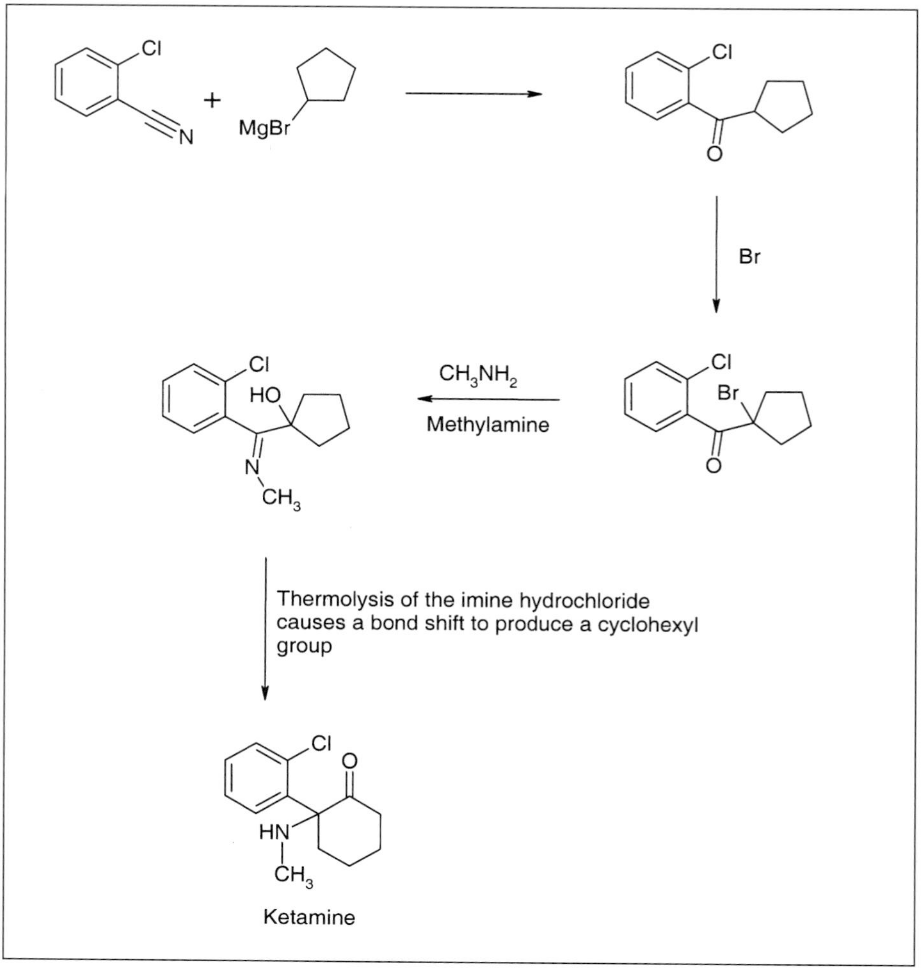

The ketone is treated with methylamine which forms an imino alcohol intermediate. Thermolysis of the hydrochloride salt of this imino alcohol causes a bond shift enlarging the cyclopentyl group resulting in the final ketamine form. This reaction is not amenable to clandestine laboratory manufacturing. Ketamine is diverted from veterinary sources where it is used as a small animal anesthetic (Anonymous, 1996). There have also been cases where bulk ketamine hydrochloride has been obtained through Asian sources.

4.1.5 CULTIVATION OF ERGOT FUNGUS AND PSILOCYBIN MUSHROOMS

Ergot fungi

Naturally occurring ergot fungi are found through a worldwide distribution in most temperate zones (Mantle, 1975). Of the ergot fungi the two of the most prolific producers of ergot alkaloids are *Claviceps paspali* and *Claviceps purpurea.* Fermentation of these fungi is a viable commercial source of these alkaloids (Adams, 1964; Arcamone *et al.*, 1970). *Claviceps purpurea* can produce up to 0.1–0.5% ergotamine while other ergot fungi such as *C. fusiformis, C. gigantea, C. sulcata* produce related alkaloids but little ergotamine and as such do not lend themselves to easy transformation into LSD precursor material. Optimization of growth media and manipulation of external factors on the production of ergot alkaloids have been extensively studied (McCrea, 1931; Taber and Vining, 1958, 1960; Arcamone *et al.*, 1960, 1961). Furthermore, work on a certain strain of ergot, *Claviceps paspali,* revealed d-lysergic acid α-hydroxyethylamide in concentrations of 1mg/ml (Arcamone *et al.*, 1961; Chain *et al.*, 1962). Kelleher *et al.* (1969) examined effects of surfactants on the production of alkaloids in *Claviceps paspali* and found that up to 1mg/ml could be produced. Similarly, Arcamone *et al.* (1970) found that up to 1mg/ml of peptide alkaloids, mainly consisting of ergotamine, were obtained from *Claviceps purpurea* strain F275 under optimal conditions. Fermentation of the ergot fungi is a viable source for the LSD precursors in a clandestine laboratory setting (Hugel, 1998).

Table 4.1

Some of the peptide ergot alkaloids

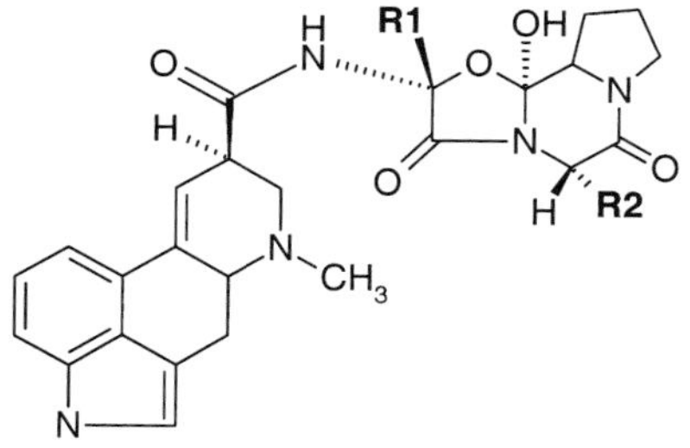

R1	R2	Peptide Ergot Alkaloid	Chemical Formula, Molecular Weight
CH3	$CH_2CH(CH_3)_2$	Ergosine / Ergosinine	$C_{30}H_{37}N_5O_5$, 547.64
$CH(CH_3)_2$	$CH(CH_3)_2$	Ergocornine / Ergocorninine	$C_{31}H_{39}N_5O_5$, 561.66
$CH(CH_3)_2$	$CH_2CH(CH_3)_2$	Ergocryptine / Ergocryptinine	$C_{32}H_{41}N_5O_5$, 575.69
CH3	$CH_2C_6H_5$	Ergotamine / Ergotaminine	$C_{33}H_{35}N_5O_5$, 581.65
$CH(CH_3)_2$	$CH_2C_6H_5$	Ergoscristine / Ergocrystinine	$C_{35}H_{39}N_5O_5$, 609.74

Other potential natural sources of ergoline alkaloids include the seeds of *Ipomoea violacea* (morning glory) where lysergamide (ergine), isolysergamide (isoergine) and clavines can be obtained in usable quantities. In the Heavenly Blue variety up to 3.42mg of lysergamide and 0.57mg of isolysergamide per 100 seeds have been reported by Genest (1966). Taber *et al.* (1963) observed that other members of the convolvulaceous family contained lysergic acid amide including Pearly Gates, Royal Blue, Crimson Rambler, Scarlet O'Hara, Darling, Cambridge Blue and Lavender Rosette where total alkaloidal content varied from 0.011–0.057% fresh weight. Most quantification work and total alkaloidal concentrations were determined using Van Urk's reagent (acidified p-dimethylaminobenzaldehyde).

The tropical wood rose, *Argyreia nervosa*, also produces ergoline alkaloids in concentrations up to 0.04% of lysergamide and 0.3% total alkaloids by weight (Miller, 1970). Yields of up to 555μg of isolysergamide and 780μg were recovered per gram of seed (Hylin and Watson, 1965).

Psilocybin fungi

Psilocybin containing mushrooms are widely dispersed throughout the world and can be found in various genus; *Panaeolus*, *Conocybe*, *Inocybe* and of course *Psilocybe* (Beug and Bigwood, 1982; Wurst *et al.*, 1992; Gartz, 1994). Some of the *Amanita* species contain the related compound bufotenin and other minor substituted hallucinogenic tryptamines. While a medium dose of psilocybin in humans is approximately 4–8mg (Hofmann, 1971) this can equate to varying quantity of dried mushroom depending upon genus and species (Table 4.2).

The harvesting of wild hallucinogenic mushrooms occurs during certain seasons, particularly duing humid spring and fall times, depending upon the species. Cultivation, on the other hand allows for year-round harvesting.

The life-cycle of a mushroom starts when spores germinate to become mycelium. (See Plate 4.1.)

As the mycelium expands it forms a mat. This mat thickens as the mycelium branches and produces sclerotia. (See Plate 4.2.)

The sclerotia serves two functions; one for survival by remaining dormant during inclement weather, and second, serving as the foundation from which the mushrooms emerge. Once a growing medium or substrate is fully encompassed by the mycelium, mat fruiting bodies form. In the early stage of fruiting, primordia form, bearing little resemblance to the final mushroom shape, which develop into fruiting bodies. The fruiting bodies will develop into full mushrooms releasing spores and starting the life cycle over (Stamets, 1993). Cultivation follows this cycle. The spores, available commercially, are grown on a substrate, such as a Petri dish of potato-dextrose agar, to form the mycelium.

Table 4.2

Some species of mushrooms containing psilocybin and psilocin or bufotenin

Mushroom	Psilocybin [c]	Psilocin [c]	Bufotenin [c]
Amanita citrina [b]	-	-	10.1
Amanita porphoria [b]	-	-	2.6
Amanita rubescens [b]	-	-	0.2
Panaeolus subbalteatus [a]	3.9	-	
Psilocybe stuntzii [a]	2.5	0.3	
Psilocybe semilanceata [a]	9.8	-	
Psilocybe semilanceata [b]	9.1	1.0	
Psilocybe pelliculosa [a]	4.1	-	
Psilocybe cyanescens [a]	3.2	5.1	
Psilocybe cyanescens Wakefield [a]	11.6	2.1	
Psilocybe cyanescens [b]	5.0	4.6	
Psilocybe cubensis [a]	6.3	1.1	
Psilocybe bohemica [a]	8.5	0.2	
Psilocybe bohemica [b]	7.5	1.9	
Psilocybe baeocystis [a]	3.0	1.9	
Inocybe aeruginascens [a]	4.0	-	
Inocybe aeruginascens [b]	2.7	0.2	
Gymnopilus purpuratus [a]	3.4	2.9	
Conocybe cyanopus [a]	9.3	-	

[a] Methanol extraction for 12 hours

[b] 70% Methanol saturated with potassium nitrate for 10 min, psilocin and bufotenin extracted with 75% ethanol for 160 minutes

[c] values given as mg/gram of dried plant material

The mycelium is propagated by transferring it to a sterilized grain substrate enriched with nutrients. As the mycelium grows it can either be enticed to bear fruiting bodies as is, or it can be transferred to another growing medium, often called a "log." The composition of the log is dependent upon growth requirements of the mushroom, and in the case with *Psilocybe cubensis* sterilized straw is an excellent medium (Plate 4.3).

The quantity of mushrooms produced is dependent upon several factors, of which perhaps most importantly is the amount of available nutrients. Gross (2000) cultivated *Psilocybe cyanescens* from spores ordered through a mail-order company. While psilocybin and psilocin were detected in the primordia stage, neither was detected in the earlier stages. However, Laing (2000) found psilocybin in the mycelium mat of *Psilocybe cubensis* on innoculated grain in 1 quart Mason jars using LC-MS. In previous experiments psilocin had been detected in mycelium cultivated on petri dishes. The difficulty in the analysis of mycelia lies with obtaining enough material to extract analyzable quantities of psilocin and psilocybin.

When cultivation employs large quantities of growth substrate high yields of mushrooms are possible.

Table 4.3

Cultivation of seized straw log cultures of Psilocybe cubensis *providing yield of mushroom to starting log weight*

Sample	Log Weight (g)	End Weight (g)	Weight Loss (g)	Wet Mushroom weight (g)	% of Mushroom to Weight Loss	% of Mushroom Log Weight
C1	3700.00	1300	2400	1553.03	64.7	42.0
C2	4000.00	1400	2600	1969.01	75.7	49.2
C3	3800.00	1100	2700	1559.44	57.8	41.0
C4	4800.00	2200	2600	1322.30	50.9	27.5
C5	5200.00	3500	1700	996.98	58.6	19.2
A1	3800.00	2100	1700	1109.73	65.3	29.2
A2	4600.00	3200	1400	580.49	41.5	12.6
A3	5000.00	2800	2200	1586.86	72.1	31.7
A4	3251.00	3000	251	214.92	85.6	6.6
Totals	38151.00	20600.00	17551.00	10892.76	63.6	28.8

In Table 4.3 the growth characteristics of nine seized cultures of *Psilocybe cubensis* are presented. The logs of inoculated straw ranged in weight from 3700g to 5200g at time of seizure. The logs all produced at least one flush of mushrooms, with six cultures producing a second flush and three cultures yielding yet a third flush. Each log was weighed after the final flush and the end weight ranged from 1300g to 3500g. The difference between the initial log weight and end weight is found under the column weight loss. The mushrooms accounted for 41.5% to 85.6% of the weight loss of the logs. All cultures were eventually over run by infection of opportunistic molds.

Typical street samples consist of the dried stalk and cap of the cultivated mushrooms. Approximately 92% reduction in weight is observed during the drying process.

4.2 WORLDWIDE DISTRIBUTION

The 1997 United Nations Drug Control Program World Drug Trend Report estimated that 25.5 million people worldwide abuse hallucinogens (UNDCP, 1997, p. 32). Surprisingly, this number exceeds the combined estimate of heroin and cocaine users, 8 and 13.3 million respectively. Typically, the hallucinogens are synthetically produced in illicit drug laboratories; however, some are derived from natural sources. In the 1990s in Europe an increase in the number of synthetic laboratories was observed. By the mid-1990s 55% of all Ecstasy laboratories detected worldwide were found in Europe, whereas in the pre-1990s, Ecstasy laboratories were largely confined to North America.

The major source for the hallucinogenic phenethylamines and amphetamines are the Benelux countries of northern Europe: the Netherlands, Belgium, and Luxemburg (Franzosa, 2001). (See Plate 4.4.)

Coupled with a large chemical and pharmaceutical industry many of the required precursor materials are available or easily diverted (Table 4.4).

Table 4.4

Commercial applications of some of the commonly encountered precursors and essential chemicals

Chemical	Legimate Use
Isosafrole	Manufacture of piperonal (heliotropin), perfumes, soap perfumes, flavours (root beer and sarsaparilla), pesticides
Safrole	Perfumes, soaps, manufacture of piperonal (heliotropin), pharmaceuticals, insecticides (not used as a flavouring agent)
Oil of Sassafras	Oil from the root of *Sassafras albidum, S. variifolium, S. officinale*, containing 80-90% Safrole; aromatherapy, masks disagreeable odors of medicaments, topical anti-infective, pediculicide
Piperonal	Pharmaceuticals (pediculicide), perfumery, suntan preparations, mosquito repellent, laboratory reagent, flavors (cherry and vanilla)
Nitroethane	Solvent for celluloids (cellulose derivatives), solvent for many resins, chemical synthesis, propellant
Nitromethane	Solvent for celluloids (cellulose derivatives), polymers, chemical synthesis, rocket fuel, gasoline additive
Hydrobromic Acid	Analytical chemistry, chemical synthesis, pharmaceuticals, solvent for ore minerals
Methylamine	Tanning, intermediate for accelerators, dyes, pharmaceuticals, insecticides, fungicides, solvent, photographic developer, rocket propellant
Indole	Perfumery, chemical reagent
Oxalyl	Military poison gas, manufacture of piperonal (heliotropin),

Table 4.4 (continued)

chloride	pharmaceuticals, insecticides
Diethylamine	Rubber chemicals, textile specialties, solvent, dyes, resins, pesticides, polymerization inhibitors, pharmaceuticals, petroleum chemicals, electroplating, corrosion inhibitors
Lysergic Acid	Medical research, synthesis of ergonovine
Hydrazine hydrate	Chemical intermediate, reducing agent, rocket propellant
Piperidine	Solvent and intermediate, curing agent for rubber and epoxy resins, catalyst for some reactions, ingredient in oils and fuels

[a] Condensed Chemical Dictionary 9th Ed (1977), Van Nostrand Reinhold Company, New York, ed Gessner G. Hawley; The Merck Index 11th Ed (1989), Merck & CO., Inc., Rahway, N.J., ed. Susan Budavari

Mexico in the mid-1990s was a source country for tablet preparations intended primarily for North America, but no longer seems to play a major role. Some former Warsaw pact countries, including Poland, are becoming more significant in the production of the Ecstasy-type drugs and is the suspected source of paramethoxyamphetamine and paramethoxymethamphetamine tablets which have caused overdose deaths in North America. The Netherlands, nonetheless, is still the major producer of Ecstasy tablets, and seizures of large to semi-industrial scale illicit laboratories are not uncommon. As an example, in the late 1990s, 750kg of MDMA was seized from a rural site, which also had approximately 750kg of Ecstasy in various stages of processing. While the major production is found in Western Europe, other lab seizures have been reported in North America, along with significant seizures of precursors.

Smuggling via domestic airline traffic is very popular, and primary and secondary European cities are commonly used by couriers with the contraband being either in their baggage or as "body packs." In the Toronto International Airport, Canada, almost once a month a major shipment was interdicted, and for the 2001 calendar year, ten seizures (flight origin: two from Germany, five from the Netherlands, two from France and one from Portugal – originating in the Netherlands) totaling almost 290,000 tablets were made. For the first quarter of 2002 there have been four seizures totaling over 76,000 tablets (Kyle, 2002). Shipping worldwide by means of courier companies, however, is more efficient and is less risky. Many companies through their Internet websites provide parcel tracking. A smuggler can track any of their packages and can

Table 4.5

Worldwide MDMA clandestine laboratory seizures

Clandestine Labs				
Country	1998	1999	Mid-2000	2001
Belgium		4	10	
Netherlands			23	
Estonia		1		
Ukraine		2		
DEA & State and Local		19	8	
Canada	5	11	11	9
Tablets				
Reporting Agency	1998	1999	Mid-2000	2001
Europe	5,000,000	14,100,000	8,400,000	
Interpol	5,600,000	22,000,000*		
Germany		1,470,000	719,300[t]	
Netherlands		3,660,000	510,000	
Belgium	171,000	585,000	532,000	
Luxembourg	145,000	1,400,000		
Sweeden			135,000	
Norway		4700	24,500	
Spain			169,200	
Italy		286,200	413,000	
Greece			52,300	
United Kindom	304,500	3,994,000		
Romaina	4200	10,500	10,100	
United States	11,900 (1996)		>3,000,000	
Canada	70,000	400,000	> 2,000,000	

* at is suspected at least 1,600,000 of these seized tablets were destined to the US and 1,000,000 to other world markets

[t] many seizures were to be transhipped to the US

predict which ones the authorities detain. The use of private mailbox companies also provides anonymity. Larger shipments of both tablets and precursors have been found on ocean-faring container ships. Not only is the trade in MDMA significant but also in precursor chemicals. Significant seizures have been made in Europe, Asia and North America. Plate 4.5 shows some of the reported seizures worldwide.

LSD, on the other hand, is presumed to be manufactured by a few groups in the United States of America and is distributed worldwide. LSD is easily concealed when small sheets of blotter paper are inserted in magazines, cards or small envelopes. As in the case of Ecstasy shipping, the use of private mailboxes is common, as well as payment via national and international cheque cashing and money transfer businesses. By using passwords and pseudonyms, along with keeping the transfers under $10,000, the drug traffickers need not worry about the transaction being reported to the authorities.

4.3 LAW ENFORCEMENT: AN INTERVIEW WITH AN LSD CLANDESTINE CHEMIST

Illicit LSD manufacturing has occurred since the early 1960s when the popularity of this drug was increasing as more young individuals discovered its psychedelic powers. At the epicenter of this "self discovery" movement stood a small group of individuals, some of whom preached the power of self-awareness, while others prepared the "psychedelic sacraments." It is worthy to note that few LSD clandestine laboratory seizures have been made since the early to mid-1970s and it is presumed that the world's production of LSD is still controlled by a small group of individuals.

In September 1996 in suburban Vancouver, Canada, a Royal Canadian Mounted Police investigation culminated in the seizure of a clandestine laboratory. The significance of this seizure was unknown at the time. After extensive fingerprint database search, all of the seized identification and passport documents were either forgeries or of deceased individuals, the identity of the principal character was determined. (See Plate 4.6.)

The chemist, an American fugitive, had jumped bail in 1974 after appealing a conviction for manufacturing LSD in California in 1973. He had been at large for more than 20 years and continued to ply his skill in the production of the "psychedelic sacraments." He granted a debriefing interview (Laing, 1999) just prior to his deportation to California. The chemist started cooking hallucinogens and LSD in 1962–63 using the Pinoche method – trifluoroacetic anhydride on lysergic acid. He discussed his role in the "League of Spiritual Discovery" in the preparation of the "sacraments" and as a guide for enlightenment (he perceived himself to still be in this role). Timothy Leary founded this group in the early 1960s. He discussed how another underground chemist taught him how to make LSD using the DMF-SO_3 complex which he said was a difficult reaction due to the toxic nature of sulfur trioxide and the critical stoichiometry of the reagents. In the 1996 seizure the chemicals and reagents for three different reactions were found: the phosphorus oxychloride reaction, CDI reaction and the DCC with HOBT reaction. When asked, he said that his main synthetic route used at the time was the CDI reaction since his diethylamine was in short supply. His preferred method, however, was the phosphorus oxychloride reaction, although it required too much diethylamine: 1 mol lysergic acid monohydrate to 2 mol $POCl_3$ to 9mol of diethylamine. He stated that the CDI method required the removal of the dimethylformamide solvent for which a two-stage vacuum stripping with a rotary evaporator, chiller and a dry ice condenser were needed. When asked about ergotamine, of which he and his co-conspirators had smuggled 24kg into Canada, he said he would obtain 375g of lysergic acid monohydrate from 1kg of ergotamine tartrate and that he only had 2kg of ergotamine left but no lysergic acid. He would purchase ergotamine in kilogram lots for $3–$7 per gram, depending upon the source. The epimerization of iso-LSD was accomplished using his own method in which potassium metal is dissolved in anhydrous ether and then refluxed with anhydrous methanol. After returning to room temperature the iso-LSD was added to the potassium solution and allowed to sit 24 hours at room temperature in the dark, providing a "98.5%" conversion. He argued that the old standard method of refluxing in potassium or sodium hydroxide provided only yields of 75% or less. Purifying by means of recrystallization from methanol worked best and the purity was estimated by means of the "flash test" – the piezoilluminesence characteristic of pure LSD when two crystals were struck together. Seized at the lab were numerous drugs in varying quantities: LSD 47g, MDA 3kg, MDMA 3kg, MDEA, DMT 4kg, diethyl and dipropyl tryptamine, 2CB (4-bromo-2,5-dimethoxyphenethylamine), mescaline, trimethoxyamphetamine, dimethoxyamphetamine, psilocybin and numerous sets of standards. (See Plate 4.7.)

Also found were 2g each of lysergic acid sec-butylamide and iso-lysergic acid sec-butylamide. When asked about the "designer" LSD, he said that he did not have enough time to try it out and was not sure as to the dosage level.

In the end he claimed to have made 900g of LSD that year.

The chemist said that his "first love" was dimethyltryptamine (DMT) and that he was synthesizing this compound before he started making LSD in the early 1960s. The DMT was made using a large 72L reaction set-up and reacting indole with oxalyl chloride and then with anhydrous dimethylamine. The DMT required distillation and recrystallization to obtain pure product. Distillation would be set up on the laboratory lattice (Figure 4.6) and 1 or 2kg would be processed at a time. A kilogram of DMT dissolved in 2L ether would recrystallize when added to 8L hexane. Rhomboid crystals would spontaneously form in the mother liquor.

Reference documents and hand written notes were found describing the use of monoamineoxidase (MAO) inhibitors, namely harmine and harmaline, with DMT to make it orally active. When asked about other tryptamines such as 5-methoxytryptamine he said he did not enjoy it but liked psilocybin. He went on to say that he did not care for the harshness of the hallucinogenic amphetamines (substituted phenylisopropylamines) but preferred the substituted phenethylamines such as 2C-B and mescaline which provided "enlightenment."

The interview ended with the chemist being escorted in to a waiting van for the trip to the airport and then on to California to serve the remainder of his original sentence.

ACKNOWLEDGEMENTS

We wish to acknowledge the support provided by Health Canada, Québec Region, Canadian clandestine laboratory seizure data from Benoit Archambault, E-Division RCMP and the Toronto International Airport RCMP.

REFERENCES

Adams, R. A. (1964) US Patent 3,117,917.

Allen, A., Carr, S., Cooper, D., Fransoza, E., Koles, J., Kram, T. and Solon, E. (1980) *Microgram,* 13, pp. 44–46.

Allen, A. C., Robles, J., Dovenski, W. and Calderon, S. (1993) *For. Sci. Intl*, 61, pp. 85–100.

Angelos, S. A., Raney, J. K., Skowronski, G.T. and Wagenhofer, R. J. (1990) *J. For. Sci.*, 35, pp. 1297–1302.

Anonymous (1996). *Microgram*, 24, p. 202.

Arcamone, F., Bonino, C., Chain, E. B., Ferretti, A., Pennella, P., Tonolo, A., Vero, L. (1960) *Nature*, 187, pp. 238–239.

Arcamone, F., Chain, E. B., Ferretti, A., Minghetti, A., Pennella, P., Tonolo, A. and Vero, L. (1961) *Proceedings of the Royal Society*, Series B, 187, pp. 26–54.

Arcamone, F., Cassinelli, G., Ferni, G., Penco, S., Pennella, P. and Pol, C. (1970). *Can. J. Microbiol.*, 16, pp. 923–931.

Bailey, K. and Legault, D. (1979) *J. Assoc. Off. Anal. Chem.*, 62, pp. 1124–1137.

Bailey, K., Gagne, D. R. and Pike, R. K. (1976) *J. Assoc. Off. Anal. Chem.*, 59, pp. 81–89.

Baker W., Penfold, A. R. and Simonsen, J. L. (1939) *J. Chem. Soc.*, pp. 439–443.

Baker, W., Jukes, E. H. T. and Subrahmanyam, C. A. (1934) *J. Chem. Soc.*, pp. 1681–1684.

Baker, W., Montgomery, L. V. and Smith, H. A. (1932) *J. Chem. Soc.*, pp. 1281–1283.

Benington, F., Morin, R. D., Clark, L. C. and Fox, R. P. (1958) *J. Org. Chem.*, 23, pp. 1979–1983.

Beug, M. W. and Bigwood, J. (1982) *J. Ethnopharm.*, 5, pp. 271–285.

Biniecki, S. and Krajewski, E. (1961) *Chem. Abstr.*, 55, p. 14350e.

Boissannas, R. A. (1951) *Helv. Chim. Acta.*, 34, pp. 874–879.

Boissannas, R. A. (1952) *Chem. Abstr.*, 47, p. 2493i.

Bourne, E. J., Henry, S. H., Tatlow, C. E. M. and Tatlow, J. C. (1952) *J. Chem. Soc.*, pp. 4014–4019.

Braun, U., Shulgin, A. T. and Braun, G. (1980) *J. Pharm. Sci.*, 69, pp. 192–195.

Budaveri, S. (ed.) (1996) *The Merck Index*, 12th edn, Whitehouse Station, NJ: Merck & Co.

Campbell, K. N., Hopper, P. F. and Campbell, B. K. (1951) *J. Org. Chem.*, 16, pp. 1736–1741.

Cerny, A. and Smonsky, M. (1954) *Chem. Abstr.*, 60, p. 8082b.

Cerny, A. and Smonsky, M. (1963) *Chem. Abstr.*, 58, p. 1501h.

Chain, E. B., Bonino, C. and Tonolo, A. (1962) US Patent 3,038,840.

Cladingboel, D. E.and Parsons, P. J. (1990) *J. Chem. Soc., Chem. Commun.*, pp. 1543–1544.

Clark, J. H., Holland, H. L. and Miller, J. M. (1976) *Tetrahedr. Lttrs.*, 38, pp. 3361–3364.

Cordell, G. A. (1981) in *Introduction to the Alkaloids: A Biogenic Approach*, NY:John Wiley & Sons, pp. 623–655.

Cowie, J. S., Holtham, A. L. and Jones, L. V. (1982) *J. For. Sci.* 27, pp. 527–540.

Crossley, F. S. and Moore, M. L. (1944) *J. Org. Chem.*, 9, pp. 529–536.

Dal Cason, T. A. (1990) *J. For. Sci.*, 35, pp. 675–697.

Elks, J. and Hey, D. H. (1943) *J. Chem. Soc.*, pp. 15–16.

Feugas, C. (1964) *Bull. Soc. Chim. Fr.*, 8, pp. 1892–1895.

Franzosa E. S. (2001) Private communication.

Gairaud, C. B. and Lappin, G. R. (1953) *J. Org. Chem.*, 18, pp. 1–3.

Garbrecht, W. L. (1956) US Patent 2,774,763.

Garbrecht, W. L. (1959) *J. Org. Chem.*, 24, pp. 368–372.

Gartz, J. (1994) *J. Basic Microbiol.*, 34, pp. 17–22.

Genest, K. (1966) *J. Pharm. Sci.*, 55, pp. 1284–1288.

Godefoi, E. F., Maddox, V. H. and Parcell, R. F. (1963) US Patent 3,097,136, pp. 1–10.

Green, M. (1962) US Patent 3,062,884.

Gross, S. T. (2000) *J. Forensic. Sci.*, 45, pp. 527–537.

Hamlin, K. E. and Fischer F. E. (1951) *J. Am. Chem. Soc.*, 73, p. 5007.

Hamlin, K. E. and Weston, A. W. (1949) *J. Am. Chem. Soc.*, 71, pp. 2210–2212.

Heacock, R. A. and Hutzinger O. (1963) *Can. J. Chem.*, 41, pp. 543—545.

Heacock, R. A., Hutzinger, O. and Nerenberg, C. (1961) *Can. J. Chem.*, 39, pp. 1143–1147.

Heinzelman, R. V. (1963) in *Organic Synthesis Collective: Volume 4*, NY: John Wiley and Sons.

Heinzleman, R. V., Anthony, W. C., Lyttle, D. A. and Szmuszkovicz, J. (1960) *J. Org. Chem.*, 25, pp. 1548–1558.

Hey, P. (1947) *Quart. J. Pharm. Pharmacol.*, 20, pp. 129–134.

Hofmann, A. (1971) *Bulletin on Narcotics*, 23, pp. 3–14.

Hofmann, A., Frey A., Ott, H., Petrzilka,T. and Troxler, F. (1958) *Experientia*, 14, p. 3975.

Hugel, J. (1998) *J. Clan. Lab. Invest. Chem. Assoc.*, 8, pp. 27–28.

Hylin, J. W. and Watson, D. P. (1965) *Science*, 148, pp. 499–500.

Johnson, F. N, Ary, I. E., Teiger, D. G. and Kassel, R. J. (1973) *J. Med. Chem.*, 16, pp. 532–537.

Julia, M., Le Goffic, F. and Baillarge, M. (1969) *Chem. Abst.*, 96, p. 69274s.

Kalbhen, D. A. (1971) *Angew. Chem.*, 10, pp. 370–374.

Kalir, A., Edery, H., Pelah, Z., Balderman, D. and Porath, G. (1969) *J. Med. Chem.*, 12, pp. 473–477.

Kawanishi, M. (1957) *Chem. Abstr.*, 51, p. 15574g.

Kelleher, W. J., Kim, B. K. and Schwarting, A. E. (1969) *Lloydia*, 32, pp. 327–333.

Kenner, G. W. and Stedman, R. J. (1952) *J. Chem. Soc.*, pp. 2069–2076.

Kornfeld, E. C., Fornefeld, E. J., Kline,G. B., Mann, M. J., Morrison, D. E., Jones, R. G. and Woodward, R. B.(1956) *J. Am. Chem. Soc.*, 78, pp. 3078–3114.

Kurihara, T., Terada, T., Harusawa, S. and Yoneda, R. (1987) *Chem. Pharm. Bull.*, 35, pp. 4793–4802.

Kyle, L., Cst. (2002) *Toronto International Airport RCMP*, Private communication.

Laing, R. R. (1999) Presented at 9th Annual Clandestine Laboratory Investigating Chemists Association, Toronto, Canada.

Laing, R. R. (2000) Presented at 10th Annual Clandestine Laboratory Investigating Chemists Association, Brisbane, Australia.

Lednicer, D. and Mitscher, L. A. (1977) *The Organic Chemistry of Drug Synthesis, Vol. 1*, NY: John Wiley & Sons, p. 57.

Lerner, O. M. (1958) *Chem. Abstr.*, 52, p. 18271g.

Lodge, B. A., Duhaime, R., Zamecniuk, J., MacMurray, P. and Brousseau, R. (1992) *For. Sci. Intl.*, 55, pp. 13–26.

Losse, G. and Mahlberg, W. (1978) *Chem. Abstr.*, 88, p. 191493n.

Maddox, V. H., Godefoi, E. F. and Parcell, R. F. (1965) *J. Med. Chem.*, 8, pp. 230–234.

Mantle, P. G. (1975) *The Filamentous Fungi*, in J. E. Smith and D.R. Berry (eds), NY: John Wiley & Sons, pp. 281–300.

McCrea, A. (1931) *Am. J. Botany*, 18, pp. 50–79.

Merck Index (The) (2001) Rahway, NJ: Merck & Co., Inc.

Milas, N. A. (1937) *J. Am Chem. Soc.*, 59, pp. 2342–2344.

Miller, M. D. (1970) *J. AOAC*, 53, pp. 123–127.

Nakai, M. and Enoralya, T. (1987) US Patent 4,638,094.

Novelli, A. (1939) *J. Am. Chem. Soc.*, 61, pp. 520–521.

Parke Davis (1960) British Patent 836,083, pp. 1–5.

Patelli, B. and Bernardi, L. (1962) US Patent 3,141,887.

Pioch, R. P. (1956) US Patent 2,736,728.

Rebek, J., Tai, D. F. and Shue, Y. K. (1984) *J. Am. Chem. Soc.*, 106, pp. 1813–1819.

Schenk, H. P. and Lamparsky, D. (1981) *J. Chromatog.*, 204, pp. 391–395.

Sheehan, J. C. and Hess, G. P. (1955) *J. Am. Chem. Soc.*, 77, pp. 1067–1068.

Shulgin, A. T. (1963) *Nature*, 197, p. 379.

Shulgin, A. T. (1967) *J. Chromat.*, 30, pp. 54–61.

Shulgin, A. T. (1968) *Can. J. Chem.*, 46, pp. 75–77.

Shulgin, A. T. and MacLean, D. E. (1976) *Clin. Toxic.*, 9, pp. 553–560.

Shulgin, A. T. and Shulgin A. (1991) in *PIHKAL, A Chemical Love Story*, Berkeley, CA: Transform Press.

Shulgin, A. T. and Shulgin A. (1997) in *TIHKAL, A Chemical Love Story*, Berkeley, CA: Transform Press, pp. 49 –493.

Speeter M. E. and Anthony, W. C. (1954) *J. Am. Chem. Soc.*, 76, p. 6209.

Stamets, P. (1993) in *Growing Gourmet and Medicinal Mushrooms*, Berkeley, CA: Ten Speed Press, p. 554.

Stoll, A. and Hofmann, A. (1937) US Patent 2.090,430.

Stoll, A. and Hofmann, A. (1941) US Patent 2,265,207.

Stoll, A. and Hofmann, A. (1941) US Patent 2,265,217.

Stoll, A. and Hofmann, A. (1943) *Helv. Chim. Acta*, 26, pp. 944–964.

Stoll, A. and Hofmann, A. (1944) *Chem. Abstr.*, 38, pp. 1500–1502.

Stoll, A. and Hofmann, A. (1955). *Helv. Chim. Acta*, 38, pp. 421–433.

Sy, W-W. and By, A. W. (1985) *Tetrahedr. Lett.*, 26, pp. 1193–1196.

Taber, W. A., Vining L. C. and Heacock, R. A. (1963). *Phytochem*, 2, pp. 65–70.

Taber, W. A. and Vining, L. C. (1957) *Can. J. Microbio.*, 3, pp. 55–60.

Taber, W. A. and Vining, L. C. (1958). *Can. J. Microbio.*, 4, pp. 611–626.

Taber, W. A. and Vining, L. C. (1960). *Can. J. Microbio.*, 6, pp. 355–365.

Tiffaneau, M. (1907) *Ann. Chim. Phys.*, 8, pp. 322–378.

Tindall, J. B. (1954) *Chem. Abstr.*, 48, p. 8259b.

Tomita, M., Fujitani, K., Aoyagi, Y. and Kajita, Y. (1968) *Chem. Pharm. Bull.*, 16, pp. 217–226.

Trikojus, V. M. and White, D. E. (1948) *J. Chem. Soc.*, pp. 436–439.

UNDCP World Drug Trend Report (1997) Northampton, England: Oxford University Press, p. 32.

Van Haeren, C., Vanbeckevoort, Y. and Coppens, H. (2000) Presented at 10th Annual Clandestine Laboratory Investigating Chemists Association, Brisbane, Australia.

Weaver, K. and Yeung, E. (1995) in *An Analyst's Guide to the Investigation of Clandestine Laboratories*, Toronto, Ontario, Canada: Health Canada.

Wurst, M., Kysilka, R. and Koza, T. (1992) *J. Chrom.*, 593, pp. 201–208.

Young E. H. P (1958) *J. Chem. Soc.*, p. 3493.

Ziegler, R. and Stuetz, P. (1983) *Chem. Abstr.*, 99, p. 71069j.

CHAPTER 5

ANALYSIS OF THE HALLUCINOGENS

John Hugel
John Meyers
David Lankin

PART I: INFRARED (IR) SPECTROSCOPY

Since the 1960s, infrared spectroscopy has been used as a powerful confirmatory test for the identification of organic compounds, including the hallucinogens and other drugs of abuse. As with the application of any technique to illicit drug analysis, the forensic chemist must understand how the technique works and how it can be applied, and its strengths and weaknesses.

5.0 THEORETICAL BASIS

An organic molecule is characterized by the composition of its atoms and the bonds they form. These bonds can absorb wavelengths of infrared light causing their vibration, stretching, bending and wagging. The Infrared (IR) Spectrometer is an analytical instrument in which organic compounds are irradiated with infrared light typically with wavelengths of 4000–400cm^{-1}. When the molecule absorbs at a certain wavelength, depending upon the type of bond, it creates a peak at that wavelength in its infrared spectrum. Slight variations in a molecule such as the positions of bonds (as with isomers), different composition of atoms, and even salt or crystalline forms will change the manner in which the molecule absorbs the IR radiation. With this, the IR spectrum can be used to distinguish small differences between two similar molecules. It also follows that the same compound will yield the same infrared spectrum ostensibly unchanged by time and instrument constraints. This further means that if one obtains an infrared spectrum of an analyte which matches that of a known molecule, then the analyte is that known molecule. As can be expected, there are pitfalls to be avoided in making such sweeping statements, but with the application of knowledge and care, the statement will hold when applied to the identification of hallucinogens.

Isomers yield different IR spectra. One exception to this rule is optical isomers which cannot be differentiated. Diastereomers (optical isomers with more than one chiral center) do, however, exhibit different spectra. For example, the spectra of D-lysergic acid diethylamide (LSD) is identical to

L-LSD, but the differences between the infrared spectra of LSD and iso-LSD are remarkable. The same applies to lysergic acid sec-butylamide (LSB) which is a structural isomer of LSD and iso-sec-LSB. The structures and infrared spectra of LSD/iso-LSD and LSB/iso-sec-LSB are shown in Figure 5.1.

Occasionally the spectrum of the optically pure isomer in the solid phase will not be the same as that of a racemic mixture. This is caused by intermolecular interactions between the two racemates changing the way they vibrate, stretch, bend or wag. This has been examined as a way of determining the optical purity of phenethylamines (CND Analytical (1994)). It, however was concluded that it is not a reliable technique.

A problem with infrared spectroscopy, which appears occasionally in the identification of hallucinogens, is polymorphism. That is, some analytes in the solid phase, can crystallize in different ways. Since the different crystal forms will change the intermolecular interactions, there can be significant changes in the infrared spectrum. It is important to remember that even if an analyte is polymorphic, as long as the spectrum of the analyte matches the spectrum of the known compound, its identity is established. If the spectrum does not match, it does not necessarily rule out that the analyte is, in fact, the known compound.

A powerful feature of infrared spectroscopy is that it can be used in elucidating the structure of a molecule since functional groups absorb at characteristic wavelengths. For example, carbonyls (-C=O) exhibit a strong absorption in the 1780–1630cm^{-1} region of the infrared spectrum. This feature means that analytes with the same functional groups will present an absorption at similar wavelengths. In addition, information about which functional groups are present or absent can be gained from the infrared spectrum. This is particularly helpful when dealing with an analyte whose identity is unknown. For a list of texts which include tables of absorption frequency versus functional group, see the further reading heading at the end of the IR spectroscopy section.

One general principle that holds true is that solid phase spectra of the same analyte will have more features than the analyte's liquid phase spectrum, which in turn will have more features than the analyte's vapor phase spectrum. Figure 5.2 illustrates the infrared spectrum of 3,4-methylenedioxymethamphetamine hydrochloride (MDMA.HCl)–solid phase, MDMA base–liquid phase, and MDMA vapor phase at 225°C.

Figure 5.1 (opposite)

Infrared spectra of LSD related diastereomers:
(a) LSD amorphous;
(b) iso-LSD amorphous;
(c) sec-LSB amorphous;
(d) iso-sec-LSB amorphous

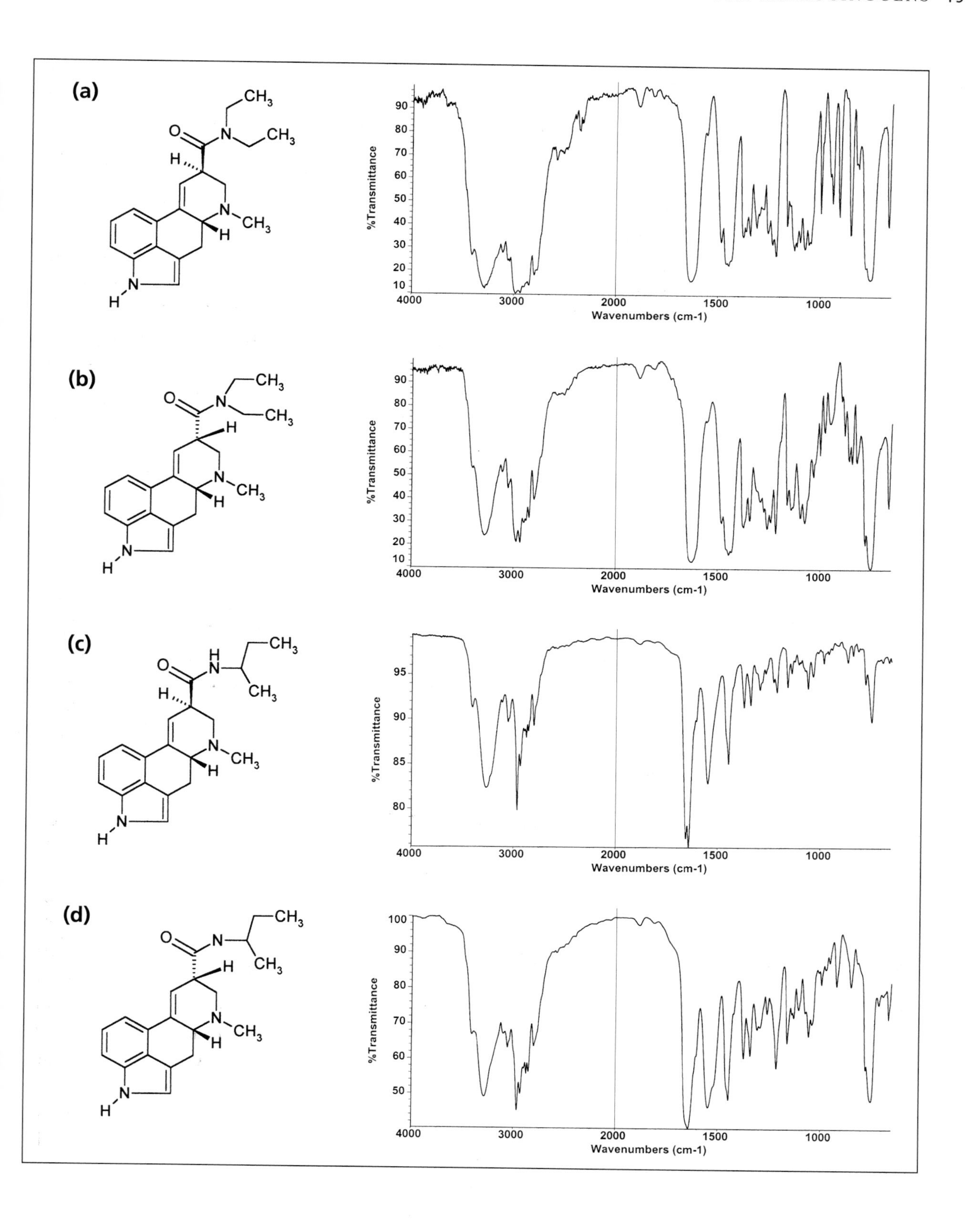
(a)
CH3
O
N
CH3
H
N
CH3
H
N
H
%Transmittance
Wavenumbers (cm-1)
(b)
(c)
(d)
4000
3000
2000
1500
1000

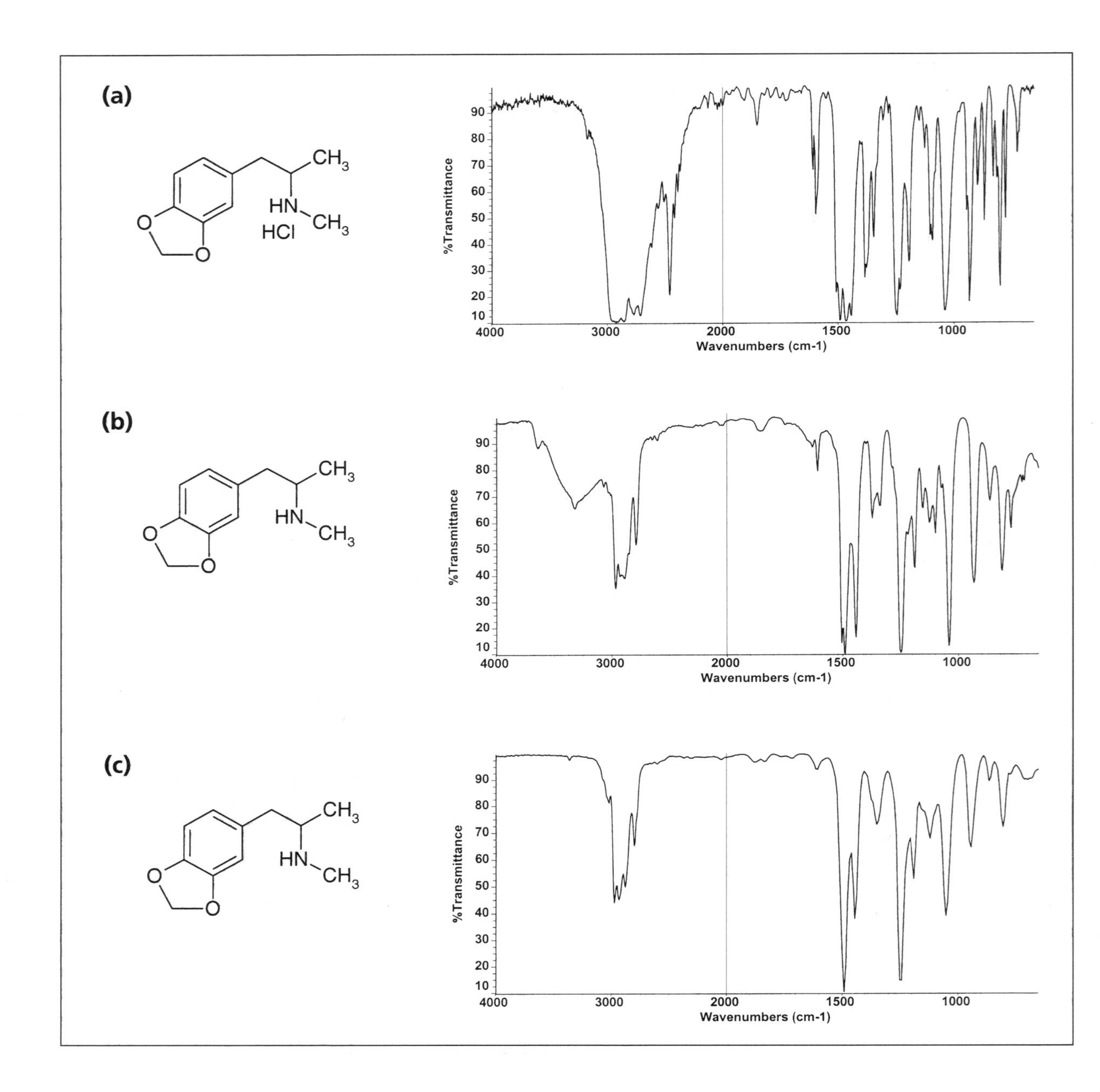

Figure 5.2

IR spectra of MDMA: (a) solid – MDMA.HCl; (b) liquid – MDMA base; (c) vapor MDMA at 225°C

5.1 INFRARED SPECTROMETER INSTRUMENTATION

Infrared spectrometers are divided into two groups – dispersive and Fourier transform instruments. The former scans each wavelength sequentially and plots the absorption on a chart. The latter obtains an interferogram of all infrared wavelengths simultaneously and then performs a mathematical function called a Fourier transform to obtain the plot of absorption versus wavelength. Performing the Fourier transform requires some computing power which, nowadays, is easily obtained. Since Fourier transform instruments are faster and more sensitive, they now dominate the market. Fourier transform instruments also have the advantage of having only one moving part. The optical benches of Fourier transform infrared (FTIR) spectrometers usually last much longer than any controlling computer system.

Dispersive instruments are often double beam, which means that the infrared spectrum of the analyte is automatically ratioed against a background infrared spectrum. This automatically eliminates the effects of ambient infrared absorbing water vapor and carbon dioxide. Fourier transform instruments, on the other hand, are single beam instruments which measure the spectrum of the analyte and the ambient interferences together. To eliminate the water vapor and carbon dioxide spectrum, a background spectrum is obtained and ratioed mathematically against the analyte spectrum. A further measure, which can be taken to reduce this effect, is to purge the spectrometer with dry, carbon dioxide-free air. This will also lengthen the life of any moisture-sensitive parts in the infrared spectrometer, namely the beam splitter. In the case of a purged sample compartment, whenever adding an analyte to the sample compartment, one must wait for the purge to be re-established. This is usually a matter of a minute or so.

Infrared spectrometers, dispersive or Fourier transform, measure the analyte that is placed in the infrared beam. Contaminants in an analyte will affect the obtained infrared spectrum. The extent of the effect depends on the infrared spectra of the analyte and the contaminant. In the worst case, where the contaminant has a strong infrared spectrum and the analyte a weak spectrum, a few percent of the contaminant will badly affect the analyte spectrum. In the best case, where the reverse is true, seemingly unaffected spectra are observed even though the analyte might be only 80% pure. An effective way of getting around this problem is the use of the gas chromatograph/Fourier transform infrared (GC-FTIR) spectrometer.

The biggest problem with applying infrared spectroscopy to the analysis of hallucinogens is ensuring that the analyte being placed in the infrared beam is sufficiently pure that the analyte spectrum is free of interferences from contaminants. It is for this reason that the following discussion on the applications of infrared spectroscopy to the analysis of hallucinogens includes outlines of extraction procedures to purify samples.

5.1.1 OBTAINING CONDENSED PHASE INFRARED SPECTRA

For the purposes of IR spectroscopy, analytes that are solids or liquids are considered to be in the condensed phase.

The classical spectrum of a solid phase analyte is obtained by making a paste of the analyte and a mulling agent and spreading the paste between two salt (usually NaCl or KBr) plates. The most common mulling agent is nujol (mineral oil). Alternatively, the analyte is mixed with a salt (usually KBr), ground and pressed into a pellet which is placed into the infrared beam. Both techniques have their pros and cons, but for hallucinogen identification, either will function well. Terry Gough (1991) includes a discussion on the two techniques. Other techniques which do not require the dilution of the sample in a mulling agent include attenuated total reflectance (ATR) and diffuse reflectance. ATR uses an IR inert crystal whose internal reflectance does not allow the IR radiation to exit the crystal but rather to bounce within the crystal and sample the analyte which is in contact with the surface. This technique is perfectly suited for films, liquids and polymers. It is also used in IR microscopy and with "diamond" cells which enable crystalline materials or powders to be sampled under pressure as a film. Diffuse reflectance, on the other hand, can be used to collect the spectra of powders directly through the focusing of the IR radiation using ellipsoid mirrors onto the sample and processing the resulting reflected radiation (Griffiths and de Haseth, 1986).

A spectrum of a (non-aqueous) liquid phase analyte can be obtained by placing the analyte between two salt (usually NaCl or KBr) plates and placing the analyte in the infrared beam. The two cells can be, but do not necessarily have to be, separated by a spacer of known width. Specialized cells with a cell cavity of known volume can also be used. By knowing the width of the spacer and the volume of the cell, quantitative experiments can be performed. Using two cells with no spacer that are pressed together are effective for identification work.

Both solids and liquids can be dissolved in appropriate solvents and the spectra obtained. The spectrum of the solvent will be superimposed on that of the analyte but can sometimes be removed using a reference spectrum and through the data manipulation of a subtraction algorithm. For routine hallucinogen identification work, this is normally unnecessary.

5.1.2 VAPOR PHASE INFRARED SPECTROSCOPY

Vapor phase IR spectroscopy can be applied to the identification of analytes which have a sufficient vapor pressure at room temperature such that the headspace of the analyte in a container can be introduced into an evacuated gas cell, and a useful spectrum obtained. Almost all solvents can be identified by this method without difficulty. This technique can be easily and successfully applied to those solvents which have a boiling point of less than 100°C. With the use of multi-pass gas cells, the technique can be applied to less volatile analytes. Since solvents are always used in the production of hallucinogens, this technique can be useful in clandestine laboratory investigations.

Another method of obtaining vapor phase spectra of hallucinogens is the use of gas chromatography – Fourier transform infrared (GC-FTIR) spectroscopy. (Dispersive IR spectrometers cannot scan quickly enough to obtain infrared spectra of peaks eluting from capillary columns.) The technique combines the separating power of the gas chromatograph with the identifying power of the FTIR spectrometer. The analyte along with contaminants is injected into a gas chromatograph which separates the components based upon physical characteristics. As the components elute from the GC, they enter a light pipe through which the infrared beam has been focused. The light pipe consists of a gold-lined heated tube with IR inert crystals on either end permitting the IR radiation to traverse through the vaporized sample. This technique requires the use of a liquid nitrogen cooled mercury-cadmium-telluride (MCT) detector for the required sensitivity and speed also to obtain the spectra of the fast eluting components from a capillary column.

Vapor phase FTIR spectra have the characteristics of condensed phase spectra such as same functional group absorptions, except that they generally show fewer features.

Subtle differences in vapor phase spectra become very significant when comparing homologous series of IR spectra of hallucinogens. Figure 5.3 illustrates the structures and vapor phase infrared spectra of 3,4-methylenedioxymethamphetamine (MDMA) versus that of N-methyl-1-(3,4-methylenedioxyphenyl)-2-butanamine (MBDB).

Minor, but reproducible, differences can be seen in the spectra in the 2950, 1350, and 1100cm^{-1} areas of the spectrum. On the other hand, structural isomers exhibit remarkably different spectra. Figure 5.4 illustrate the vapor phase IR spectra of 2,3-methylenedioxyamphetamine versus 3,4-methylenedioxyamphetamine (MDA).

Figure 5.5 demonstrates the differences between the homologous series related to MDA (Figure 5.4):

- 3,4-methylenedioxyphenethylamine;
- 1-(3,4-methylenedioxyphenyl)-2-butanamine; and
- 1-(3,4-methylenedioxyphenyl)-2-pentanamine.

As the homologue side chain grows in length, the differences become more subtle.

A skilled and knowledgeable forensic chemist can make use of the vapor phase FTIR spectra to distinguish among isomers and homologous series of hallucinogens.

Figure 5.3 (p. 198, top)

Vapor phase infrared spectra of homologues: (a) MDMA at 225°C; (b) MBDB at 225°C

Figure 5.4 ((p. 198, bottom)

Vapor phase IR spectra of structural isomers: (a) 2,3-methylenedioxyamphetamine at 225°C; (b) 3,4-methylenedioxyamphetamine (MDA) at 225°C

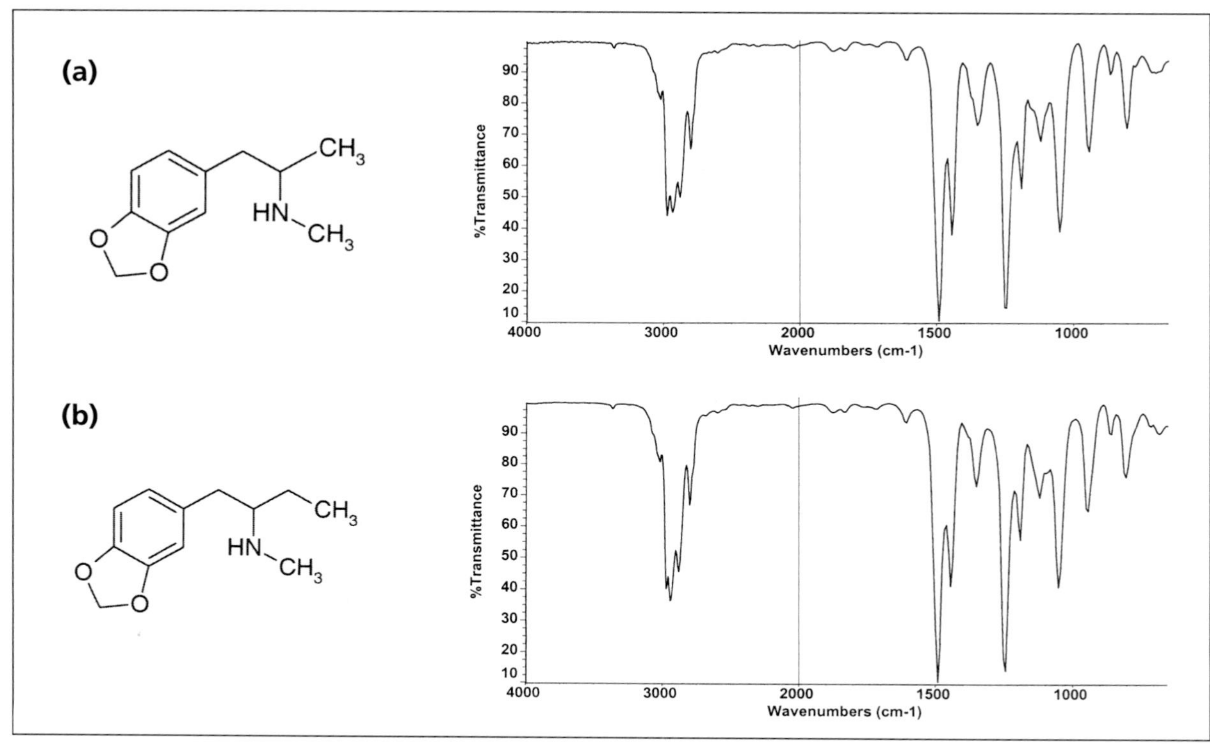
(a)
CH3
HN
CH3
O
O
%Transmittance
Wavenumbers (cm-1)
(b)
CH3
HN
CH3
O
O
%Transmittance
Wavenumbers (cm-1)

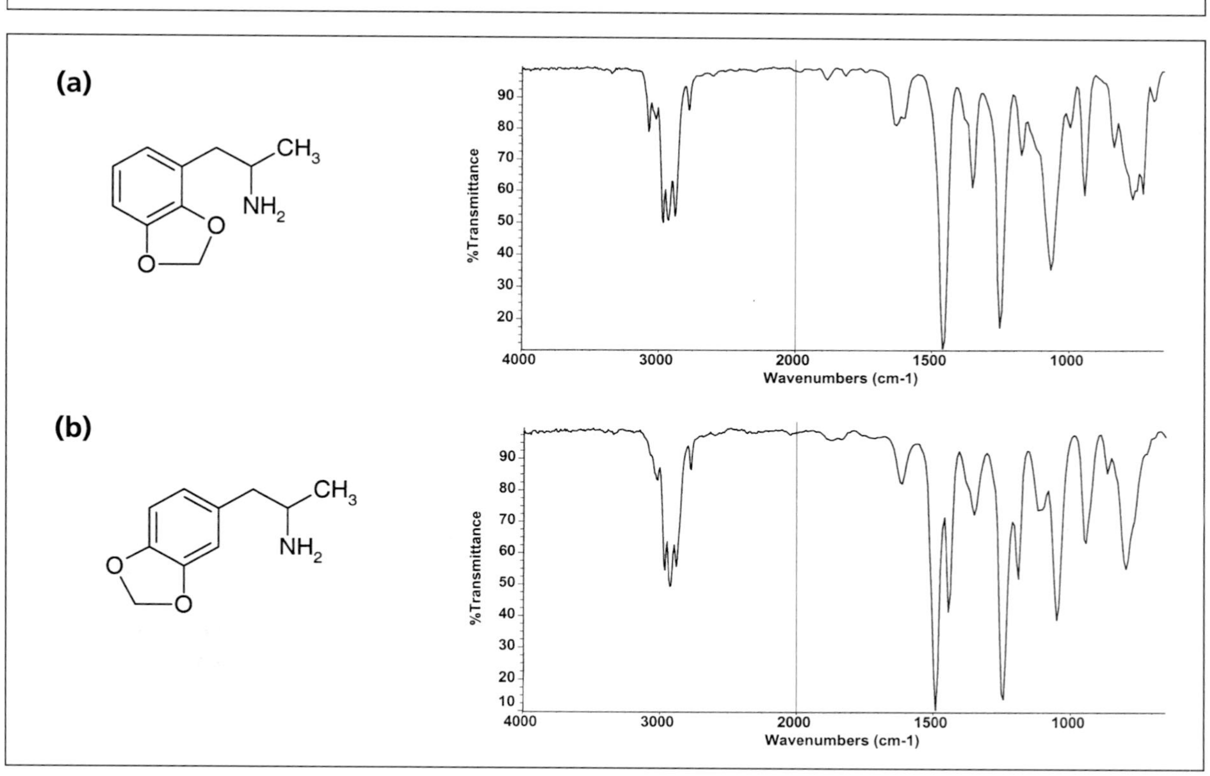
(a)
CH3
NH2
O
O
%Transmittance
Wavenumbers (cm-1)
(b)
CH3
NH2
O
O
%Transmittance
Wavenumbers (cm-1)

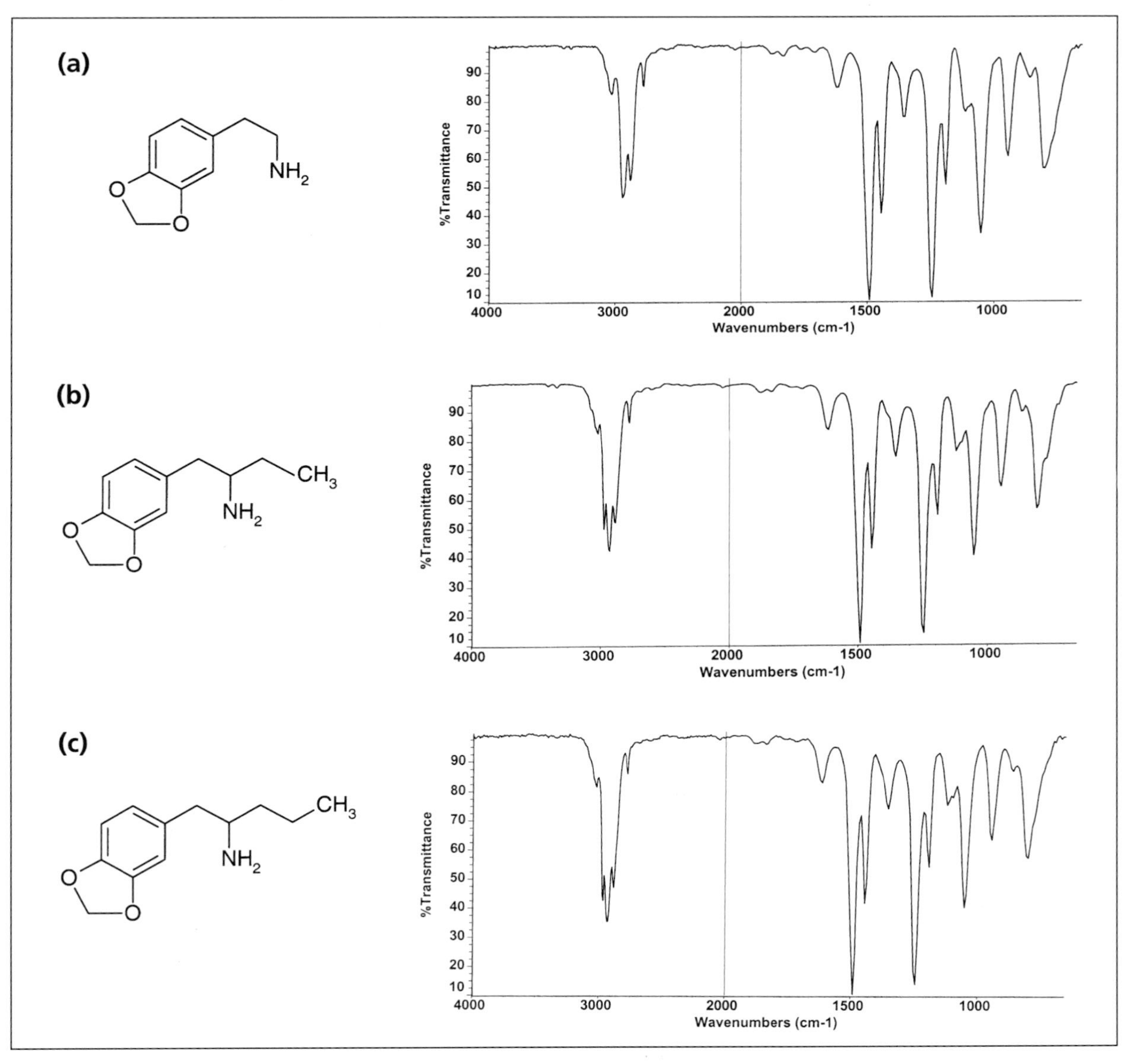

Figure 5.5 (above)

Vapor phase IR spectra of the homologous series related to MDA: (a) 3,4-methylenedioxyphenethylamine at 225°C, (b) 1-(3,4-methylenedioxyphenyl)-2-butanamine at 225°C; (c) 1-(3,4-methylenedioxyphenyl)-2-pentanamine at 225°C

Another problem which arises when using GC-FTIR is the effect that temperature can have on vapor phase spectra. Since the differentiation and identification of vapor phase infrared spectra is often performed on the basis of subtle differences in the spectra, and since temperature can subtly affect the infrared spectrum, the temperature at which the vapor phase spectrum is obtained is a crucial parameter when identifying hallucinogens. Figure 5.6 illustrates the spectra of MDMA at 150°C.

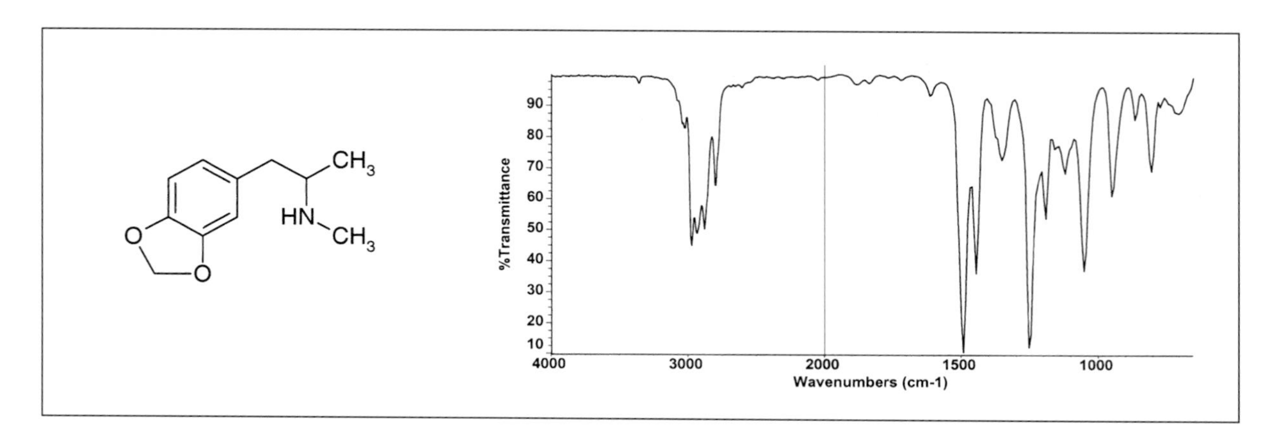

Figure 5.6

Vapor phase IR spectrum of MDMA at 150°C

The spectrum of MDMA at 225°C is found in Figure 5.3. The key to working around this problem is to have a spectrum of the standard and the sample obtained at the same temperature.

The sensitivity of GC-FTIR is typically in the 10ng range dependent on the instrument's characteristics and the intensity of the analyte's IR spectrum. The gas chromatograph should be equipped with a capillary column. The use of wide bore (about 0.3mm ID diameter) or mega-bore (about 0.5mm ID diameter) columns are recommended for GC-FTIR work. The common capillary dimethylpolysiloxane and (5%phenyl)methylpolysiloxane columns will work well.

Unlike gas chromatography-mass spectrometry, there is no need to turn off the FTIR spectrometer at any time during the GC run. This means that low boiling contaminants and solvent mixtures can be identified in or near the solvent front. Another characteristic of vapor phase FTIR spectrometry is that all wavelengths are being measured concurrently. The implication of this is that as the concentration of the analyte increases as on the up slope of an eluting chromatographic peak or decreasing as on the down slope of an eluting chromatographic peak, the obtained spectrum is unaffected. This is an important feature which leads to vapor phase IR spectra which are remarkably reproducible.

Since the FTIR spectrometer collects the GC-FTIR data as spectra, an algorithm called a Gram-Schmidt orthogonalization is used to generate an analyzable chromatogram where the peak height and area are representative of the analyte's concentration.

Since GC-FTIR spectroscopy is less sensitive than gas chromatography-mass spectrometry (GC-MS), GC-MS is much more popular. The mass spectra of structural isomers are often very similar (see section in this chapter on GC-MS). On the other hand, the mass spectra of a homologous series of compounds are usually different. The two techniques, GC-FTIR and GC-MS, are therefore complimentary.

5.1.3 SPECTRAL COMPILATIONS

In order to identify the infrared spectrum of various hallucinogens, the easiest way is to search a library of infrared spectra. This has led to compilations of infrared spectra which are available in electronic or hard copy (book) form. Electronic libraries have the advantage that they can be searched by FTIR instrument manufacturer's software. A good match when working with condensed phase spectrum is one which has a significantly better match value than the next. The absolute value of the obtained numbers is not necessarily a good indication of the identity of the unknown. It is always important to remember that, regardless of how "good" a match is, visual comparison of the analyte spectrum against that of the appropriate standard is an absolute requirement.

As discussed earlier, analytes which include the same functional groups will give similar infrared spectra. Infrared search results will reflect this similarity. This means that, if the analyte's spectrum is not in the IR spectral library, the best search results will be of compounds containing similar functional groups.

5.2 INFRARED SPECTROSCOPY OF LYSERGIC ACID DIETHYLAMIDE (LSD)

Lysergic acid diethylamine (LSD) is well known as a powerful hallucinogen. The dose of LSD necessary to have a hallucinogenic effect is about 100μg. Tablets and pieces of blotter will normally contain between 20 and 100μg of LSD tartrate. The common sample preparation techniques for condensed phase infrared spectroscopy, that is pellet formation and mulling, normally require a few milligrams. This means that extraordinary means are needed to obtain usable infrared spectra of LSD. Techniques such as preparing a micro-pellet – 1 to 3mm in diameter – for insertion into a beam condenser are often needed. The placing of a small spot of LSD amorphous solid on a small silver chloride disk and inserting it into the focus of a beam condenser also works well. With these techniques quantities as low as a single dose of LSD should be identifiable.

In addition, LSD when isolated as the base, does not crystallize well. It forms what it is termed an amorphous solid. The one saving grace of this formation is that the spectrum amorphous solid is reproducible. Mesley and Evans (1969) characterized the infrared spectrum of the amorphous solid LSD.

The isomers of LSD are all quite distinguishable by infrared spectroscopy. Iso-LSD as well as lysergic acid methyl propyl amide (LAMPA) are all readily distinguishable by the infrared spectra of their amorphous solids. Figure 5.1 illustrates the differences between LSD and iso-LSD. Bailey *et al.* (1973)

published the infrared spectra of iso-LSD and LAMPA. Neville *et al.* (1992) included the infrared spectra of LSD and LAMPA.

Infrared spectroscopy is an excellent method for the identification of LSD. The problem to be overcome is isolating the LSD in a sufficiently pure form that a suitable infrared spectrum is obtained. LSD acts as a base and is extracted from aqueous basic solutions by organic solvents. The use of very strong bases such as sodium or potassium hydroxide is not, however, recommended as LSD will decompose in their presence. For LSD soaked or spotted onto papers the following method will often work to isolate the LSD:

- cut the paper up into small pieces;
- soak the paper in dilute sulfuric acid solution;
- extract the sulfuric acid solution with an organic solvent such as chloroform to remove any acidic or neutral dyes;
- make the aqueous solution basic with ammonia;
- extract with chloroform;
- evaporate the solvent.

For LSD in tablets, crushing the tablet and following the same procedure will often work. In some cases, however, an emulsion will result. The alternative is a basic celite column which is described as follows:

- crush tablets and triturate with dilute sulfuric acid;
- add sodium bicarbonate and mix;
- add acid washed celite and mix;
- pack dry mixture is packed in column stoppered with chloroform washed cotton or glass wool;
- elute column with water washed chloroform;
- extract chloroform twice with dilute sulfuric acid (chloroform is discarded);
- make dilute sulfuric basic with ammonia;
- extract basic solution twice with chloroform;
- evaporate chloroform to yield LSD.

Because LSD is present in such small doses, either technique is very sensitive to sources of contamination. All glassware should be carefully inspected and rinsed and care be used in handling solvents in that a small contamination of plasticizer or grease will mask the LSD spectrum.

If the above technique does not yield LSD of sufficient purity that the spectrum is contaminated, preparatory thin layer chromatography using a silica gel plate in a solvent system of chloroform:methanol 90:10 will often work.

There are other approaches on the use of infrared spectroscopy to identify LSD. Harris and Kane (1991) outlined the use of a microscope sampling device attached to an FTIR instrument in conjunction with thin layer chromatography to distinguish among LSD, iso-LSD and LAMPA. Kovar *et al.* (1995) and Pfiefer and Kovar (1995) discussed the application of High Performance Thin Layer Chromatography-Ultra Violet (HPTLC-UV)/ FTIR on line coupling to LSD identification. Kempfert (1988) included the identification of LSD and its isomers in his discussion of the applications of GC-FTIR.

5.3 INFRARED SPECTROSCOPY OF PHENYLALKYLAMINES

The phenylalkylamines are good examples of several of the general principles discussed earlier in the chapter. For instance, the rule that the number of spectral features decreases from the condensed phase to the vapor phase is demonstrated in Figure 5.2 on the phenylalkylamine 3,4-methylenedioxymethamphetamine (MDMA). The power of FTIR spectroscopy to distinguish among structural isomers is illustrated in Figure 5.4 using the example of 2,3-methylenedioxyamphetamine versus 3,4-methylenedioxyamphetamine (MDA). Distinguishing among homologues is demonstrated with the example of the vapor phase spectra of MDMA versus N-methyl-1-(3,4-methylenedioxyphenyl)-2-butanamine (MBDB) in Figure 5.3 and MDA and its homologues in Figure 5.5. It is important to note that while the vapor phase spectra of MDMA and MBDB show few differences, the liquid phase spectra, also termed a film of the base extract, of MDMA and MBDB show more (Figure 5.7).

The hydrochloride spectra of the two isomers are even more distinctly different, as Figure 5.8 illustrates.

It follows from the previous discussion that the easiest way of identifying phenylalkylamines is to obtain an uncontaminated spectrum of their hydrochloride salts. Liquid and vapor phase spectra can be used quite effectively in identifying phenylalkylamines as long as the techniques are applied with care and knowledge. The problem then becomes how to isolate the phenylalkylamine from the sample matrix. One of the best ways is by extraction.

A dry extraction technique which will isolate the salt of the phenylalkylamine directly is as follows:

- grind the tablet or other matrix into a fine powder;
- extract the dry powder with diethyl ether to extract any tablet lubricant such as methyl stearate. The ether can usually be discarded;
- extract the powder with chloroform and filter the chloroform;

(a)

CH3
HN
CH3
O
O
%Transmittance
90
80
70
60
50
40
30
20
10
4000
3000
2000
1500
1000
Wavenumbers (cm-1)

(b)

CH3
HN
CH3
O
O
%Transmittance
90
80
70
60
50
40
30
20
4000
3000
2000
1500
1000
Wavenumbers (cm-1)

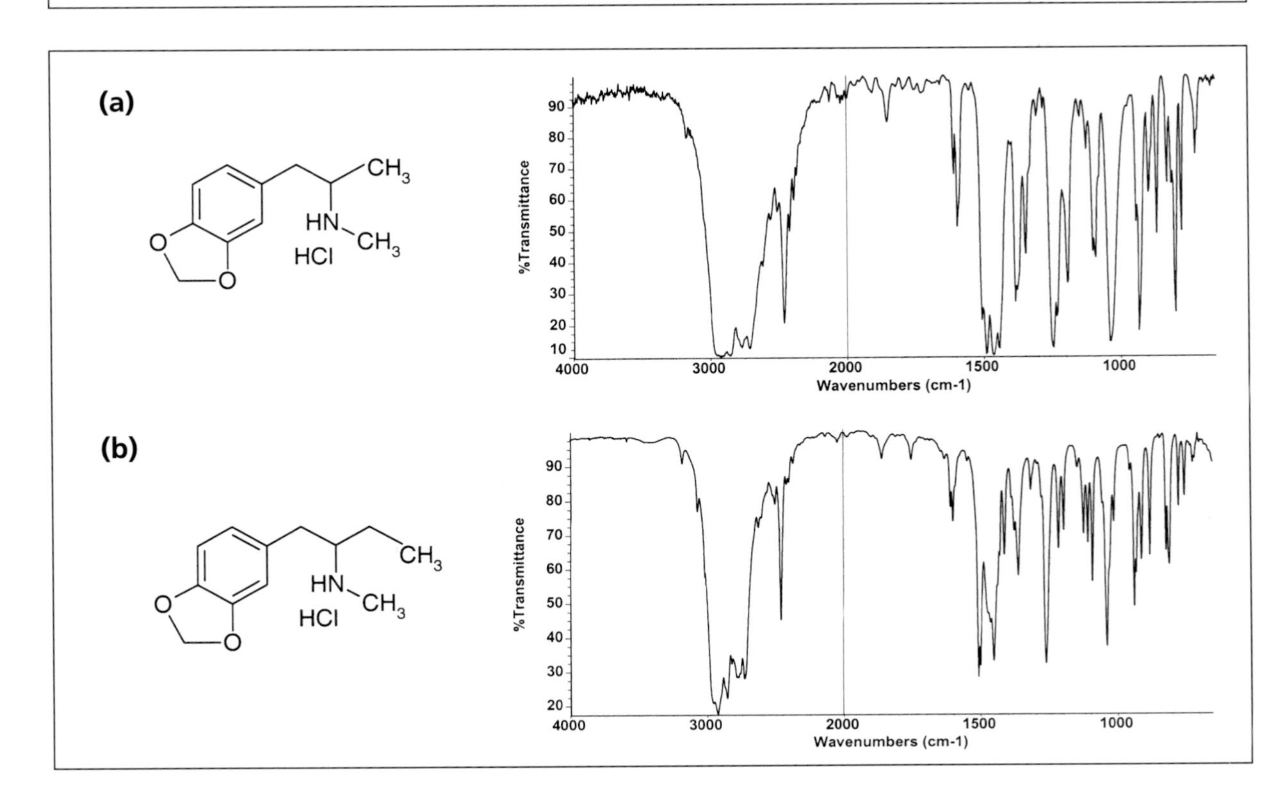

- evaporate the chloroform;
- in some cases, methanol is a better choice than chloroform.

The above technique will work using chloroform with most phenylalkylamines including MDA.HCl and MDMA.HCl. Methanol can be used in place of chloroform, but some of the tablet excipients may contaminate the spectrum.

A second extraction technique which is a liquid-liquid extraction is as follows:

- grind the tablet or other matrix into a fine powder;
- add the powder to a separatory funnel;
- add dilute sulfuric acid;
- extract at least twice with an organic solvent (chloroform works well);
- render solution basic with sodium hydroxide solution or ammonia;
- extract twice with an organic solvent;
- evaporate the solvent;
- add a few drops of concentrated hydrochloric acid before all the solvent has evaporated.

The second technique can be used to obtain the base in the liquid phase by not adding hydrochloric acid. The base can also be diluted with an organic solvent and injected into a GC-FTIR spectrometer. It is important to evaporate the solvent carefully; some of the phenylalkylamines are sufficiently volatile that they can be lost if heated unduly after all solvent has been evaporated.

It is also important to not use methanol or ethanol as the solvent when injecting onto a gas chromatograph. Clark *et al.* (1992) detailed the formation of condensation products of phenylalkylamines with methanol or ethanol.

There are several articles on the infrared spectra of the base and hydrochloride forms of the phenylalkylamines. Dal Cason (1989) published several spectra of MDA and MDMA and their analogues. Bailey *et al.* (1975) published the spectra of the N-methylated analogues of the methoxyamphetamines and MDA. Noggle *et al.* (1986) published the spectra of MDA, MDMA, and 3,4-methylenedioxy-N-ethylamphetamine (MDEA) and their hydrochloride salts. Hugel and Weaver (1988) published the infrared spectra of 2-methoxy-3,4-methylenedioxyamphetamine, N,N-dimethyl-MDA and MDEA. Veress *et al.* (1994) published the solid phase and vapor phase spectra of MDEA. CND Analytical (1988) and (1991) published many of the infrared spectra of the phenylalkylamines related to MDA.

Figure 5.7 (opposite, top)

Liquid phase infrared spectra of: (a) MDMA base; (b) MBDB base

Figure 5.8 (opposite, bottom)

Solid phase IR spectra of: (a) MDMA.HCl; (b) MBDB.HCl

5.4 INFRARED SPECTROSCOPY OF PHENCYCLIDINE AND ANALOGUES

The approach and comments outlined in the phenylalkylamines hold true for phencyclidine and its analogues with a few notable differences. The generalization that the number of spectral features decreases from the condensed phase to the vapor phase is demonstrated by Figure 5.9 where the spectra of phencyclidine HCl-solid phase, phencyclidine base-liquid phase, and phencyclidine base at 225°C-vapor phase are presented.

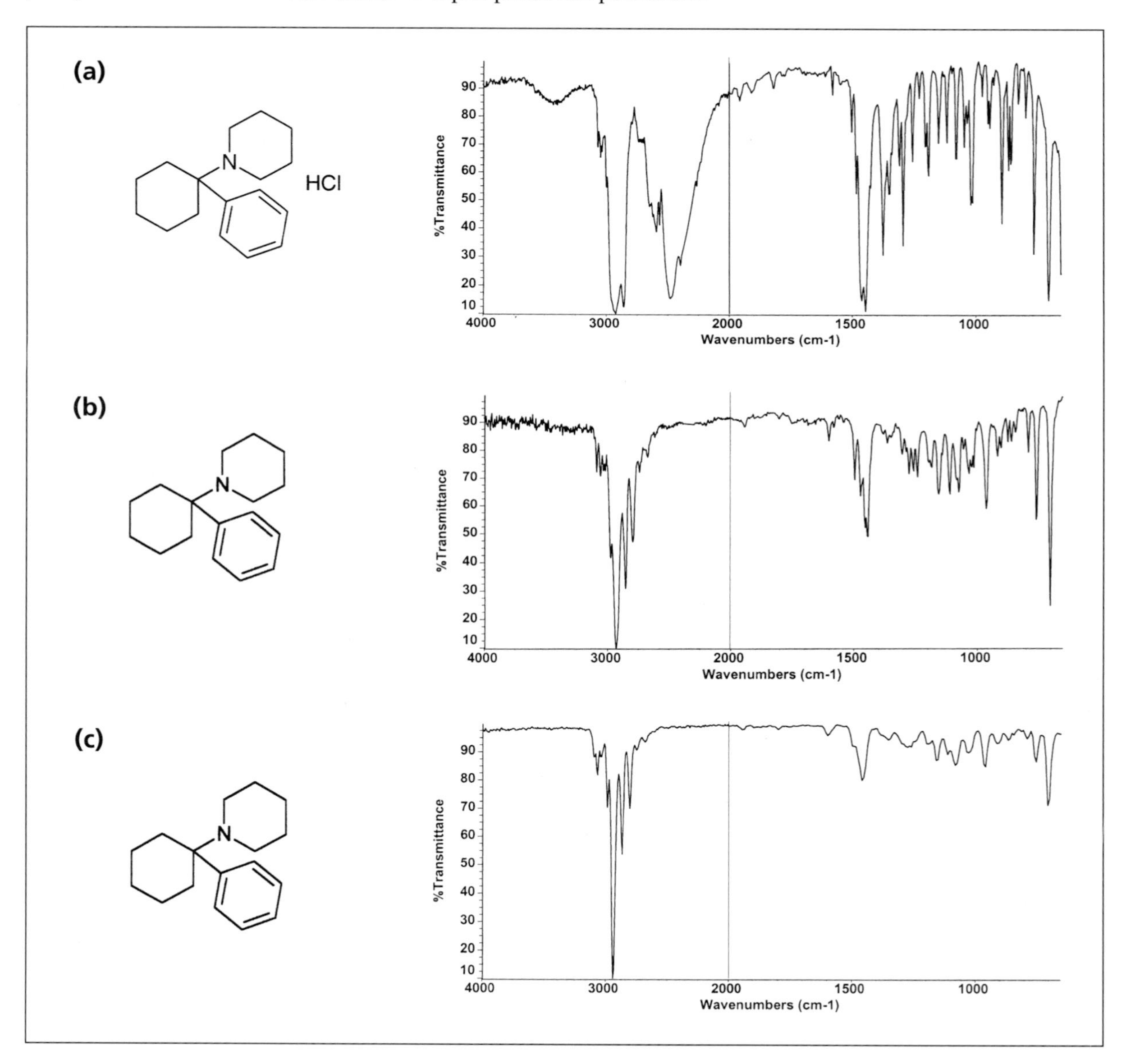

Figure 5.9

Infrared spectra of phencyclidine (PCP): (a) solid – phencyclidine HCl; (b) liquid – phencyclidine base; (c) vapor – phencyclidine at 225°C

Some methods for the synthesis of phencyclidine involve the use of hydrobromic acid near the end of the reaction. This will sometimes result in the occurrence of phencyclidine hydrobromide (PCP.HBr) in street samples in place of the more common phencyclidine hydrochloride (PCP.HCl). Figure 5.10 illustrates the spectra of PCP.HBr in comparison to the spectrum of PCP.HCl in Figure 5.9.

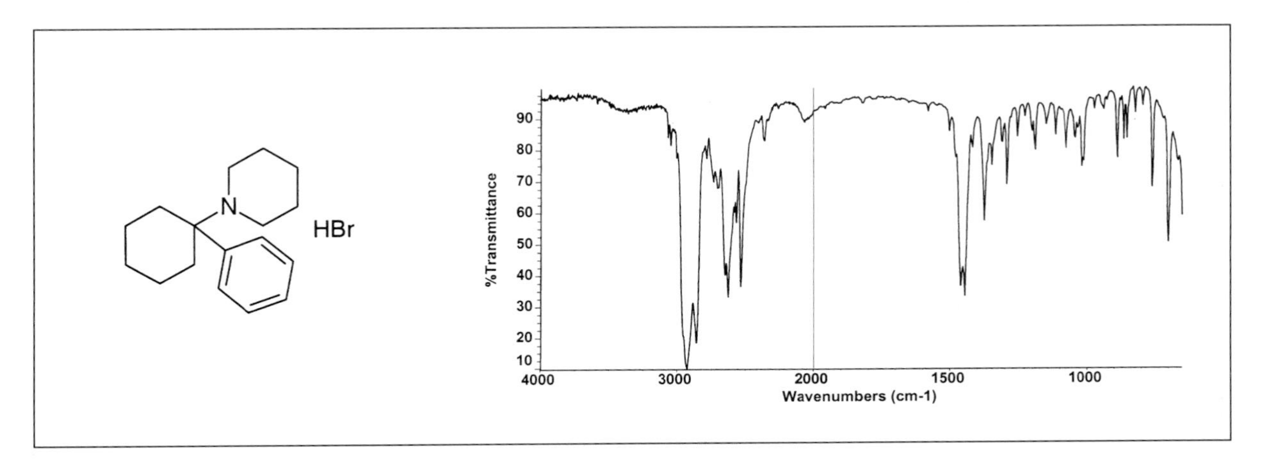

Figure 5.10
IR spectra of phencyclidine HBr

Note the differences in the spectra in the 2600 to 2900cm^{-1} region of the spectrum.

The spectra of the base films of the following homologous series are presented in Figure 5.11:

- N-(1-phenylcyclohexyl)methylamine;
- N-(1-phenylcyclohexyl)ethylamine;
- N-(1-phenylcyclohexyl)propylamine;
- N-(1-phenylcyclohexyl)isopropylamine.

The series of spectra demonstrate why IR spectroscopy is an excellent technique in distinguishing phencyclidine and its analogues.

Phencyclidine and analogues can be extracted as outlined in the phenylalkylamines section with a few exceptions. Phencyclidine HCl is so soluble in chloroform that it will extract from the aqueous acid solution in the liquid-liquid extraction method. 1-Phenylcyclohexene is a common gas chromatographic artifact when phencyclidine is injected into the gas chromatograph.

There are several articles on the identification of phencyclidine and its analogues by infrared spectroscopy. Bailey *et al.* (1976) published the base and base hydrochloride spectra of six phencyclidine analogues observing that the base or hydrochloride spectra were all distinguishable. Bailey *et al.*

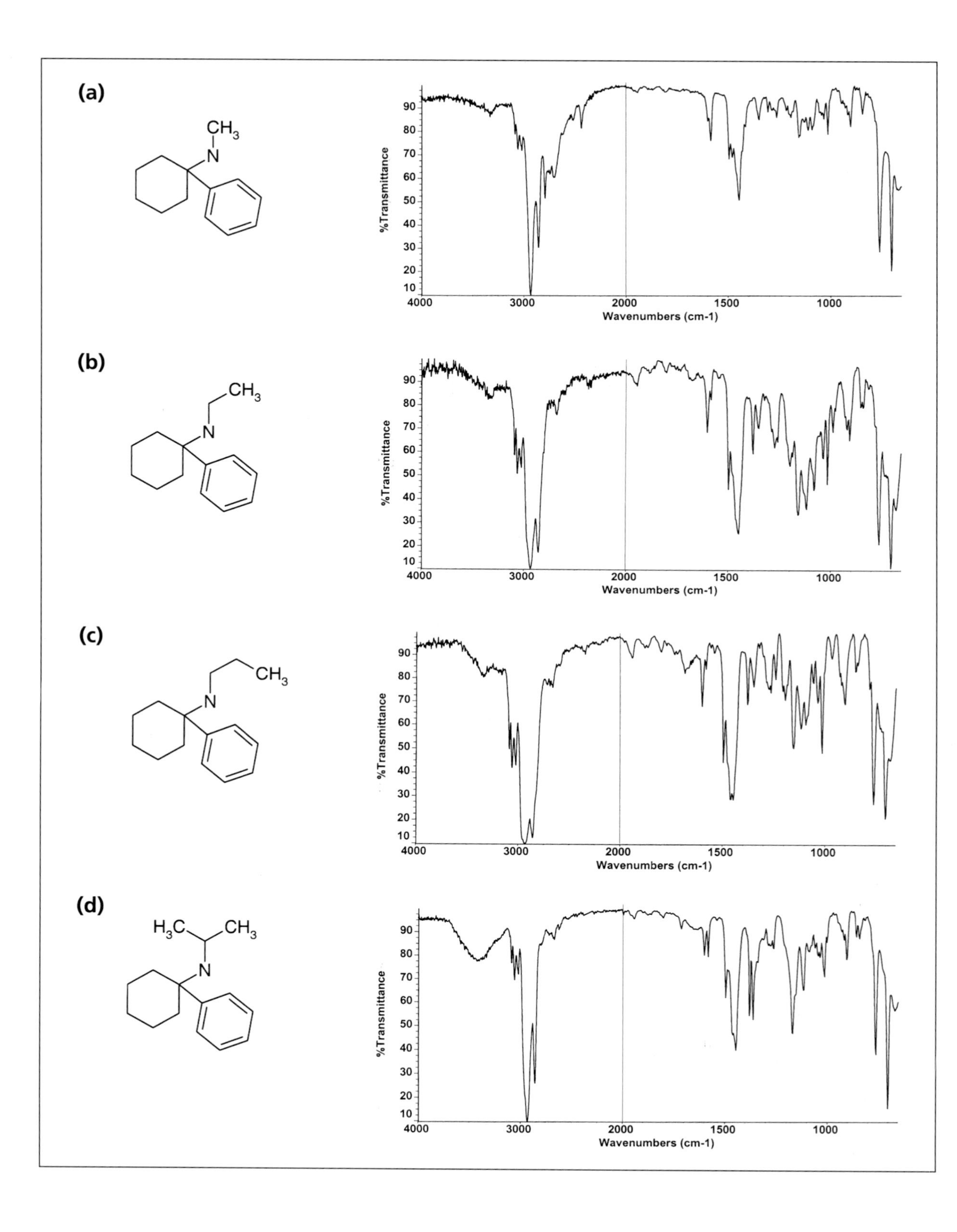
(a)
CH_3
N
%Transmittance
Wavenumbers (cm-1)
(b)
CH_3
N
%Transmittance
Wavenumbers (cm-1)
(c)
CH_3
N
%Transmittance
Wavenumbers (cm-1)
(d)
H_3C
CH_3
N
%Transmittance
Wavenumbers (cm-1)

(1979) argued that while the base spectra of some of the analogues of N-(1-phenylcyclohexyl)ethylamine (cyclohexylamine) were similar they could be distinguished. Allen *et al.* (1980) presented infrared spectra of 1-(1-phenylcyclohexyl)-4-methylpiperidine. Lodge *et al.* (1992) presented the hydrochloride spectra of analogues of phencyclidine where the phenyl group had been replaced by a benzyl, 2-methylphenyl, 3-methylphenyl, and 4-methylphenyl. The spectra were readily distinguishable.

Figure 5.11 (opposite)

IR spectra of phencyclidine homologues in liquid form are: (a) N-(1-phenylcyclohexyl)methylamine; (b) N-(1-phenylcyclohexyl)ethylamine; (c) N-(1-phenylcyclohexyl)propyl amine; (d) N-(1-phenylcyclohexyl)isopropylamine

5.5 INFRARED SPECTROSCOPY OF TRYPTAMINES

The techniques outlined in the phenylalkylamine section also hold for the identification of the tryptamines. Spectra of N,N-dimethyltryptamine, N,N-diethyltryptamine and N,N-dipropyltryptamine are illustrated in Figure 5.12 and are easily distinguishable.

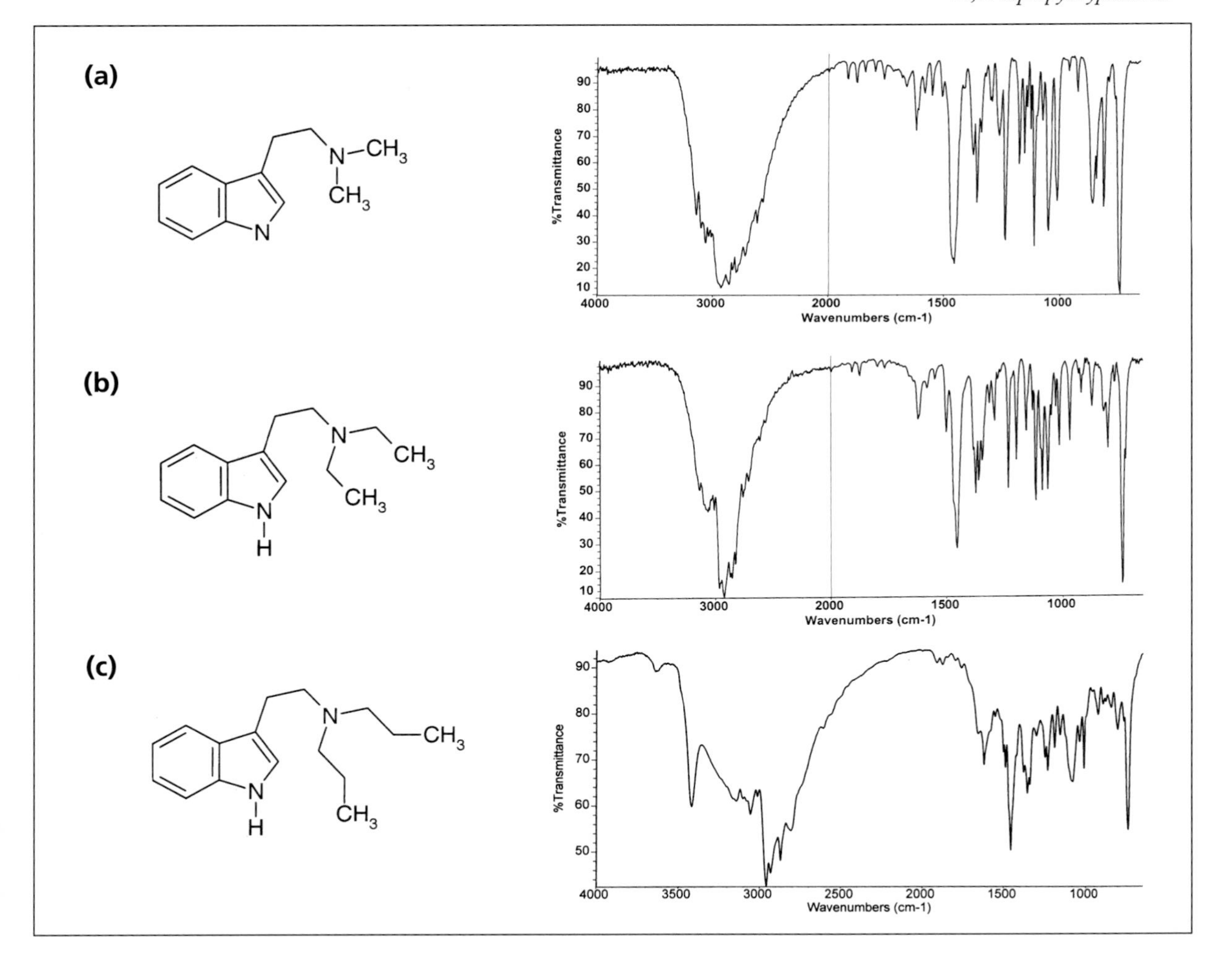

Figure 5.12

IR spectra of tryptamine homologues: (a) solid – N,N-dimethyltryptamine; (b) solid – N,N-diethyltryptamine; (c) liquid – N,N-dipropyltryptamine

The newest member of the tryptamine family, 5-methoxy-N,N-diisopropyl-tryptamine is easily identified using the infrared spectrometer (Figure 5.13).

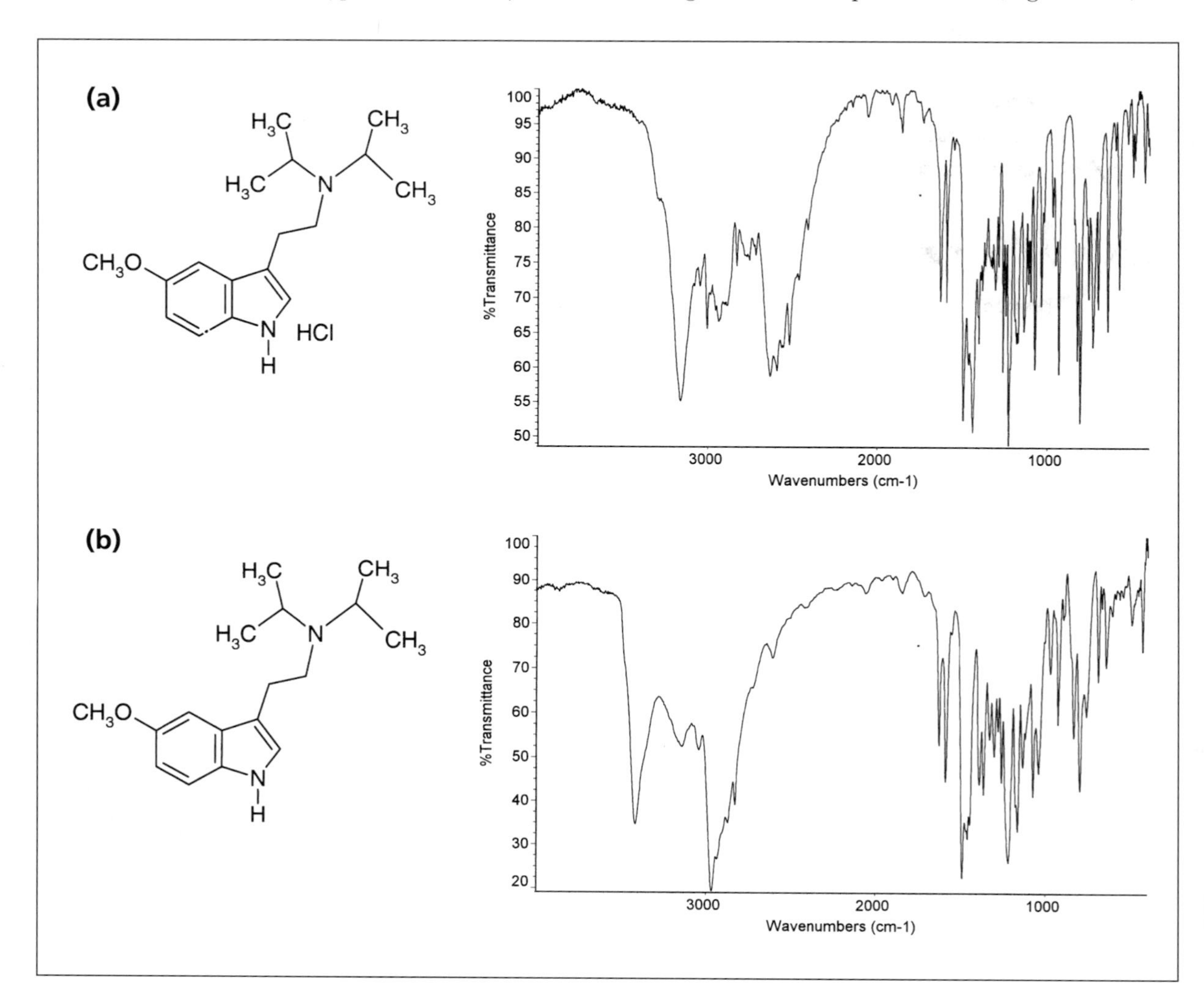

Figure 5.13

IR spectra of 5-methoxy-N,N-diisopropyl tryptamine: (a) solid – as HCl salt; (b) liquid – freebase

Most tryptamines will extract as described in the phenylalkylamine section. A notable exception is psilocybin which has a labile phosphate group. The phosphate group is easily removed from the psilocybin molecule to make psilocin. The phosphate group will hydrolyze under acidic, basic, and high temperature treatment. Psilocybin is isolated from its matrix (usually mushrooms or mycelia) by preparatory HPLC, Hugel (1984). The spectrum of the purified analyte can be obtained as an amorphous solid on a silver chloride disk or as a KBr micropellet. Spectra of psilocin, psilocybin standards as KBr pellets as well as a spectrum of psilocybin as an amorphous solid are presented in Figure 5.14.

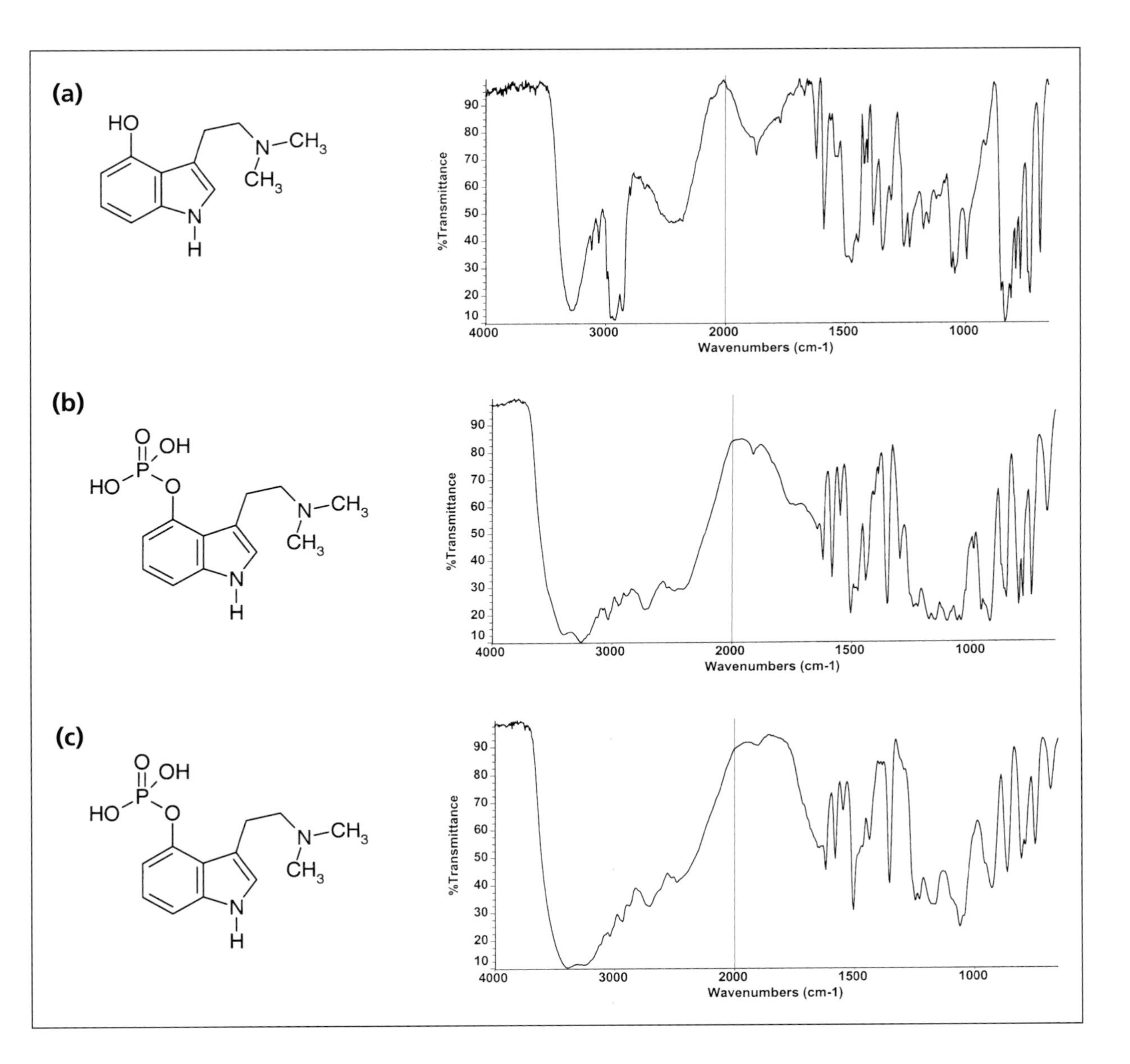

Figure 5.14

Infrared spectra of psilocin and psilocybin: (a) solid – psilocin; (b) solid – psilocybin; (c) solid – amorphous psilocybin

5.6 FURTHER READING

Detailed discussions on the theoretical basis of infrared spectroscopy can be found in many university level textbooks. In particular, Fifield and Kealey (1995) and Ege (1999) discussed the theoretical basis of infrared spectroscopy and its application to the identification of organic molecules. The assignment of absorption frequencies to particular functional groups was also outlined. The text edited by Gough (1991) similarly contained a chapter which includes the theoretical basis of infrared spectroscopy in the context of illicit drug analysis. CND Analytical (1994) provided a similar discussion in the context of the analysis of phenylalkylamines.

Texts which outline the absorption frequencies of functional groups include:

- *The Analysis of Drugs of Abuse* (1991) by T. Gough (editor).
- *Handbook of Spectrophotometric Analysis of Drugs* (1981b) by I. Sunshine (editor).
- *Spectroscopic Methods in Organic Chemistry* (1966) by D. H. Williams and I. Fleming.
- *Applications of the Absorption Spectroscopy of Organic Compounds* (1965) by J. R. Dyer.

Less common ways of obtaining condensed phase spectra are specular reflectance and photoacoustic spectroscopy. All are discussed in the text by Gough (1991). These techniques can be effective for the analysis of unusual analytes, but are not in general use for hallucinogen analysis.

Electronic compilations of condensed phase infrared spectra which include hallucinogens and / or their precursors include:

- Georgia State Crime Laboratory Drug Library containing 1900 spectra which includes abused drugs, isomers, and precursors;
- Toronto Forensic FTIR Library containing 3400 spectra which focuses on abused drugs, pharmaceuticals, precursors, and reagents;
- Nicolet/Aldrich Spectral Library containing 18,500 spectra which includes precursors, essential chemicals, and solvents used in the clandestine production of hallucinogens;
- Nicolet/Sigma Library containing 10,400 spectra which includes many biochemicals as well as spectra of many abused drugs;
- Sadtler/Bio-Rad Condensed Phase IR Standards containing 75,500 spectra which includes precursors, essential chemicals, and solvents used in the clandestine production of hallucinogens.

Electronic compilations of vapor phase infrared spectra which include hallucinogens and / or their precursors include:

- Sadtler/Bio-Rad Vapor Phase IR Standards containing 9100 spectra which includes precursors, essential chemicals, and solvents used in the clandestine production of hallucinogens;
- Nicolet Vapor Phase Library containing 8600 spectra which includes pre-

cursors, essential chemicals, and solvents used in the clandestine production of hallucinogens;
- Aldrich FTIR Vapor Phase Library containing 5000 spectra which includes precursors, essential chemicals, and solvents used in the clandestine production of hallucinogens;
- EPA Vapor Phase Library containing 3300 spectra which includes essential chemicals and solvents used in the clandestine production of hallucinogens.

Hard copy (text form) compilations of condensed phase infrared spectra which include hallucinogens and their precursors include:

- *Instrumental Data for Drug Analysis Volumes 1 to 4* (1987) by Mills and Roberson;
- *Instrumental Data for Drug Analysis Volume 5* (1992) by Mills *et al.*;
- *Instrumental Data for Drug Analysis Volumes 6 and 7* (1996) by Mills *et al.*;
- *Clarke's Isolation and Identification of Drugs* (1986) by Moffat *et al.* (editors);
- *Sigma Library of FT-IR Spectra* (1987) by Keller;
- *The Aldrich Library of FT-IR Spectra* (1985) by Pouchert;
- *Analytical Profiles of Amphetamines and Related Phenethylamines* (1989) by CND Analytical;
- *Analytical Profiles of Designer Drugs Related to the 3,4-Methylenedioxyamphetamines (MDA's)* (1991) by CND Analytical;
- *Analytical Profiles of Substituted 3,4-Methylenedioxyamphetamines: Designer Drugs Related to MDA* (1988) by CND Analytical;
- *Analytical Profiles of Precursors and Essential Chemicals* (1990) CND Analytical
- *Analytical Profiles of the Hallucinogens* (1991) by CND Analytical;
- *Forensic and Analytical Chemistry of Clandestine Phenethylamines* (1994) by CND Analytical.

Hard copy (text form) compilations of vapor phase infrared spectra which include hallucinogens and their precursors include:

- *The Aldrich Library of FT-IR Spectra Volume 3* (1989) by Pouchert;
- *Instrumental Data for Drug Analysis Volumes 6 and 7* (1996) by Mills *et al.*

PART II MASS SPECTROMETRY

The application of gas chromatography-mass spectrometry (GC-MS) has become a common tool in most forensic labs charged with the identification of illicit drugs seized by law enforcement agencies. Being a sensitive technique, only small quantities of analyte are required in order to obtain a usable spectrum. It is also a quick technique and, with advances in computer technology, easy to operate. As with the application of any technique in illicit drug analysis, the forensic chemist must understand how GC-MS works, how it can be applied, and the strengths and weaknesses of the technique.

5.7 THEORETICAL BASIS

Classical electron impact mass spectroscopy works by introducing the analyte to the mass spectrometer at a very low pressure. At this low pressure, inside the mass spectrometer's source, the analyte is bombarded by an electron beam. The high energy electrons when passing in close proximity to the molecule, can cause it to lose an electron. The molecule, carrying a positive charge, then becomes unstable (termed metastable) and fragments. The positively charged fragments are repelled from the source into the mass analyzer. The mass analyzer, based upon the generation of electro-magnetic or similar fields, allows only one fragment at a time to enter the detector, an electron multiplier. The mass analyzer rapidly scans through the atomic mass range of interest. In this process, the ions that are being selected reach the electron multiplier and their mass to charge ratio can be calculated by knowing what electro-magnetic field strength allowed them to reach the detector. The resulting data is plotted as ion intensity, or abundance, versus the mass to charge ratio of the fragment. For simple organic molecules normally only one positive charge can be accommodated and therefore the mass to charge ratio can be referenced as mass only.

The formation of fragment ions is dependant upon several factors including the stability and amount of energy imparted upon the molecular ion (M^+). The molecular ion degrades through various pathways dependant upon the formation of energetically favored intermediates. Structural differences from one analyte to another means that different pathways and fragment ions form. This in turn means that the mass spectrum will be different. In the application of GC-MS to hallucinogen analysis, this description holds true for most, but not all applications. In particular, distinguishing isomers by mass spectra is often problematic. Optical isomers are not distinguishable. Diastereomers are sometimes distinguished by their mass spectra. In some cases, structural isomers are not distinguished. For instance, the mass spectra of lysergic acid diethylamine (LSD) and iso-LSD, lysergic acid methyl-propylamide (LAMPA), and lysergic acid sec-butylamide (LSB) are virtually identical as illustrated in Figure 5.15.

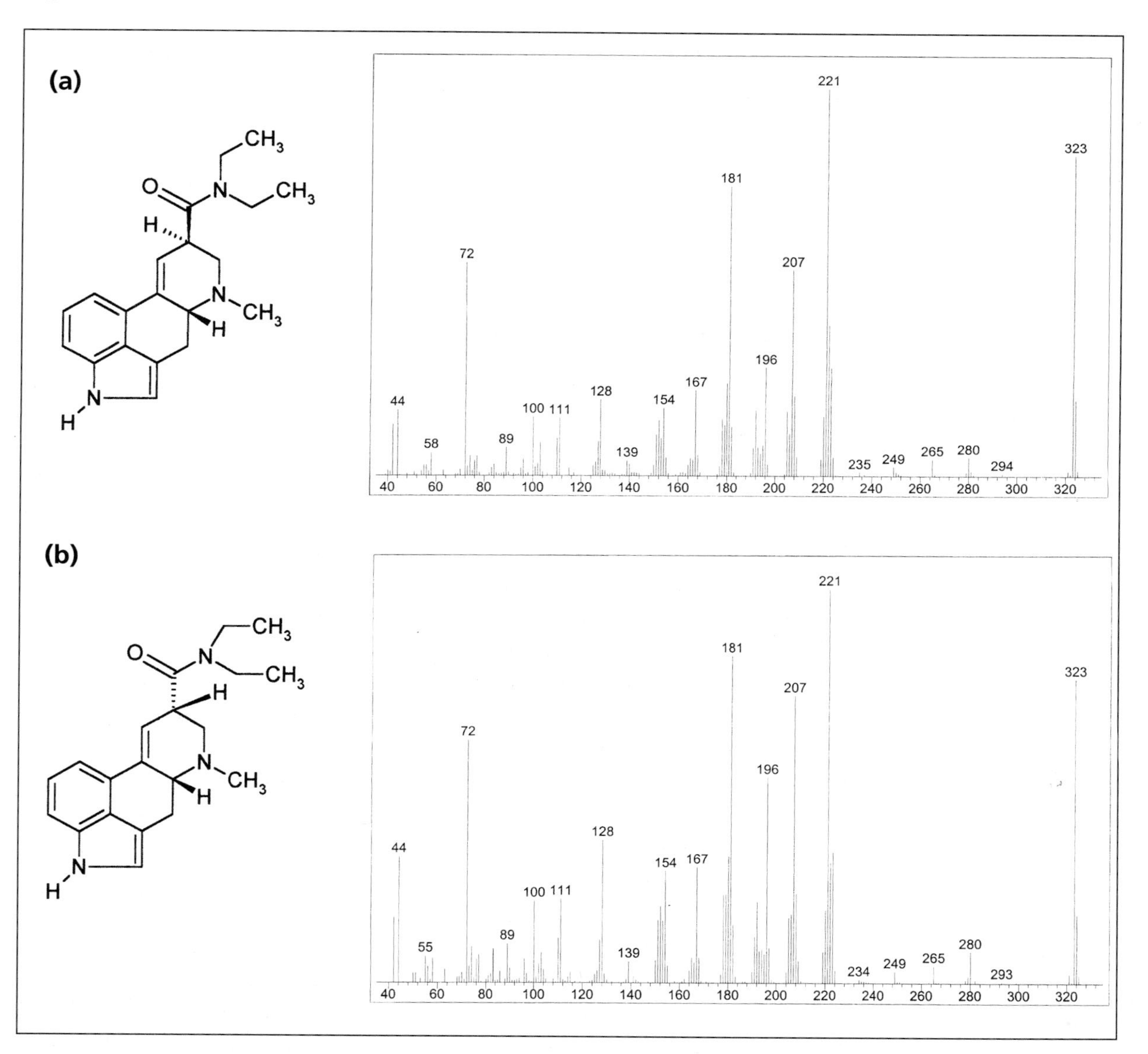

Figure 5.15

Mass spectra of LSD related compounds: (a) LSD; (b) iso-LSD; (c) Lysergic acid N,N-methyl propylamide (LAMPA); (d) Lysergic acid sec-butylamide (sec-LSB)

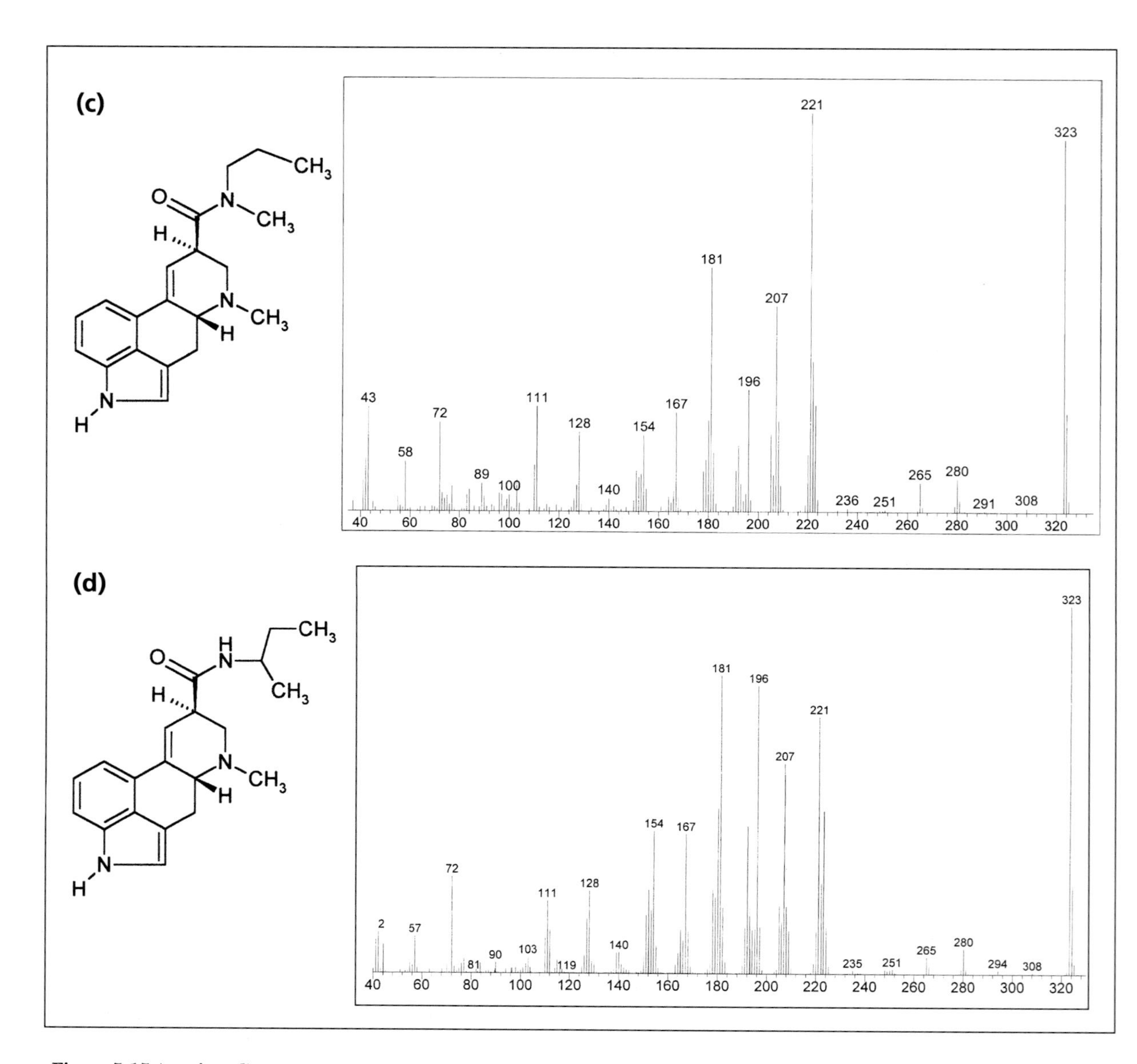

Figure 5.15 (continued)

On the other hand, the addition of a methylene group to an analyte will invariably change its mass spectrum relative to its non-methylated homologue. For example, as shown in Figure 5.16 the easily distinguished mass spectra of the following homologous series is presented:

- 3,4-methylenedioxyphenethylamine
- 3,4-methylenedioxyamphetamine (MDA)
- 1-(3,4-methylenedioxyphenyl)-2-butanamine
- 1-(3,4-methylenedioxyphenyl)-2-pentanamine.

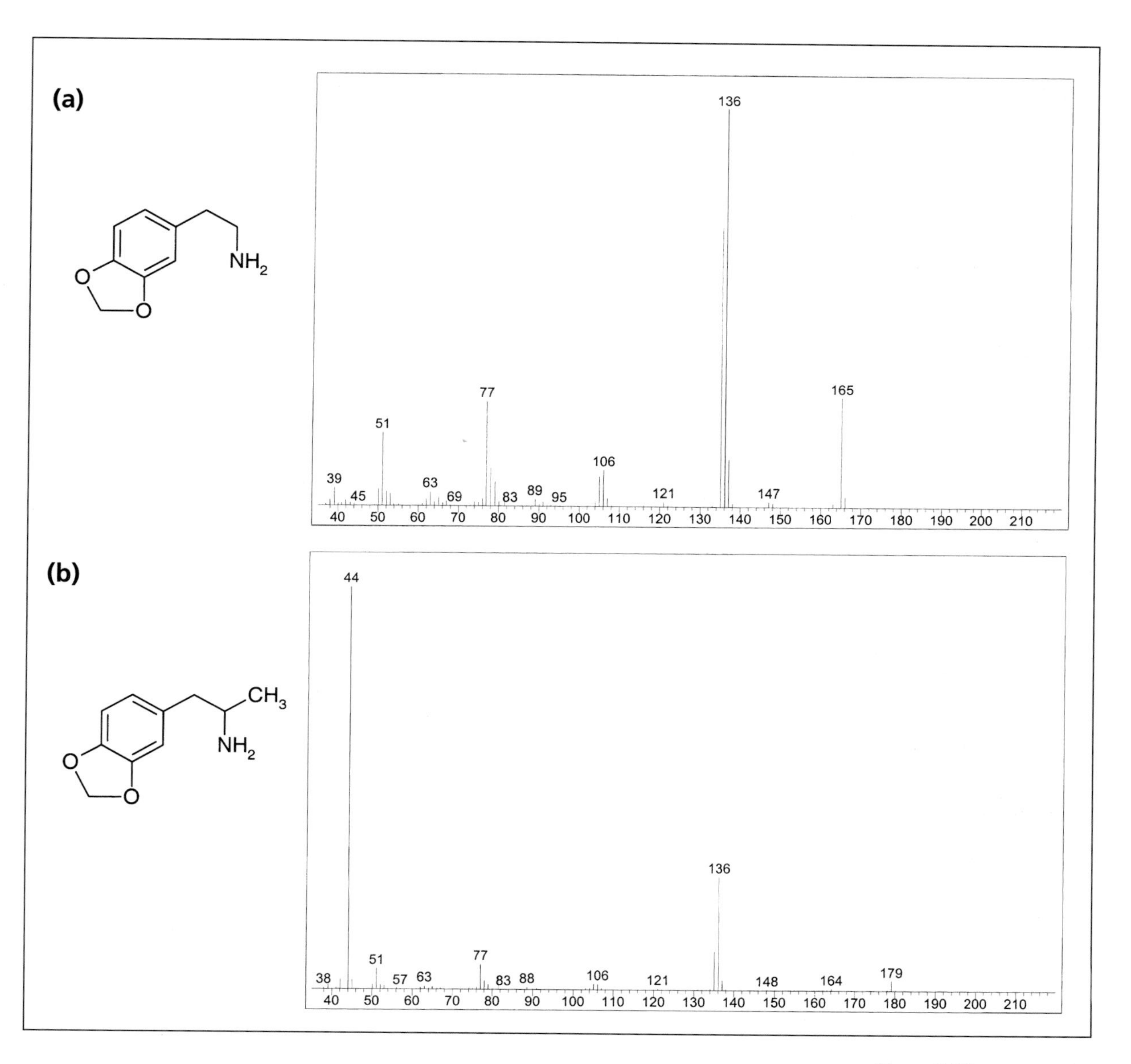

Figure 5.16

Mass spectra of the homologous series based on MDA: (a) 3,4-methylenedioxyphenethylamine; (b) MDA; (c) 1-(3,4-methylenedioxyphenyl)-2-butanamine; (d) 1-(3,4-methylenedioxyphenyl)-2-pentanamine

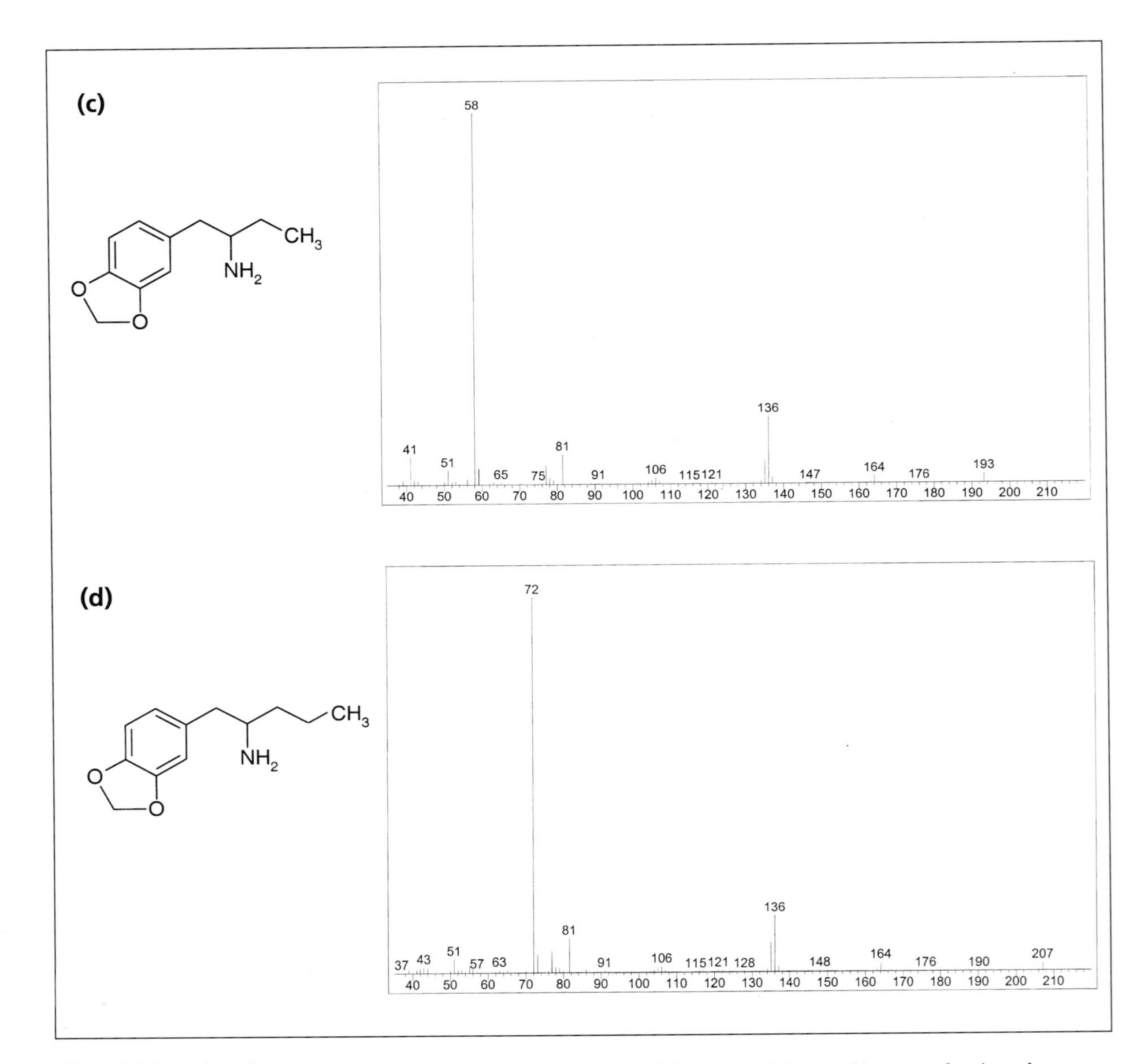

Figure 5.16 (continued)

One common way of dealing with the problems of isomers having the same mass spectra is the use of GC retention times. Examples abound where the mass spectra are identical, but the GC retention time is different enough that the two isomers can be distinguished. For example, the mass spectra of LSD, iso-LSD and lysergic acid methyl propyl amide (LAMPA) are all identical. It is, however, a relatively simple matter to develop a method using either a dimethylpolysiloxane or (5% phenyl)methylpolysiloxane capillary column to separate the three isomers.

The sensitivity of mass spectrometers is nothing short of remarkable. Some instruments are more sensitive than others and some analytes intrinsically provide more intense and more featured mass spectra than others. To obtain full scans of hallucinogens where the features of the mass spectrum are clear enough to allow conclusive identification are generally in the nanogram range. The downside to this sensitivity is that the analyst must be sure that the small amount of drug that is being detected was not caused by inadvertent contamination.

The resolution of a mass spectrometer is a measure of the accuracy of the mass that is measured by the mass analyzer. Unit resolution means that the mass analyzer can distinguish between a mass of 323 and 322 or 324. It cannot distinguish between a mass of 323.2 and 323.5. If the mass of a peak is 43.0184 daltons, then the peak is that of $C_2H_3O^+$ and not $C_3H_7^+$ (43.0547 daltons) or $C_2H_5N^+$ (43.0421 daltons) (McLafferty, 1980). A unit resolution mass spectrometer cannot make this distinction. However, this differentiation is beyond the capabilities of many common GC-MS instruments. There are mass analyzers which can obtain resolution in the tens of thousands range. Popular mass analyzers that are used in forensic laboratories are normally unit resolution.

5.7.1 MASS ANALYZERS

A magnetic sector instrument deflects the ion fragments based on their mass to charge (m/z) ratio by means of a magnetic field. By varying the magnetic field, only one m/z will reach the detector at a time. Magnetic field instruments when coupled with an electrostatic field analyzer are capable of much better than unit resolution. These double focusing instruments are not typically used in the identification of routine drug exhibits.

The mass analyzer in quadrupole instruments consist of four precisely aligned parallel rods to which radio frequency and direct current voltages are applied. By varying the frequency and voltage only one mass unit can travel among the quadrupoles and hit the electron multiplier detector. These instruments have enjoyed increasing popularity over the past few years.

Ion trap mass analyzers have the sample ionization and mass analysis occur in the same region. A radio frequency is applied to a ring electrode which traps the ions. The ions are selectively expelled by varying the radio frequency. This, as with the quadrupole, is a unit resolution instrument and is commonly used for drug analysis.

In time of flight mass analyzers ion fragments are accelerated into the mass analyzer. The ions drift through a field free region and arrive at the detector at different times based on their velocity and hence their mass can be calculated. These types of instruments are typically of low resolution and, to date have not been popular.

5.7.2 IONIZATION TECHNIQUES

In electron impact (EI) ionization, electrons are accelerated to 70eV and focused on the analyte in the vapor phase in the mass spectrometer's source. The electrons hit the analyte and remove an electron from the analyte causing fragmentation. Most of the published work to date on the mass spectra of illicit drugs, including hallucinogens has been electron impact data with ionization energy of 70eV. Libraries of mass spectra consisting of several hundred thousand spectra have been accumulated from such data. This continues to be the ionization method of choice for most hallucinogen analyses. In some cases, the positively charged molecular ion fragments so readily that molecular ions are not seen. This can be a problem when attempting to identify an unknown compound using its mass spectrum.

In chemical ionization (CI), a positively charged reagent gas such as methane, isobutane, or ammonia collides with the analyte transferring a proton (H^+) from the reagent gas to the analyte (A). The $(A+H)^+$ ion is quite stable and is often the base peak in the CI spectrum. This can be an important characteristic for determining the molecular ion for those hallucinogens, such as the phenethylamines, which do not have strong molecular ions in their EI spectra.

5.7.3 ANALYTE INTRODUCTION

There are several techniques for introducing the analyte to the mass spectrometer.

Direct insertion probe

The analyte is coated onto a glass holder which is introduced directly into the mass spectrometer source through a vacuum lock. The end of the probe can be temperature programmed to selectively desorb any components that have been deposited on the glass holder. Most often, however, the technique is used with pure analytes. This technique is quite often employed when the analyte is not chromatographable and as such will decompose in the GC inlet. It does not lend itself to routine analysis or automation.

Gas chromatography (GC)

The components, including the analyte of interest are injected onto a capillary column gas chromatograph. The gas chromatograph separates the components which are sequentially introduced to the mass spectrometer. Before the commercial availability of capillary GC columns there was a need for separators to eliminate the carrier gas before the components reached the mass analyzer

(Gough, 1991). In almost all cases now, the end of the GC capillary column is inserted directly into the mass spectrometer source with the pumping system able to handle the carrier gas flow rates. Helium or hydrogen are typical GC carrier gases providing the best chromatographic separation characteristics.

The technique of gas chromatography-mass spectrometry (GC-MS) is the most common mass spectral technique used in the identification of hallucinogens. The most common columns are dimethylpolysiloxane and (5%phenyl)methylpolysiloxane. With few exceptions, either of these columns will separate the mixtures of hallucinogens and contaminants such that the analyte of interest can be identified. One consideration in determining the internal diameter of the gas chromatograph capillary column to be used is the ability of the mass spectrometer vacuum pumps to remove the carrier gas from the source. Generally, for an instrument that is capable of handling a few mL/min carrier gas and can obtain a full scan in less than a second, the internal diameter of the capillary columns should be in the 0.15 to 0.25mm range. The length of the column should be determined by the resolving power required to separate analytes of similar composition. Columns of length between 10 and 30m are quite satisfactory for normal hallucinogen identification.

The problems associated with GC-MS are that the analyte must be gas chromatographicable. Compounds which are not volatile and which are thermally labile fall into the category of analytes which are not suitable for GC-MS.

Liquid chromatography (LC)

The use of liquid chromatography permits the analysis of thermally labile and involatile analytes.

The major difficulty with LC-MS has been sample introduction where the liquid mobile phase cannot be directly introduced into the mass spectrometer. One of the first techniques for LC-MS was the moving belt system. The eluant from the LC is sprayed onto a moving belt which is heated to remove the mobile phase and which then passes through a vacuum lock to the mass analyzer source. This technique suffered from low sensitivity and the possibility of contamination of the belt.

Another early direct liquid introduction techniques involved forcing the LC column elute through a diaphragm orifice which then enters a chamber where most of the liquid is turned into vapor. The solvent vapor acts as the CI gas and ionization takes place.

A technique that has had commercial success is the thermospray technique. The eluate is introduced into the mass spectrometer through a heated capillary tube. The aqueous mobile phase contains electrolytes which are charged and in turn cause ionization of the sample molecules. This is mild form of ionization which means that there is little fragmentation. In addition, however, if the molecule is thermally labile it could decompose in the heated inlet.

Another recent technique which does not require heating the sample is Atmospheric Pressure Ionization (API). The sample and the mobile phase are vaporized through a nebulizer which then passes through a veil of nitrogen gas before entering the source where the bulk of the solvent is removed and soft-ionization of the analyte takes place.

A more recent sample introduction technique involves the use of a particle beam interface. From a conventional reverse phase LC, the low volatility analyte is enriched in concentration as it passes through the interface. Classical EI spectra can then be obtained as the analyte is introduced into the mass analyzer. The advantage of this technique is that the obtained EI spectra can be compared with the vast amount of published EI spectra.

5.7.4 TANDEM MASS SPECTROMETRY

Tandem mass spectrometers involve the use of two mass spectrometers in series. The first mass spectrometer purifies the introduced mixture by focusing only the molecular ion of the analyte of interest. That molecular ion then collides with neutral gas molecules which cause fragmentation of the molecular ion which fragments and is measured by the second mass spectrometer. The advantage of this technique is that there is no need to have a chromatograph at the front end of the mass analyzer. Mixtures can be introduced directly to the first mass analyzer. Tandem mass spectrometers are perfectly suited for studying the origin of an ion and their fragmentation pathways through MS-MS or "linked-scan" studies.

5.7.5 DATA ACQUISITION AND MANIPULATION

In order to obtain usable information from the mass spectrometer, the data must be in an easy-to-use format. With the advances in computer technology within the last few years, this has been effectively obtained. One of the techniques that is important to working with GC-MS or LC-MS data is obtaining a chromatogram from the data. The ions that are obtained by the mass spectrometer are summed by the computer, referred to as the Total Ion Current (TIC), to derive an analyzable chromatogram which is analogous to a gas chromatographic (or liquid chromatographic) run. For low level analyses where interferences from numerous extraneous sources are possible the mass spectrometer can be programmed to analyze for a small set of ions characteristic to the analyte(s) in question (Selected Ion Monitoring: SIM).

In most cases, however, for the identification of hallucinogens, it is best to obtain full scan spectra. This is accomplished by most modern systems at a rate of about one complete scan a second. This speed is quite satisfactory to identify the peaks eluting from a capillary column. Even if only a few scans are obtained per peak, there is no difficulty in obtaining good spectra providing there is enough sample.

One of the implications of the use of a capillary column gas chromatograph as a separation tool for the mass spectrometer is that the concentration of the analyte as it elutes from a column is constantly changing. As the analyte begins to appear at the mass spectrometer, its concentration is increasing. After reaching the peak maximum the concentration of the analyte begins to decrease. Since the mass spectrometer is measuring individual masses sequentially and those masses are changing with time, it follows that the mass spectrum can be skewed depending on the changing analyte concentration. For example, on the upside of a peak when the mass spectrometer scans from high to low masses, the lower masses will be stronger in relative intensity. Conversely, on the downside of the peak, the higher masses will be weaker in relative intensity. The problem is exacerbated by the narrow peak shapes that are characteristic of modern capillary columns. The problem of skewed spectra can be overcome by averaging scans on the upside, at the apex of the peak, and on the downside of the peak. Modern software will accomplish this with minimal effort on the user's part and can be automated.

5.7.6 SPECTRAL COMPILATIONS

In order to identify the mass spectra of the hallucinogens, the easiest way is to search a library of mass spectra. This has lead to compilations of 70eV electron impact mass spectra which are available in electronic or hard copy (book) form. Electronic libraries have the advantage that they can be searched by instrument manufacturer's software. A good match is one which has a significantly better hit value than the next. As with the case of the IR library searches, proper care must be taken to interpret the search results It is always important to visually compare the sample spectrum with the appropriate standard spectrum to identify an analyte. The heading on further reading details electronic and hard copy compilations of mass spectra.

5.8 MASS SPECTROSCOPY OF LYSERGIC ACID DIETHYLAMIDE (LSD)

As noted earlier in the chapter in the section on infrared spectroscopy, lysergic acid diethylamine (LSD) dosage units are typically in the 30 to 100μg range. An alternative to obtaining a condensed phase infrared spectrum is to obtain a mass spectrum using a GC-MS. In order to properly interpret the obtained mass spectrum, there are pitfalls to avoid.

LSD is particularly sensitive to contamination in the injection port. If the injection port is dirty, the mass spectrum of LSD will not be obtained. Changing the injection port insert is an excellent idea before attempting to obtain a mass spectrum of LSD using a GC-MS.

Isomers of LSD can be problematic when obtaining mass spectra. In particular, iso-LSD, lysergic acid methyl propyl amide (LAMPA), and lysergic acid sec-butylamide (sec-LSB) give essentially identical spectra. See Figure 5.15 at the beginning of the GC-MS section. The way to conclusively identify which isomer is present is to determine the retention times of the standards. As long as the isomers are well separated, the retention time and the mass spectrum form a conclusive identification. It is also important to either run a standard LSD in the same sequence as the sample or run the sample with an internal standard to ensure that the analyte retention time does indeed match that of LSD. As far as which columns are the best to use, the standard dimethylpolysiloxane and (5%phenyl)methylpolysiloxane columns both work.

The similarity of the spectra of LSD and its isomers has been discussed in several articles. Bailey *et al.* (1973) observed that the distinction between LSD, iso-LSD, and the methyl propyl amides of lysergic acid by mass spectrum alone is not satisfactory. Ardrey and Moffat (1979) published spectra of several ergot alkaloids as well as iso-LSD and LSD. Nichols *et al.* (1983) noted that while the mass spectra of LSD and LAMPA are similar, they could be separated by capillary column gas chromatography. Japp *et al.* (1987), Kessler (1988), and Boshears (1990) made similar observations. Clark (1989) published many mass spectra related to LSD and outlined reproducible minor differences in the iso-LSD and LAMPA mass spectra but went on to recommend the use of other techniques in addition to MS for positive identification. Neville *et al.* (1992) published the spectra of LSD and LAMPA. Blackwell (1998) demonstrated the retention time difference between LSD and LAMPA.

Since LSD is present in such small quantities in dosage units, it follows that the amounts of LSD to be detected in biological samples is likewise very small. This has led to the application of several innovative approaches to LSD analysis. Francom *et al.* (1988) discussed the derivatization of LSD and then the application of GC-MS to identify LSD in urine. Paul (1990) discussed a similar procedure for the quantitation of LSD in urine. Sun (1989) and Bukowski and Eaton (1993) outlined a procedure for the quantitation of LSD using trimethylsilyl derivatives. Papac and Foltz (1990) described a method for the measurement of LSD in plasma using capillary column negative ion mass spectrometry. Nelson and Foltz (1992) discussed the use of gas chromatography/ tandem mass spectrometry (MS/MS) for the identification of LSD, iso-LSD and N-demethyl-LSD. Several ionization techniques are discussed along with various derivatization techniques. Ohno and Kawabata (1988) outlined the use of direct inlet chemical ionization mass spectrometry for LSD analysis. Duffin *et al.* (1992) discussed several applications of liquid chromatography/mass spectrometry including to the identification of LSD. Rule and Henion (1992) and Cai and Henion (1996) use an immunoaffinity chromatograph before the LC/MS in their approach. Hopfgartner *et al.* (1993) used a high flow ion spray LC/MS to identify LSD. Pseudo-chemical ionization mass spectra were obtained using this method. Bogusz *et al.* (1998) discussed the LC-MS of LSD in biological samples using atmospheric pressure chemical ionization mass spectrometry.

5.9 MASS SPECTROSCOPY OF PHENYLALKYLAMINES

The spectrum of phenylalkylamines is dominated by the amine-containing fragment formed by cleavage of the bond beta to the amine group. Figure 5.17 shows this α-cleavage for a number of the phenylalkylamines.

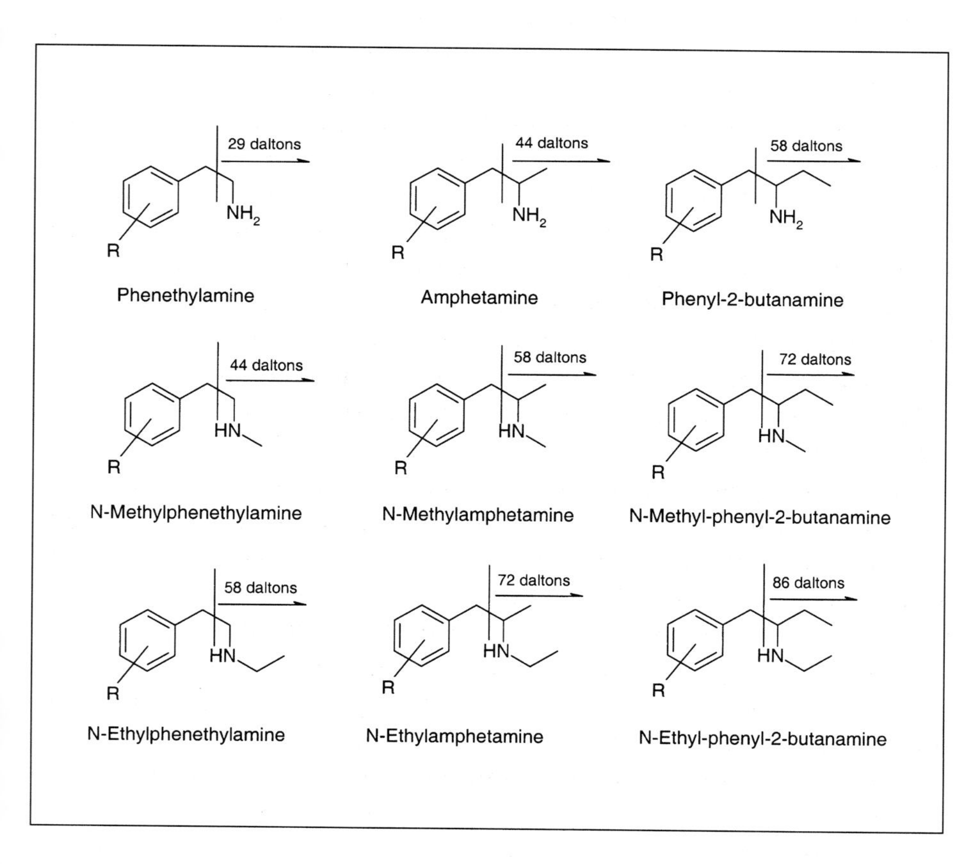

Figure 5.17

α-Cleavage that the aliphatic amine side chain of several phenylalkylamines

In Figure 5.17 the R substitution can be any functional group on the phenyl ring. One of the problems that is immediately evident from Figure 5.17 is that some of the structural isomers can generate the same base peak. This problem is further exacerbated by the possibility that other amines which are attached to the carbon alpha to the phenyl group give base peaks in an analogous manner. Figure 5.18 illustrates three other amines that would give base peaks at 58 daltons.

Figure 5.18

Other amines which give rise to 58m/z base peaks

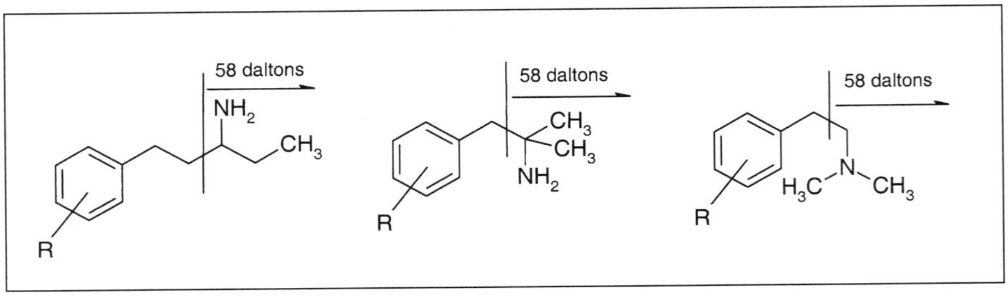

Another characteristic of the mass spectra of phenylalkylamines is that the molecular ion is weak, but usually present. Figure 5.19 illustrates the spectra of 3,4-methylenedioxyamphetamine (MDA), 3,4-methylenedioxymethamphetamine (MDMA), and 3,4-methylenedioxy-N-ethylamphetamine (MDEA).

The forensic drug chemist must determine the identification of the analyte among the possible structural isomers. In some cases, scale expansion of the mass spectrum will assist. (This would be done by making the base peak well off scale.) This is demonstrated in Figure 5.19 by comparing the spectra with the scale expanded spectra of the same three analytes in Figure 5.20.

Using this technique along with a comparison of the spectra of the possible structural isomers can sometimes be sufficient to obtain a positive identification. Literature spectra can be used with this technique, but spectra in libraries are often obtained using several different instruments, including different mass analyzers. It is best, but often impractical, to obtain or synthesize possible isomers and obtain their spectra on the forensic laboratory's instruments.

Another technique to differentiate among the mass spectra of structural isomers is to obtain the gas chromatograph retention time of the analyte and of the suspected hallucinogen. With modern capillary column gas chromatography, the separation among even closely related structural isomers is usually significant. Again the use of standards run at one time to establish the separation among isomers is necessary. As well the retention time of the standard must be established in the current sequence either by obtaining its retention time by its injection or by establishing the analyte's relative retention time by use of internal standard.

Figure 5.19 (opposite)

Mass spectra of:
(a) MDA; (b) MDMA; (c) MDEA

It is best practice to inject the bases of phenylalkylamines into gas

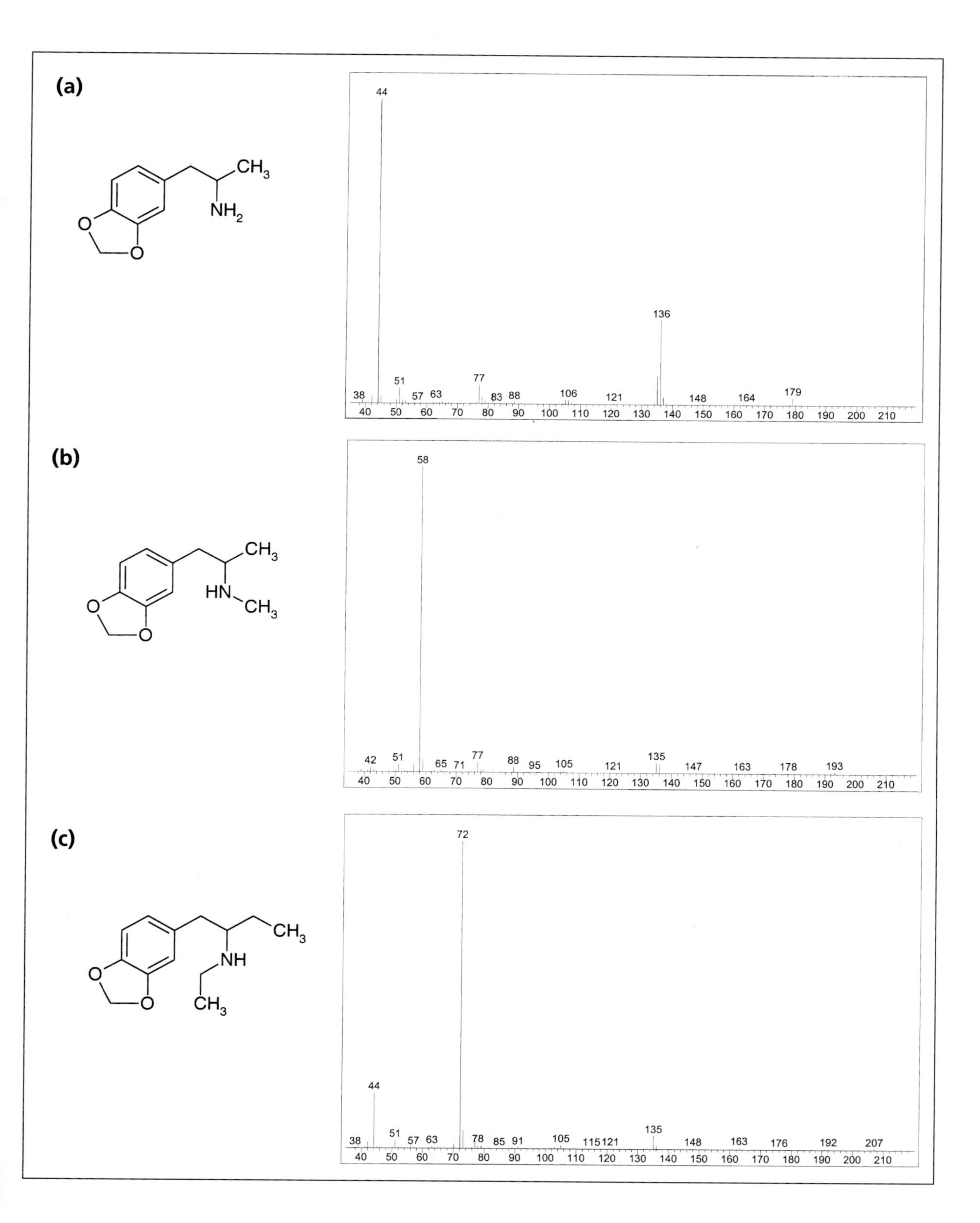
(a)
CH3
NH2
O
O
44
136
51
77
38
57 63
83 88
106
121
148
164
179
40 50 60 70 80 90 100 110 120 130 140 150 160 170 180 190 200 210
(b)
CH3
HN
CH3
O
O
58
42
51
65 71
77
88
95
105
121
135
147
163
178
193
40 50 60 70 80 90 100 110 120 130 140 150 160 170 180 190 200 210
(c)
CH3
NH
CH3
O
O
72
44
38
51
57 63
78 85 91
105
115121
135
148
163
176
192
207
40 50 60 70 80 90 100 110 120 130 140 150 160 170 180 190 200 210

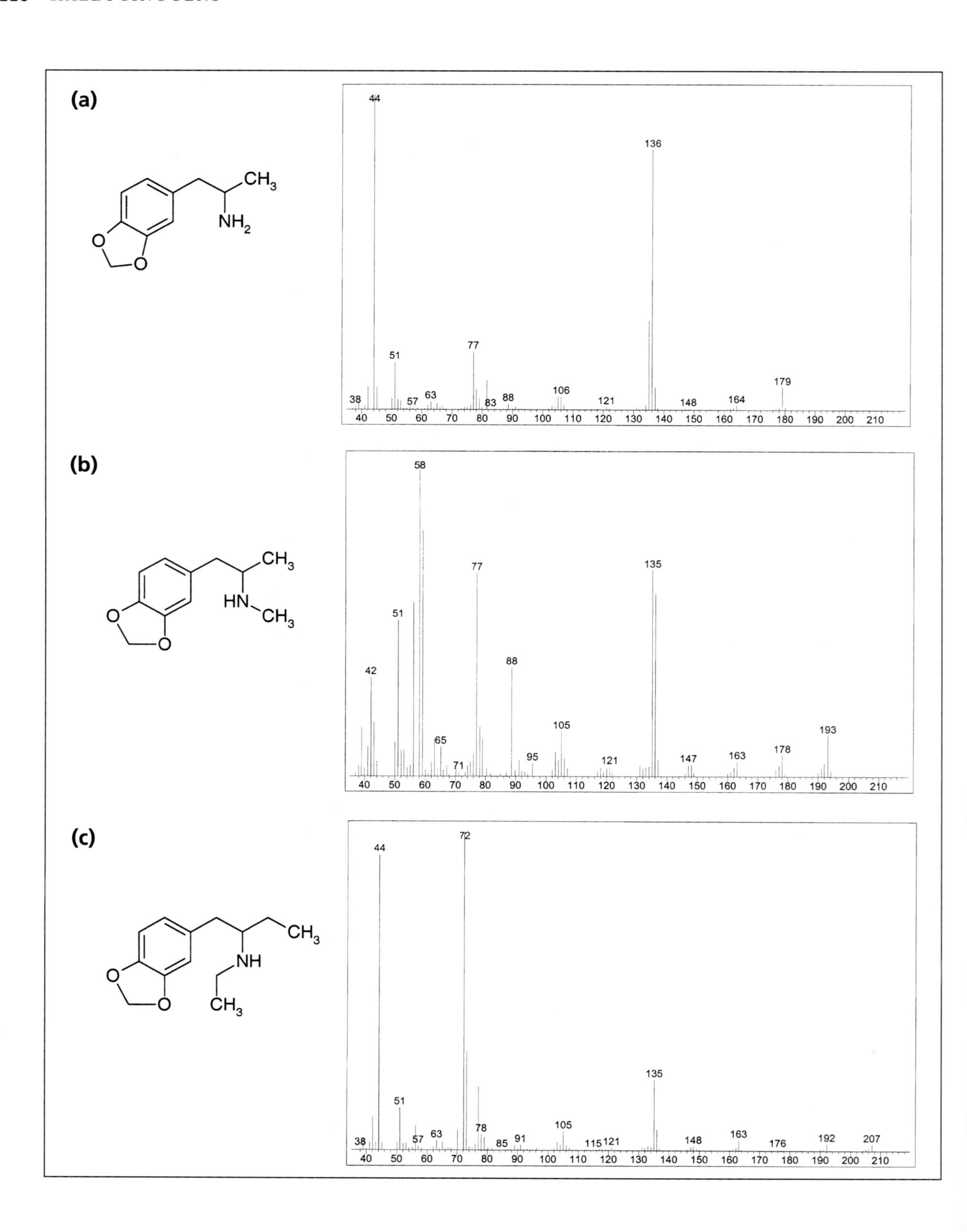
(a)
CH3
NH2
O
O
44
136
77
51
179
38
57
63
83
88
106
121
148
164
40 50 60 70 80 90 100 110 120 130 140 150 160 170 180 190 200 210
(b)
CH3
HN
CH3
O
O
58
77
135
51
88
42
105
193
65
95
121
147
163
178
71
40 50 60 70 80 90 100 110 120 130 140 150 160 170 180 190 200 210
(c)
CH3
NH
CH3
O
O
72
44
135
51
105
78
163
38
57
63
85
91
115
121
148
176
192
207
40 50 60 70 80 90 100 110 120 130 140 150 160 170 180 190 200 210

chromatographs (including GC-MSs) in order to obtain best chromatography. Phenylalkylamine hydrochlorides lose the hydrochloride when injected, the chromatography suffers, but the spectra so obtained are quite usable. The caveat mentioned in the infrared section of the chapter that alcohols are poor choices of solvent for the injection of phenylalkylamines into a GC-MS should always be kept in mind. See the article by Clark *et al.* (1992) for details.

There were several articles on the mass spectra of phenylalkylamines. Dal Cason (1989) published classical EI spectra and methane chemical ionization spectra of MDA and MDMA and its analogues. He observed that most of these analogues have weak molecular ions. CND Analytical (1994) discussed the analysis of phenylalkylamines by GC-MS. They pointed out the formation of the 44, 58, and 72 dalton ions that dominate the spectra of phenylalkylamines and the difficulty in distinguishing among regioisomeric phenethylamines. An example given in their text was the isomers of methamphetamine. Renton *et al.* presented the mass spectrum of the trifluoroacetic anhydride (TFA) derivative of MDA and N,N-dimethyl-MDA. Bailey *et al.* (1975) presented the N-methylated analogues of methoxyamphetamine and MDA and noted the similarity of mass spectra of some of the isomers. Hugel and Weaver (1988) published the infrared spectra of 2-methoxy-3,4-methylenedioxyamphetamine, N,N-dimethyl-MDA and MDEA. Veress *et al.* (1994) published the EI and CI spectrum of MDEA.

Figure 5.20 (opposite)

Mass spectra of Figure 5.19 with expanded y-scale: (a) MDA; (b) MDMA; (c) MDEA

5.10 MASS SPECTROSCOPY OF PHENCYCLIDINE AND ANALOGUES

Phencyclidine and its analogues are well distinguished by their mass spectra. Figure 5.21 illustrates the structure of phencyclidine, which is 1-(1-phenylcyclohexyl)piperidine, and its analogues where the phenyl group has been substituted.

Figure 5.22 illustrate the structure of the analogues of phencyclidine where the piperidine group has been substituted.

The mass spectra of phencyclidine, along with 1-(1-(2-thienyl)cyclohexyl) piperidine, are presented in Figure 5.23.

To further demonstrate the ability of mass spectroscopy to identify phencyclidine and its analogues, Figure 5.24 includes the following mass spectra:

- N-(1-phenylcyclohexyl)methylamine
- N-(1-phenylcyclohexyl)ethylamine
- N-(1-phenylcyclohexyl)propylamine
- N-(1-phenylcyclohexyl)isopropylamine.

1-(1-phenylcyclohexyl)piperidine (PCP)
1-(benzylcyclohexyl)piperidine (BCP)
1-(1-(4-methylphenyl)cyclohexyl)piperidine
1-(1-(2-thienyl)cyclohexyl)piperidine (TCP)

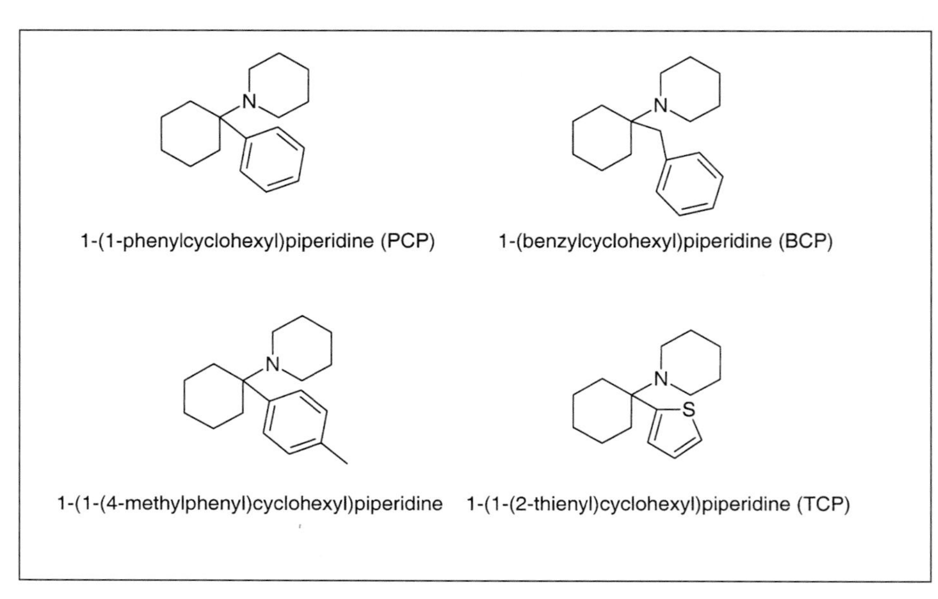

Figure 5.21
Phencyclidine and street analogues where the phenyl group is substituted

1-(1-phenylcyclohexyl)pyrrolidine (PCPy)
1-(1-phenylcyclohexyl)-4-methylpiperidine
1-(1-phenylcyclohexyl)ethylamine (PCE)
1-(1-phenylcyclohexyl)propylamine

Figure 5.22
Street analogues of phencyclidine where the piperidine group is substituted

There were several literature references on the mass spectra of phencyclidine and its analogues. Bailey *et al.* (1976) detailed that changing the piperidine to pyrrolidine or to morpholine in the phencyclidine molecule creates easily distinguishable spectra. The article went on to discuss replacing the phenyl group with thienyl in the phencyclidine molecule. The mass spectra so obtained were again easily distinguished. The mass spectra of these analogues exhibit moderate to strong molecular ions. Lodge *et al.* (1992), on the other hand, found that replacing the phenyl group with a benzyl moiety yielded a mass

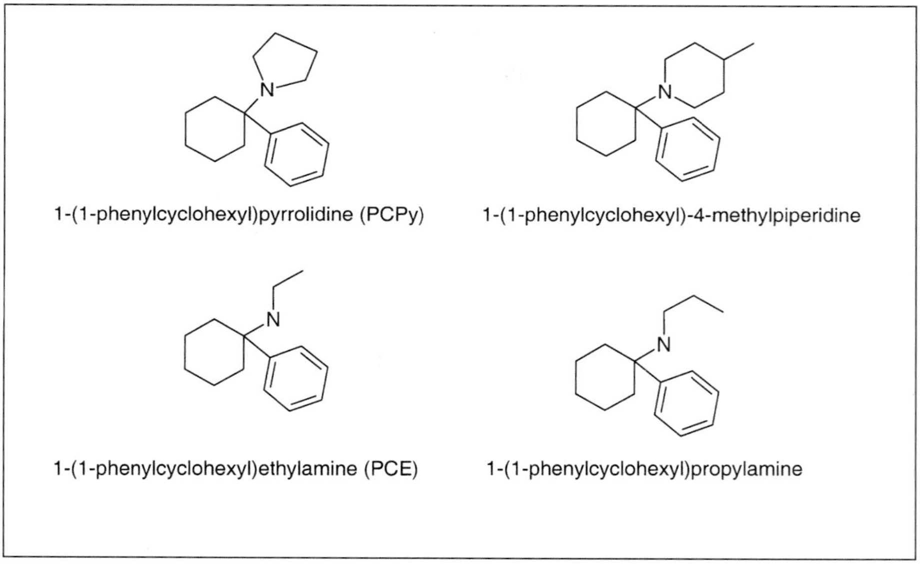

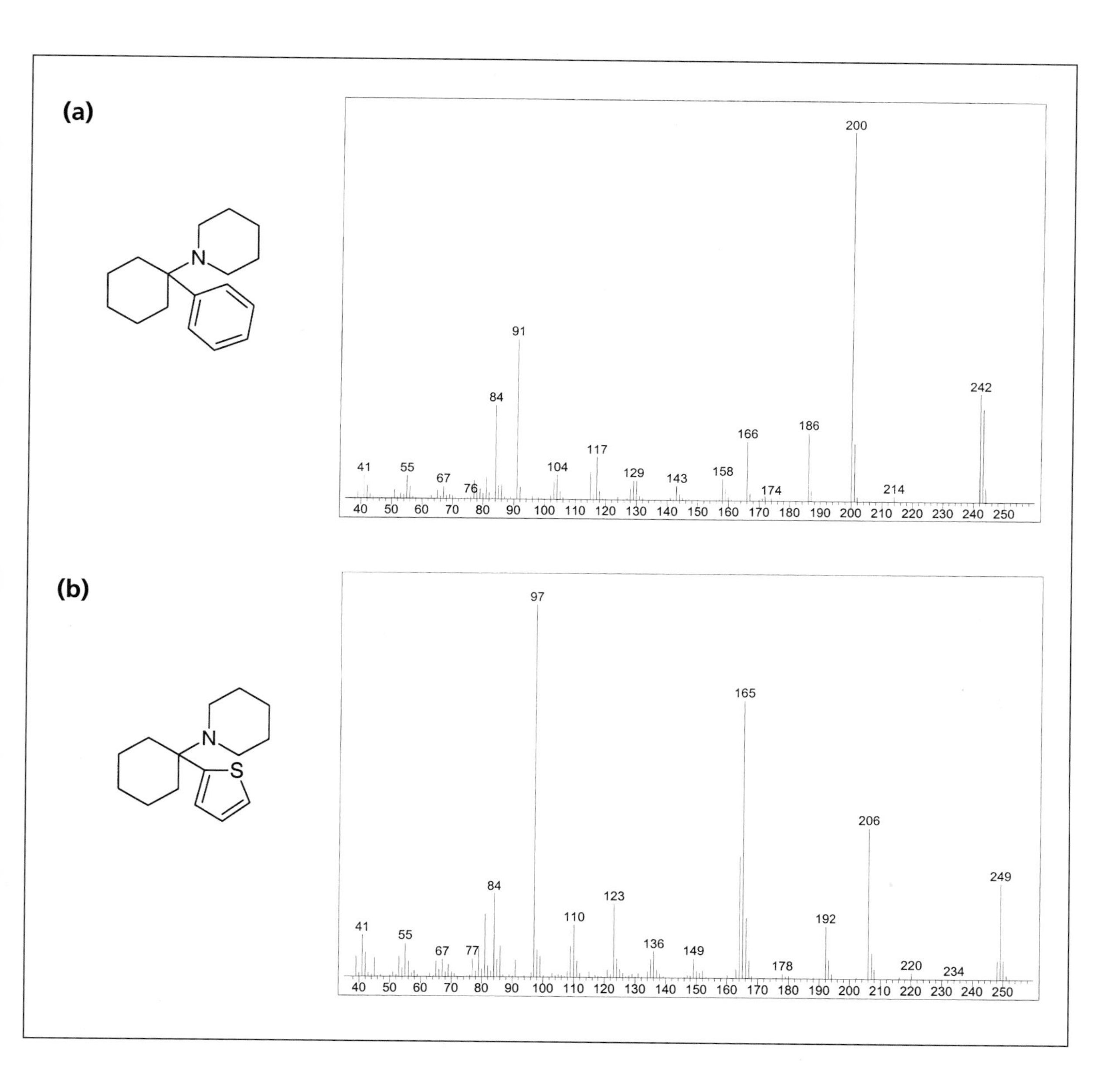

Figure 5.23

Mass spectra of: (a) phencyclidine; (b) 1-(2-thienylcyclohexyl)piperidine (TCP)

spectrum with no molecular ion. Close examination of that mass spectrum along with the mass spectra of the structural isomers where the phenyl group has been replaced by 2-, 3-, and 4-methylphenyl revealed significant differences. Bailey *et al.* (1979) investigated the mass spectra of the analogues of N-(1-phenylcyclohexyl)ethylamine and again concluded that the mass spectra were easily distinguishable and that GC-MS was a good technique to distinguish among the analogues. Allen *et al.* (1980) discussed the mass spectrum of 1-(1-phenylcyclohexyl)-4-methylpiperidine and also came to the same conclusion.

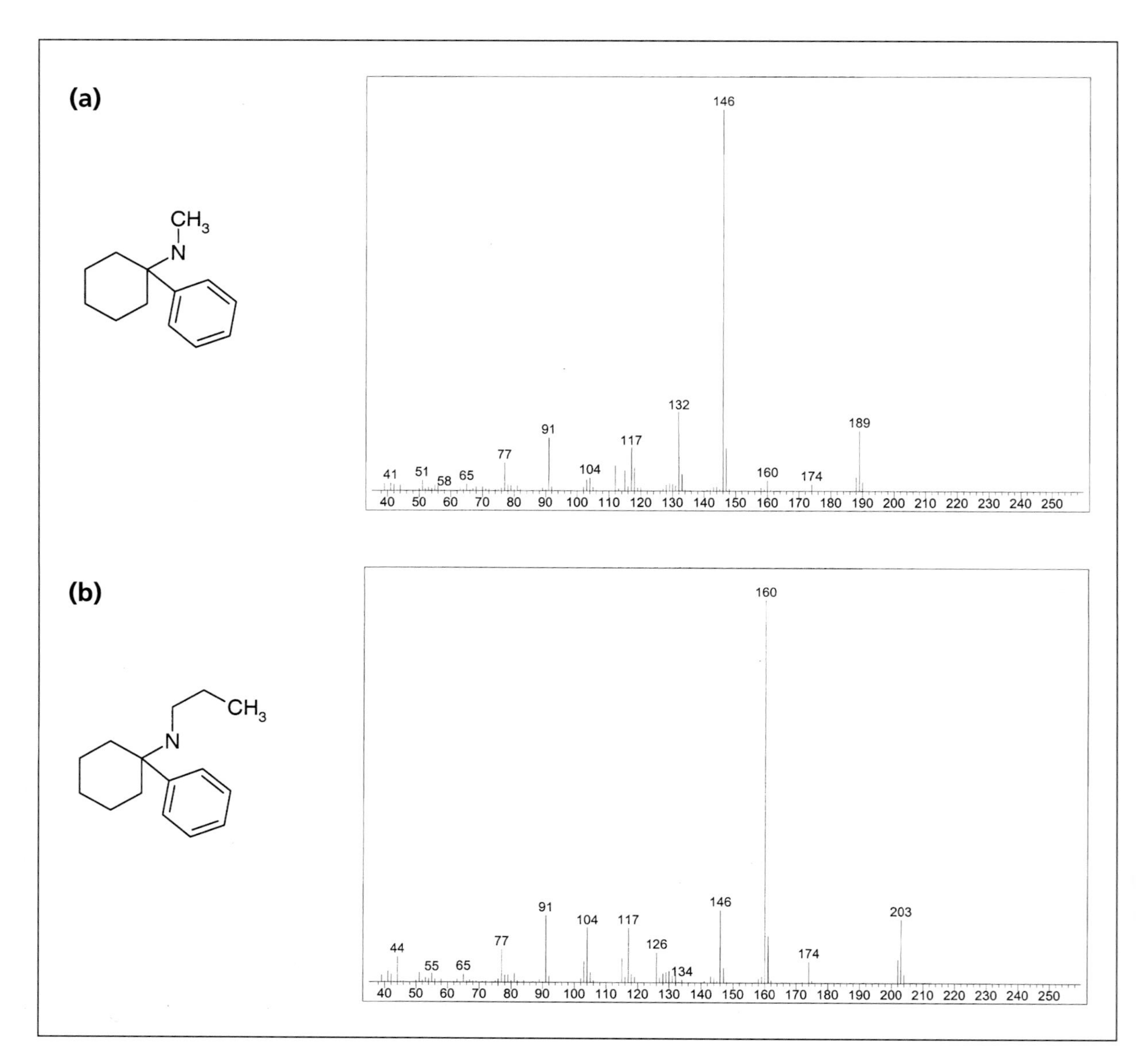

Figure 5.24

Mass spectra of a homologous series of phencyclidine related compounds: (a) N-(1-phenylcyclohexyl)methylamine; (b) N-(1-phenylcyclohexyl)ethylamine; (c) N-(1-phenyl cyclohexyl)propylamine; (d) N-(1-phenylcyclohexyl)isopropylamine

The extraction of phencyclidine and its analogues is discussed in the infrared spectroscopy section of this chapter. In general terms, the extraction is straightforward except that phencyclidine hydrochloride will extract from acid solutions. The formation of the GC artifact 1-phenylcyclohexene from the injection of phencyclidine should also be kept in mind.

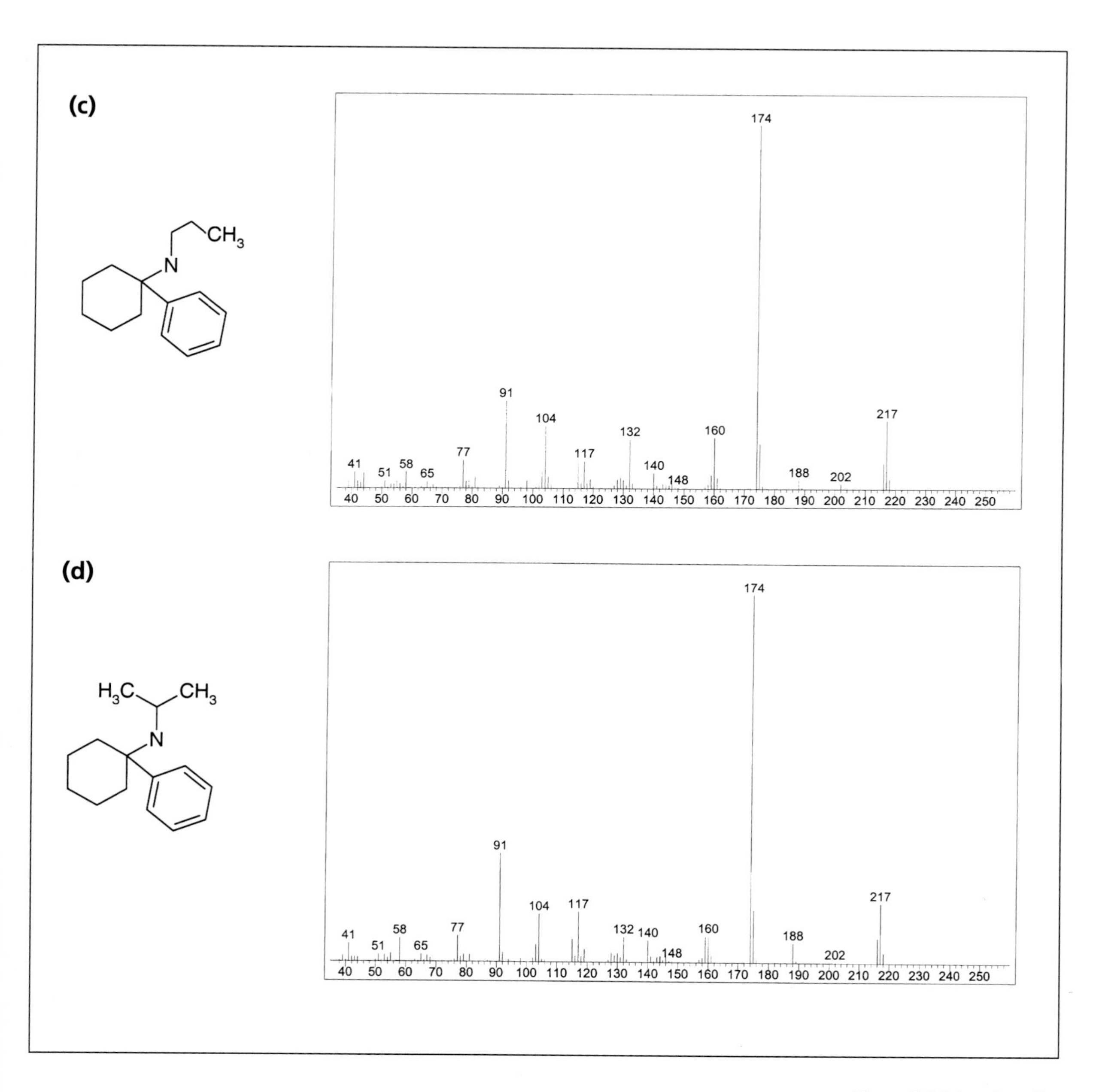

Figure 5.24 (continued)

5.11 MASS SPECTROSCOPY OF TRYPTAMINES

The mass spectra of dimethyltryptamine and diethyltryptamine are dominated by the fragment formed by cleavage beta to the amino group. Figure 5.25 illustrates this.

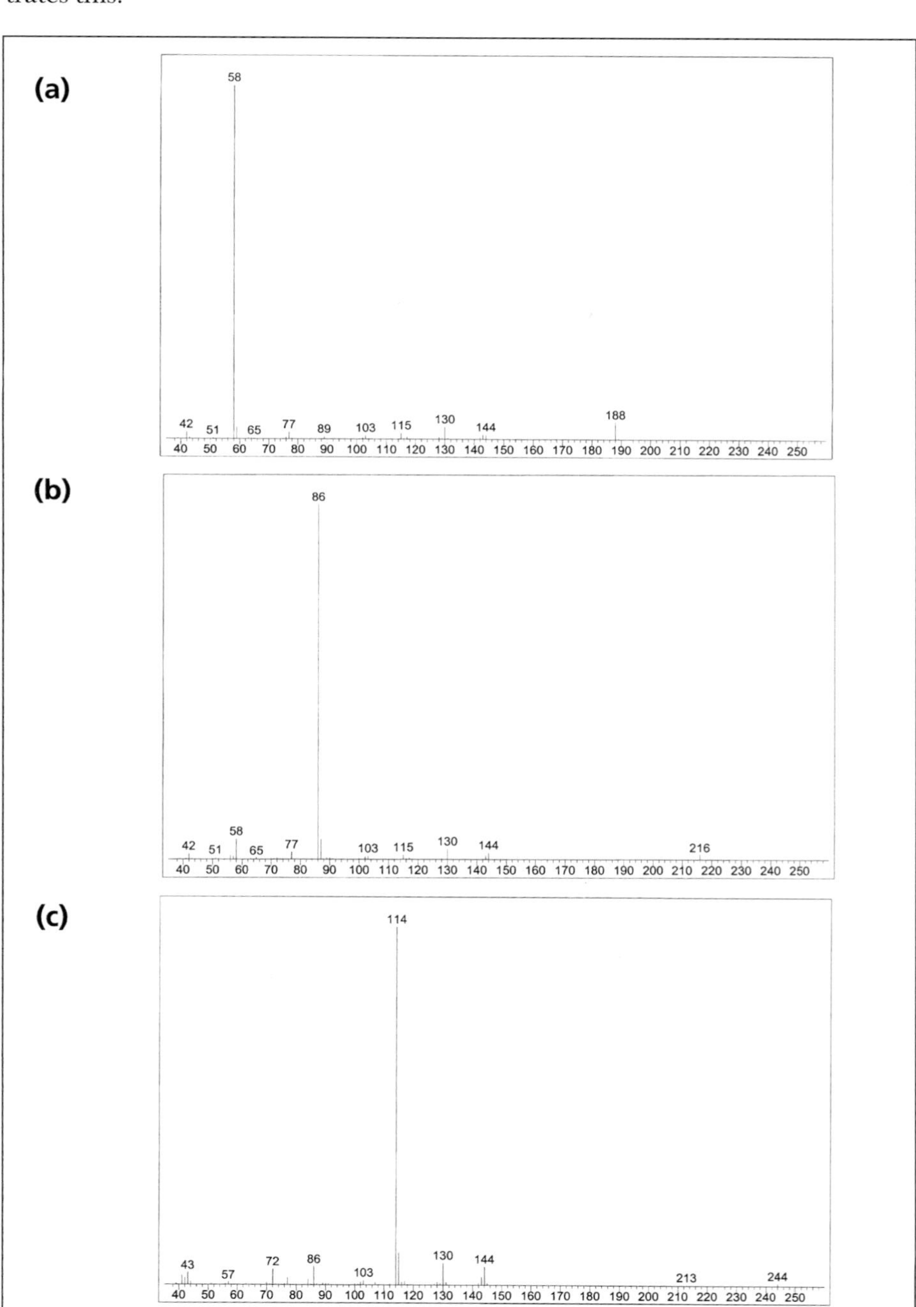

Figure 5.25

Mass spectra of:
(a) Dimethyltryptamine;
(b) Diethyltryptamine;
(c) Dipropyltryptamine;
(d) origin of base peak in the mass spectrum of DMT, DET and DPT

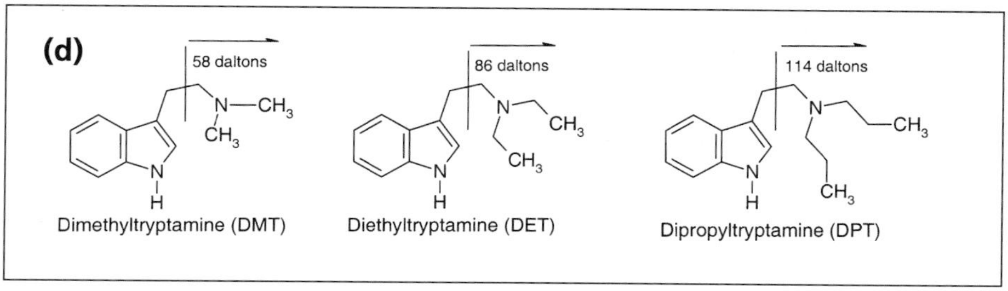

Figure 5.25 (continued)

(d)

58 daltons
Dimethyltryptamine (DMT)

86 daltons
Diethyltryptamine (DET)

114 daltons
Dipropyltryptamine (DPT)

The molecular ions are either very weak or missing. The mass spectra of dimethyltryptamine (DMT), diethyltryptamine (DET) and dipropyltryptamine (DPT) in expanded scale is presented in Figure 5.26.

In Figure 5.27 5-methoxy-diisopropyltryptamine is presented in both normal and expanded scales.

As mentioned in the section on the infrared spectroscopy of tryptamines, psilocybin easily loses its phosphate group to form psilocin. Derivatization of psilocybin with N,O-bis-(trimethylsilyl)trifluoroacetamide (BSTFA) will yield a derivatized spectrum which is easily distinguished from psilocin. Figure 5.28 presents the mass spectrum of underivatized psilocin as well as derivatized psilocybin and psilocin.

There are a few references on the identification of psilocybin. Repke *et al.* (1977) outlined the need to derivatize psilocybin to identify it by mass spectrometry. Hugel (1984) described the separation and identification of psilocybin from chocolate cookies containing psilocybin mycelia. Redhead (1984) described how the mycelia in the chocolate cookies was biologically identified. Timmons (1984) published a method using the described derivatizing technique to identify psilocybin and psilocin in mushrooms.

Most of the common hallucinogens chromatograph readily on a gas chromatograph. In Figure 5.29 a text mix of several amphetamine, phenthylamine and tryptamine based compounds were effectively separated on a 15m, DB-1 (J & W) using routine conditions.

The tryptamines and LSD analogues are easily chromatographed on Thin Layer Chromatographic systems (TLC) and will fluoresce under ultraviolet light. These compounds will also react with highly selective color reagents such as DMBA and Erelich's reagent.

(a)

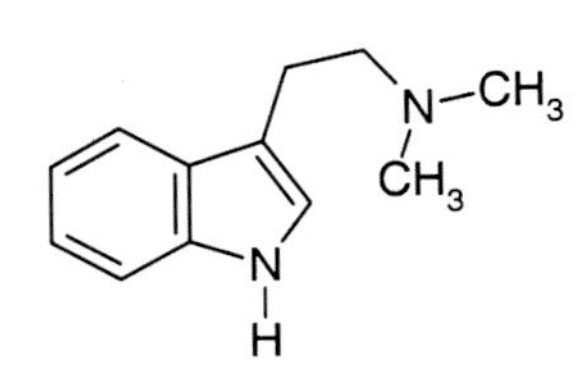
N
CH3
CH3
N
H

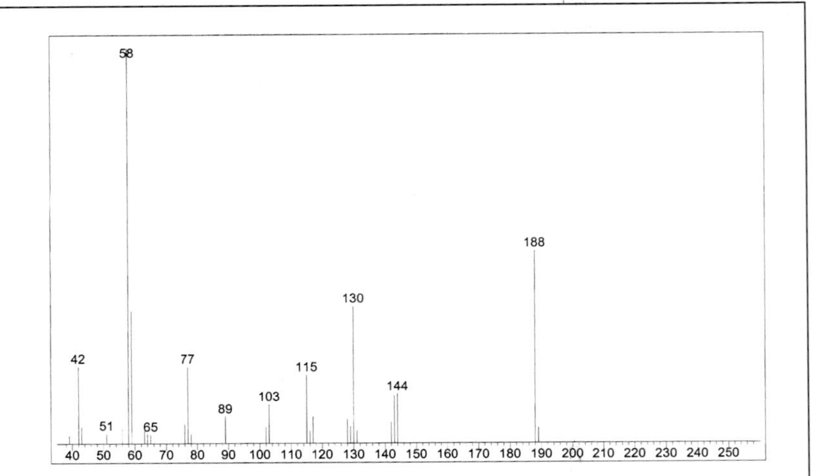
58
188
130
42
77
115
144
103
89
51
65
40 50 60 70 80 90 100 110 120 130 140 150 160 170 180 190 200 210 220 230 240 250

(b)

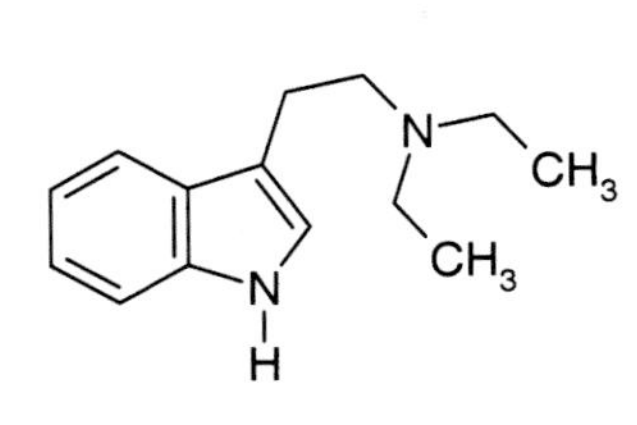
N
CH3
CH3
N
H

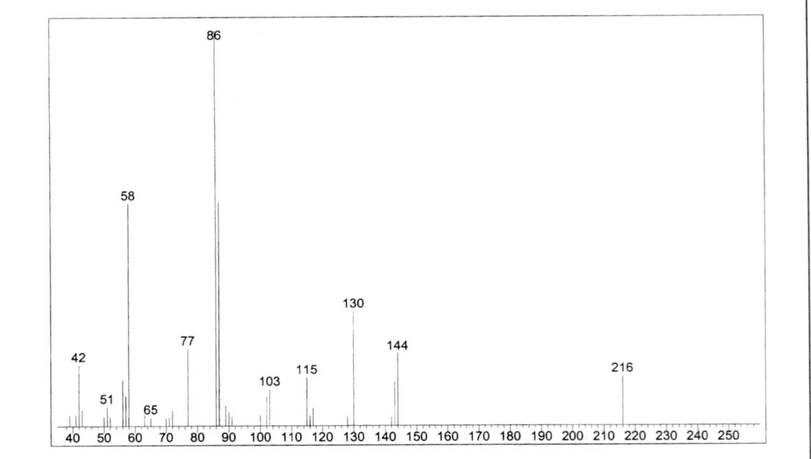
86
58
130
77
144
42
115
216
103
51
65
40 50 60 70 80 90 100 110 120 130 140 150 160 170 180 190 200 210 220 230 240 250

(c)

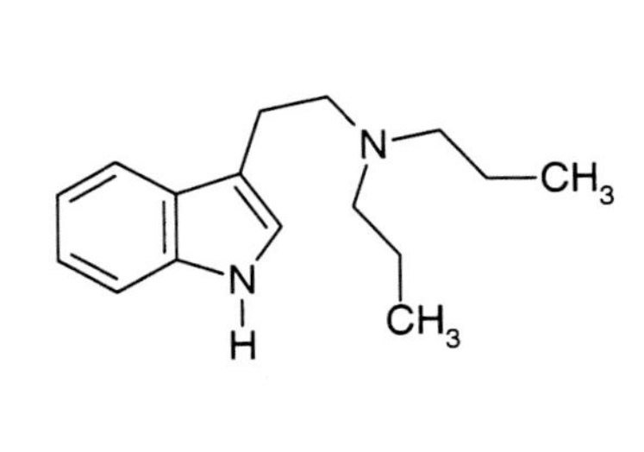
N
CH3
CH3
N
H

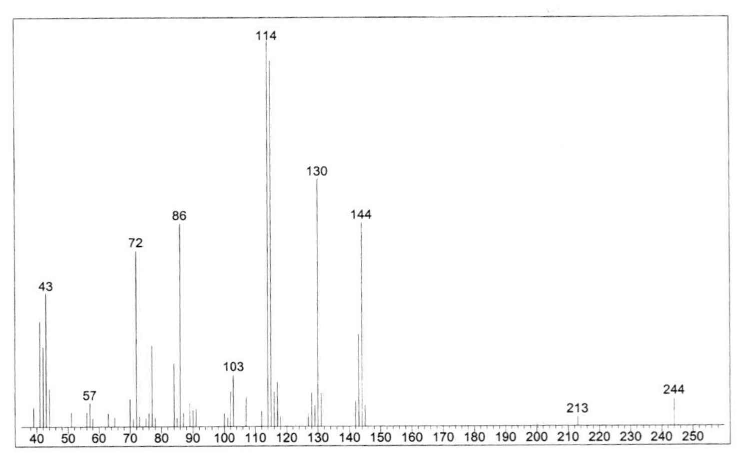
114
130
86
144
72
43
103
57
244
213
40 50 60 70 80 90 100 110 120 130 140 150 160 170 180 190 200 210 220 230 240 250

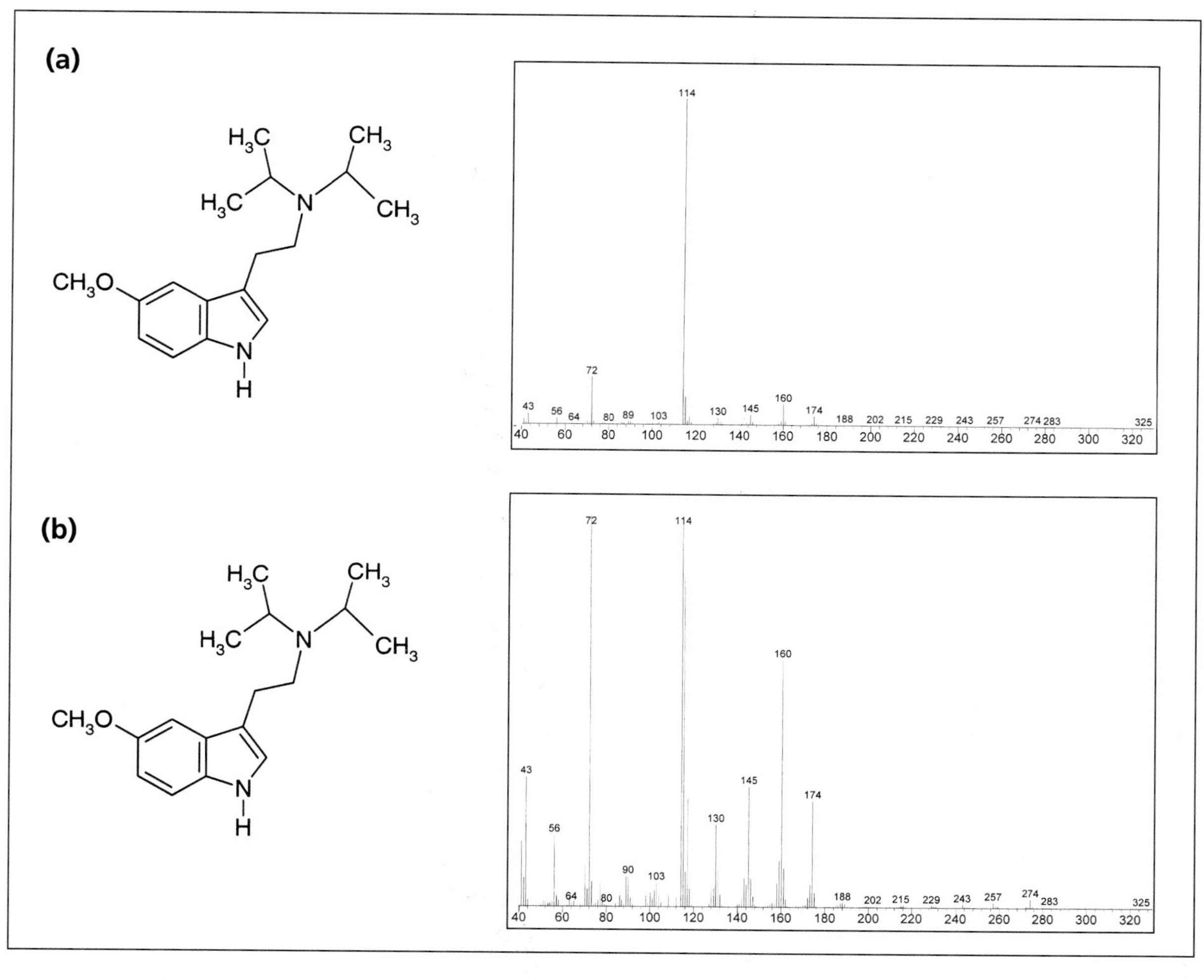

Figure 5.27

Mass spectra of: (a) 5-methoxy-N,N- diisopropyl tryptamine; (b) 5-methoxy-N,N- diisopropyl tryptamine scale expanded

Figure 5.26 (opposite)

Mass spectra of Figure 5.25 with expanded y-scale: (a) dimethyl tryptamine; (b) diethyltryptamine; (c) dipropyltryptamine scale expanded

(a)

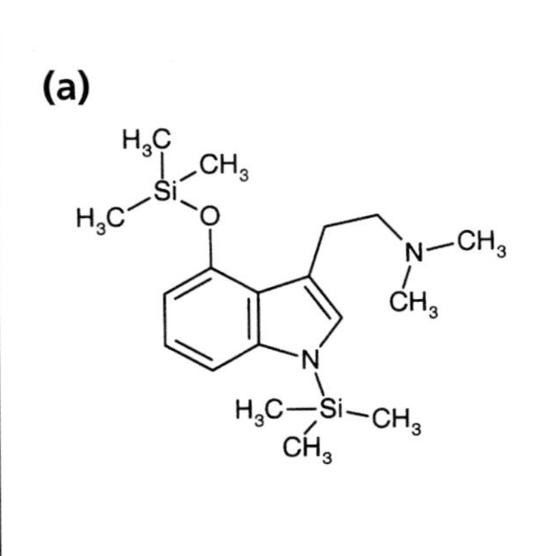
H3C
CH3
Si
H3C
O
N
CH3
CH3
N
H3C
Si
CH3
CH3

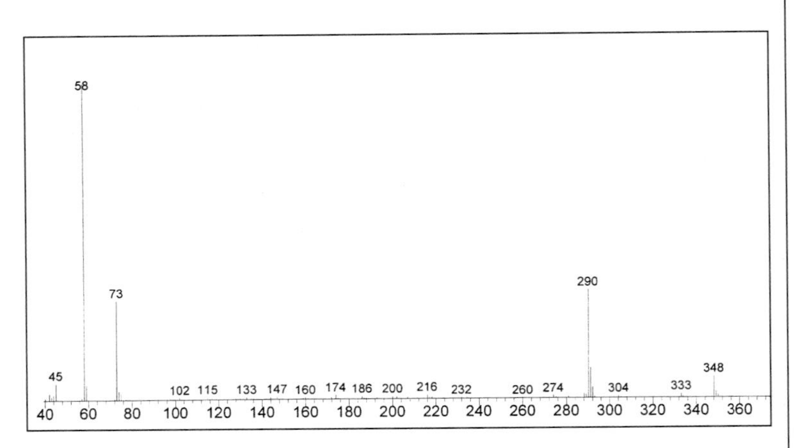
58
73
45
102 115 133 147 160 174 186 200 216 232 260 274
290
304 333
348
40 60 80 100 120 140 160 180 200 220 240 260 280 300 320 340 360

(b)

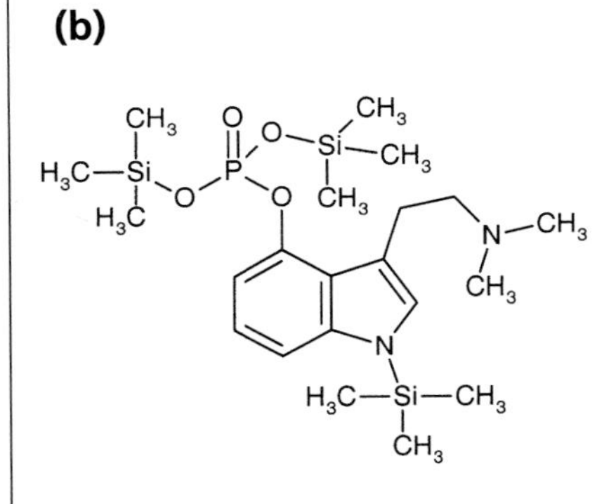
CH3
O
O
CH3
H3C
Si
O
P
O
Si
CH3
H3C
CH3
N
CH3
CH3
N
H3C
Si
CH3
CH3

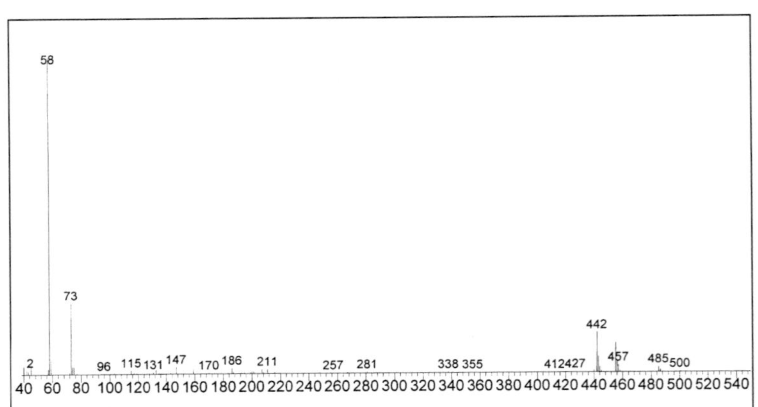
58
73
2
96 115 131 147 170 186 211 257 281 338 355 412 427
442
457
485 500
40 60 80 100 120 140 160 180 200 220 240 260 280 300 320 340 360 380 400 420 440 460 480 500 520 540

(c)

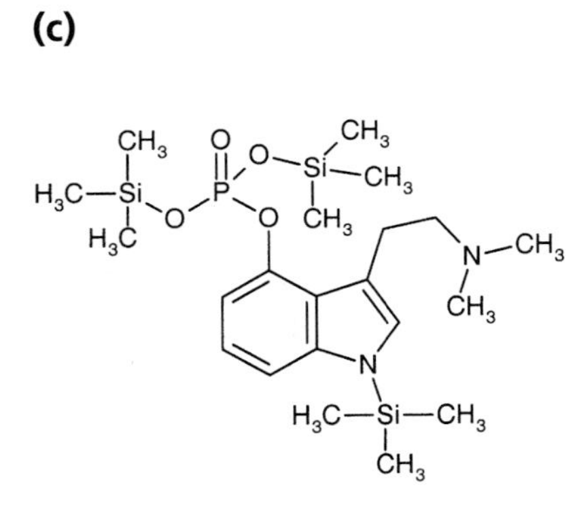
CH3
O
O
CH3
H3C
Si
O
P
O
Si
CH3
H3C
CH3
N
CH3
CH3
N
H3C
Si
CH3
CH3

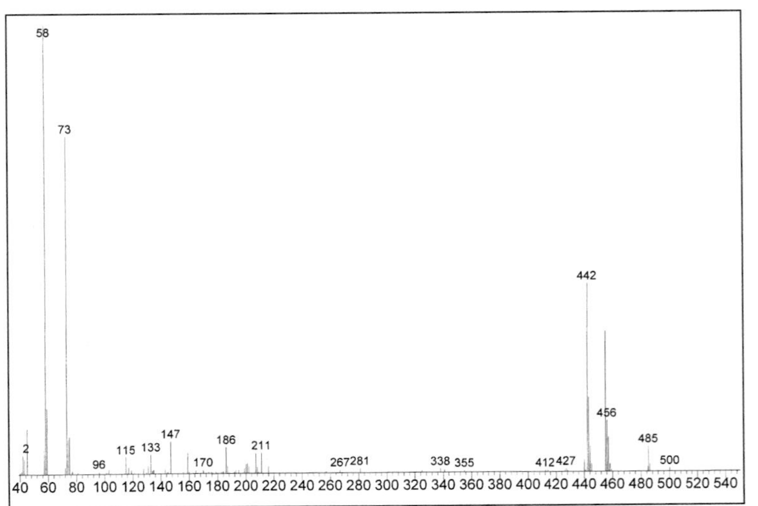
58
73
2
96 115 133 147 170 186 211 267 281 338 355 412 427
442
456
485
500
40 60 80 100 120 140 160 180 200 220 240 260 280 300 320 340 360 380 400 420 440 460 480 500 520 540

Retention Time (min.)	Compound
4.93	3,4-Methylenedioxy Amphetamine (MDA)
5.17	3,4-Methylenedioxy-N-Methylamphetamine (MDMA)
5.37	3,4-Methylenedioxy-N-Ethylamphetamine (MDEA)
5.53	2,3,4-Trimethoxy Amphetamine (2,3,4-TMA)
5.68	3,4-Methylenedioxy-N-Hydroxyamphetamine (NOH-MDA)
6.19	N,N-Dimethyl Tryptamine (DMT)
6.29	4-Bromo-2,5-Dimethoxy Phenethylamine (2C-B, Nexus)
6.62	Ketamine
6.82	N,N-Diethyl Tryptamine (DET)
8.34	5-Methoxy-N,N-Diisopropyl Tryptamine (5-OMeDIPT)
14.31	Lysergic Acid N-sec-butylamide (LSsB)
16.14	Lysergic Acid N,N-Diethylamide (LSD)
16.98	Lysergic Acid N-Methyl-N-Propylamide (LAMPA)

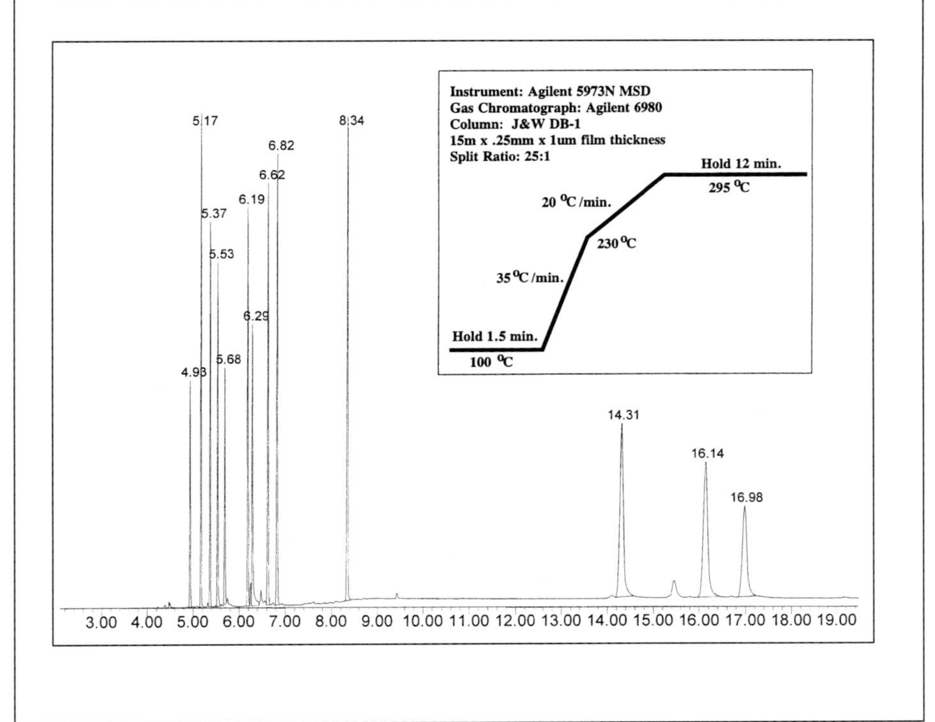

Figure 5.29

A gas chromatogram of a mixture of several hallucinogenic compounds

Figure 5.28 (opposite)

Mass spectra of derivatized psilocin and psilocybin: (a) BSTFA derivatized psilocin; (b) BSTFA derivatized psilocybin; (c) BSTFA derivatized psilocybin expanded scale

5.12 FURTHER READING

More detailed discussion on mass spectrometry and the interpretation of mass spectra can be found in the text by McLafferty (1980). The university textbook by Ege (1999) includes a description of the technique as it relates to the identification of organic compounds. The chapter in the book by Gough (1991) detailed the mass analyzers as well as other ionization techniques such as field ionization/field desorption techniques, fast atom bombardment, and atmospheric pressure ionization techniques as well as LC-MS. Willoughby *et al.* (1998) presented a modern view of the technique of LC-MS and includes a discussion on the particle beam interface.

Electronic compilations of mass spectra which include hallucinogens and their precursors are:

- Pfleger/Maurer/Weber Library consisting of 6300 spectra of drugs and metabolites;
- Wiley Mass Spectral Data 7th Edition consisting of 390,000 spectra library which include hallucinogens, precursors, essential chemicals, and reagents;
- NIST/EPA/NIH Mass Spectral Library consisting of 107,800 spectra which include hallucinogens, precursors, essential chemicals, and reagents.

Hard copy (text form) compilations of infrared spectra which include hallucinogens and their precursors are:

- *Instrumental Data for Drug Analysis Volumes 1 to 4* (1987) by Mills and Roberson.
- *Instrumental Data for Drug Analysis Volume 5* (1992) by Mills *et al.*
- *Instrumental Data for Drug Analysis Volumes 6 and 7* (1996) by Mills *et al.*
- *Clarke's Isolation and Identification of Drugs* (1986) by Moffat *et al.* (editors).
- *Handbook of Mass Spectra of Drugs* (1981a) by I. Sunshine.
- *Analytical Profiles of Amphetamines and Related Phenethylamines* (1989) by CND Analytical.
- *Analytical Profiles of Designer Drugs Related to the 3,4-Methylenedioxyamphetamines (MDA's)* (1991) by CND Analytical.
- *Analytical Profiles of Substituted 3,4-Methylenedioxyamphetamines: Designer Drugs Related to MDA* (1988) by CND Analytical.
- *Analytical Profiles of Precursors and Essential Chemicals* (1990) CND Analytical.
- *Analytical Profiles of the Hallucinogens* (1991) by CND Analytical.

- *Forensic and Analytical Chemistry of Clandestine Phenethylamines* (1994) by CND Analytical.

ACKNOWLEDGEMENTS

We wish to gratefully acknowledge permission given by Health Canada to include data generated by their Drug Analysis Service.

REFERENCES

Allen, A., Carr, S., Cooper, D., Fransoza, E., Koles, J., Kram, T. and Solon, E. (1980) *Microgram*, 13, pp. 44–46.

Ardrey, R. E. and Moffat, A. C. (1979) *Jour. For. Sci. Soc.*, 19, pp. 253–282.

Bailey, K., By, A. W., Legault, D. and Verner, D. (1975) *Jour. Assoc. Off. Anal. Chem.*, 58, pp. 62–69.

Bailey, K., Gagne, D. R. and Pike, R. W. (1976) *Jour. Off. Anal. Chem.*, 59, pp. 81–89.

Bailey, K. and Legault, D. (1979) *Jour. Assoc. Off. Anal. Chem.*, 62, pp. 1124–1137.

Bailey, K., Verner, D. and Legault, D. (1973) *Jour. Assoc. Off. Anal. Chem.*, 56, pp. 88–99.

Blackwell, T. M. (1998) *Microgram*, 31, pp. 51–61.

Bogusz, M. J., Maier, R-D., Kruger, K-D. and Kohls, U. *Jour. Anal. Toxicol.*, 22, pp. 549ff.

Boshears, F. E. (1990) *Microgram*, 23, pp. 99–100.

Bukowski, N. and Eaton, A. N. (1993) *Rapid Commun. Mass Spectrom.*, 7, pp. 106–108.

Cai, J. and Henion, J. (1996) *Anal. Chem.*, 68, pp. 72–78.

Clark, C. C. (1989) *Jour. For. Sci.*, 34, pp. 532–546.

Clark, C. R., Noggle, F. T. and DeRuiter, J. (1992) *Microgram*, 25, pp. 330–340.

CND Analytical (1988) *Analytical Profiles of Substituted 3,4-Methylenedioxyamphetamines: Designer Drugs Related to MDA*, Auburn, AL: CND Analytical.

CND Analytical (1989) *Analytical Profiles of Amphetamines and Related Phenethylamines*, Auburn, AL: CND Analytical.

CND Analytical (1990), *Analytical Profiles of Precursors and Essential Chemicals*, Auburn, AL: CND Analytical.

CND Analytical (1991) *Analytical Profiles of Designer Drugs Related to the 3,4-Methylene-dioxyamphetamines (MDA's)*, Auburn, AL: CND Analytical.

CND Analytical (1991) *Analytical Profiles of the Hallucinogens, Auburn*, AL: CND Analytical.

CND Analytical (1994) *Forensic and Analytical Chemistry of Clandestine Phenethylamines*, Auburn, AL: CND Analytical.

Dal Cason, T. (1989) *Jour. For. Sci.*, 34, pp. 928–961.

Duffin, K. L., Wachs, T. and Henion, J. D. (1992) *Anal. Chem.*, 64, pp. 61–68.

Dyer, J. R. (1965) in *Applications of the Absorption Spectroscopy of Organic Compounds*, Englewood Cliffs, NJ: Prentice-Hall Inc..

Ege, S. (1999) in *Organic Chemistry Structure and Reactivity*, 4th edn, NY: Houghton Mifflin Company.

Fifield, F. W. and Kealey, D. (1995) in *Principles and Practices of Analytical Chemistry*, 4th edn, Glasgow: Blackie Academic and Professional.

Francom, P., Andrenyak, D., Lim, H-K., Bridges, R. R., Foltz, R. L. and Jones, R. T. (1988) *Jour. Anal. Toxicol.*, 12, pp. 1–8.

Gough, Terry (ed.) (1991) in *The Analysis of Drugs of Abuse*, West Sussex, England: John Wiley & Sons Ltd.

Griffiths, P. R. and de Haseth, J. A. (1986) in *Fourier Transform Infrared Spectrometry*, NY: John Wiley & Sons, p. 191.

Japp, M., Gill, R. and Osselton, M. D. (1987) *Jour. For. Sci.*, 32, pp. 933–940.

Harris, H. A. and Kane, T. (1991) *Jour. For. Sci.*, 36, pp. 1186–1191.

Hopfgartner, G., Wachs, T., Bean, K. and Henion, J. (1993) *Anal. Chem.*, 65, pp. 439–446.

Hugel, J. (1984) *Microgram*, 17, pp. 111–119.

Hugel, J. and Weaver, K. (1988) *Microgram*, 21, pp. 681–686.

Keller, R. J. (1986) in *The Sigma Library of FT-IR Spectra*, St Louis, MO: Sigma Chemical Company.

Kempfert, K. (1988) *App. Spectrosc.*, 42, pp. 845–849.

Kessler, R. R. (1988) *Microgram*, 21, pp. 217–221.

Kovar, K. A., Dinkelacker, J., Pfiefer, A. M., Pisternick, W. and Woessener, A. (1995) *GIT Spez. Chromatog.*, 15, pp. 19–24 .

Kovar, K. A., Dinkelacker, J., Pfiefer, A. M., Pisternick, W. and Woessener, A. (1995) *Chem. Abstr.*, 123, p. 248677y.

Lodge, B. A., Duhaime, R., Zamecnik, J., MacMurray, P. and Brousseau, R. (1992) *For. Sci. Int'l*, 55, pp. 13–26.

McLafferty, F. W. (1980) *Interpretation of Mass Spectra*, Mill Valley, CA: University Science Books.

Mesley, R. J. and Evans, W. H. (1969) *Jour. Pharm. Pharmac.*, 21, pp. 713–720.

Mills, T. and Roberson, J.C. (1987) *Instrumental Data for Drug Analysis*, Volumes 1 to 4, 2nd edn, NY: Elsevier Science Publishing.

Mills, T., Roberson, J. C., McCurdy, H. H. and Hall, W. H. (1992) in *Instrumental Data for Drug Analysis*, Volume 5, 2nd edn, NY: Elsevier Science Publishing.

Mills, T., Roberson, J. C., Wall, W. H., Lothridge, K. L., McDougall, W. D. and Gilbert, M. W. (1996) *Instrumental Data for Drug Analysis*, Volumes 6 and 7, NY: CRC Press.

Moffat, A. C., Jackson, J. V., Moss, M. S. and Widdop, B. (1986) in *Clarke's Isolation and Identification of Drugs*, London: The Pharmaceutical Press.

Nelson, C. C. and Foltz, R. L. (1992) *Anal. Chem.*, 64, pp. 1578–1585.

Neville, G. A., Beckstead, H. D., Black, D. B., Dawson, B. A. and Ethier, J-C. (1992) *Can. Jour. App. Spectrosc.*, 37, pp. 149–157.

Nichols, H. S., Anderson, W. H. and Stafford, D. T. (1983) *Jour. High Res. Chromatog. & Chromatog. Comm.*, 6, pp. 101–103.

Noggle, F. T, DeRuiter, J. and Long, M. J. (1986) *Jour. Assoc. Off. Anal. Chem.*, 69, pp. 681–686.

Ohno, Y. and Kawabata, S. (1988) *Chem. Abstr.*, 108, p. 89060p.

Papac, D. I. and Foltz, R. L. (1990) *Jour. Anal. Toxicol.*, 14, pp. 189–191.

Paul, B. D., Mitchel, J. M., Burbage, R., Moy, M. and Sroka, R. (1990) *Jour. Chromatog.*, 529, pp. 103–112.

Pfiefer, A. M. and Kovar, K. A. (1995) *Jour. Planar Chromatogr. – Mod. TLC*, 8, pp. 388–392.

Pouchert, C. J. (1985) in *The Aldrich Library of FT-IR Spectra*, Volumes 1 and 2, Milwaukee, WI: Aldrich Chemical Company.

Pouchert, C. J. (1989) in *The Aldrich Library of FT-IR Spectra Volume 3 Vapor Phase*, Milwaukee, WI: Aldrich Chemical Company.

Redhead, S. A. (1984) *Microgram*, 17, pp. 120–122.

Renton, R. J., Cowie, J. S. and Oon, M. C. H. (1993) *For. Sci. Int'l*, 60, pp. 189–202.

Repke, D. B., Leslie, D. T., Mandell, D. M. and Kish, N. G. (1977) *Jour. Pharm. Sci.*, 66, pp. 743–744.

Rule, G. S. and Henion, J. D. (1992) *Jour. Chromatog.*, 582, pp. 103–112.

Sun, J. (1989) *Chem. Abstr.*, 111, p. 72627t.

Sunshine, I. (1981a) in *Handbook of Mass Spectra of Drugs*, Boca Raton, FL: CRC Press Inc.

Sunshine, I. (1981b) in *Handbook of Spectrophotometric Data of Drugs*, Boca Raton, FL: CRC Press Inc..

Timmons, J. E. (1984) *Microgram*, 17, pp. 28–32.

Veress, T., Gal, T., Nagy, G., Nagy, J. and Korosi, A. (1994) *Microgram*, 27, pp. 48–57.

Williams, D. H. and Fleming, I. (1966) in *Spectroscopic Methods in Organic Chemistry*, London: McGraw-Hill Publishing Co.

Willoughby, R., Sheehan, E. and Mitrovich, S. (1998) in *A Global View of LC/MS,* Pittsburgh, PA: Global View Publishing.

PART III NUCLEAR MAGNETIC RESONANCE SPECTROSCOPY

5.13 THEORETICAL BASIS

Infrared Spectroscopy (IR) and Gas Chromatography/Mass Spectrometry (GC/MS) are considered the preferred instrumentation used in forensic laboratories for the identification of drugs of abuse. Certain laboratories, however, have greatly enhanced their identification capabilities with the use of nuclear magnetic resonance (NMR) spectroscopy. IR spectroscopy can only provide a single element of structural information – that of functional group frequencies. However, the IR spectra taken together with GC/MS data can provide additional components of structural information – that of the mass spectrum which provides MW information, structural fragments and the GC retention time along with the IR functional group frequencies. NMR can provide multiple structural elements of information, including carbon/proton count, types of protons and carbons in the molecule, carbon framework, information deduced from carbon-13 chemical shift measurements and correlated with IR data, leading to the formal assembling of molecular structure. IR and NMR represent non-destructive analytical techniques whereas with GC/MS the samples consumed during the analysis, albeit in a low level amount.

Early NMR instrumentation suffered from low sensitivity that precluded rapid identification. Generally speaking, ^{1}H NMR results obtained in the 1960s and 1970s were reported at 60, 90, and 100MHz. During the 1980s, with the development of commercial superconducting magnets, NMR results were being reported at ^{1}H frequencies of 200, 300, 400, and 500 MHz. In the 1990s to the present day, the majority of NMR results have reported at ^{1}H frequencies of 400–600MHz and more recently at ultrahigh magnetic fields for biological NMR where proton frequencies are 700, 750, 800, and 900MHz. Carbon data was introduced as a routine technique in the late 1970s and ^{13}C frequencies of 20MHz (80MHz for proton)–125MHz (500MHz for proton) were the most common frequencies to be reported well into the 1990s. The technology has evolved dramatically over the past 30 years as much higher magnetic fields, more stable RF components and more sensitive probes have become readily available. More recently the use of cryogenic probe technology has evolved and this has permitted NMR results to be obtained on lower levels of more complex samples. Additionally, the advent of two-dimensional NMR (2-D NMR) in the late 1970s and early 1980s has become an invaluable tool for the identification and structure elucidation of complete unknowns. The introduction of Pulsed Field Gradients (PFG) has introduced new experiments probing different kinds of information and has significantly reduced the time for 2-D data collection. As a result, the use of 2-D NMR for structural problems has become almost as common as 1-D methods.

NMR spectroscopy continues to be the premiere tool for structure elucidation. It has the greatest usable information content of all of the forms of spectroscopy (Claridge, 1999). In recent years, NMR experiments have been developed which permit gross structure elucidation and stereochemical determination, and exact assignments of NMR resonance signals (^{1}H, ^{13}C, ^{19}F, ^{31}P). The principal caveat to NMR as a technique is that of sensitivity and its limitations, which are imposed by potential low sample availability. This general problem, as noted earlier, is being addressed with the use of cryogenic NMR probes as well as with the use of proton-detected experiments for 2-dimensional NMR. For the forensic area, lack of sample availability is generally not a problem. Therefore, NMR can provide rapid and *definitive* structural answers to problems, which arise. Most of the NMR analyses in the forensic area will fall into one of two categories:

1. 1-dimensional proton and carbon-13 NMR spectra which are obtained on a sample of suspected unknown or actual known structure where comparison of the resultant spectra is made to a spectral library or to spectra produced from authentic samples for the purposes of identification, both qualitatively and quantitatively; and
2. appropriate 1- and 2-dimensional NMR spectra are obtained on a complete unknown and the structure is deduced from the NMR data (as well as from the results of other complementary structural techniques, e.g. IR and MS and possibly chemical synthesis).

The purpose of this section is to:

1. provide a very brief overview of the NMR technique as it is currently practiced and
2. summarize some of the specific applications of NMR spectroscopy to forensic analysis of hallucinogenic drugs and provide key useful resources for practitioners of forensic drug analysis.

For pure structure elucidation, we will restrict the NMR discussion to the uses of proton (^{1}H) and carbon-13 (^{13}C) NMR. These nuclei represent the principle NMR active nuclei, which are traditionally employed for organic structure elucidation. Other NMR active nuclei (e.g., ^{15}N, ^{19}F, ^{31}P) can provide additional and complementary structural information to that obtained from proton and carbon-13, but they will not be discussed here.

5.14 THE NMR EXPERIMENT

The measurement of NMR spectra is generally performed in solution, although extensive applications using solid-state NMR (principally ^{13}C) are well known (Fyfe, 1983; Mehring, 1983). The sample is usually dissolved in a solvent, which is generally deuterated to provide a deuterium (^{2}H) source for internal field frequency lock of the spectrometer and is placed in a strong magnetic field. The sample is then pulsed with a burst of radio frequency (rf). The wavelength of the frequency (MHz) corresponds to the radio frequency of the nucleus at a given magnet field. The result is a "free-induction decay" or FID, which is a time domain response to the rf excitation. A FID records the free precession of nuclei after the pulse and represents the change in voltages induced in the receiver coil in the probe versus time. The FID information is digitized and stored in a computer. The digitized FID is then converted to a frequency domain spectrum by applying a Fourier transformation to the time domain FID information. The result is the typical NMR spectrum we are used to viewing for interpretative purposes. The result of a typical proton (^{1}H) NMR experiment is shown in Figure 5.30.

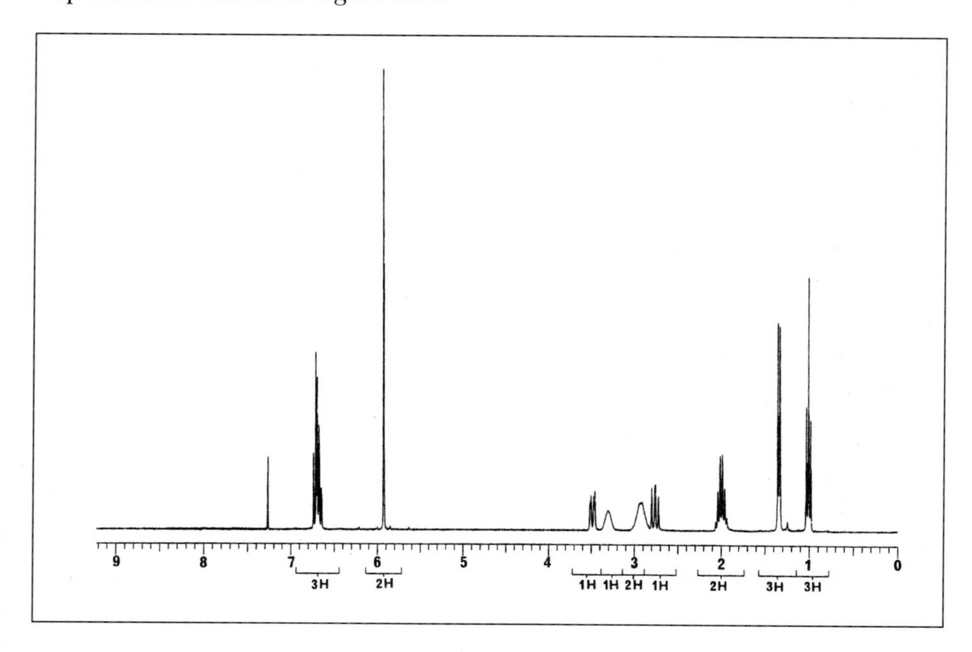

Figure 5.30

Proton NMR spectrum (300 MHz, CDCl3) of N-(n-propyl) MDA hydrochloride

The information contained in a survey spectrum, in this instance a 1-dimensional proton NMR spectrum, provides three pieces of basic information:

1. the position of the various resonance patterns corresponds to the different chemical environment that the protons find themselves;

2. spin-spin splitting (coupling), the multiplicity of lines in the spectrum, relates to how the protons are bonded within the molecule; and
3. electronic integration of the spectrum, in particular the resonance patterns, provides a quantitative relationship of the relative numbers and types of protons in the molecule.

Thus the proton NMR spectrum and the pattern(s), which are observed, are characteristic of the structure of a pure chemical compound or perhaps a mixture of chemical compounds found in the sample.

The patterns observed in proton NMR are sensitive to solvent effects and temperature. The same compound dissolved in different deuterated solvents or obtained at different temperatures may give slightly different proton NMR spectra due to shifting of the resonance positions and/or broadening of the lines. Thus, when making comparative analysis by proton NMR it is important to be sure that the same solvent is used in all cases and the same data acquisition parameters (instrument parameters) are used and that the temperature be regulated. Obtaining a survey proton NMR spectrum is usually the first logical step in using NMR for structural identification. All proton NMR spectra run in organic deuterated solvents, e.g. deuterochloroform, deuteromethanol, and deuteroacetone are internally referenced to tetramethylsilane (TMS). This internal reference is assigned 0.00 ppm (the ppm scale is used for NMR spectra and the resonance positions, or chemical shifts as they are referred to, is expressed relative to internal TMS). For deuterium oxide trimethylsilylpropionate sodium salt (TSP) is used as the internal reference.

The next logical step would be to obtain a survey carbon-13 (^{13}C) NMR (Wehrli, 1988) spectrum of the sample in question. Carbon-13 NMR spectra are more difficult to obtain than proton NMR spectrum. The natural abundance of natural NMR active nucleus of carbon (i.e., ^{13}C) in a molecule is 1.1% (as compared to 99.6% for proton). NMR also more easily detects the hydrogen nucleus than the carbon-13. These two factors make carbon-13 less sensitive than proton to NMR detection by a factor of 5600. Carbon-13 spectra are obtained under conditions where the spin-coupling effects from the protons directly attached to the carbons as well as long-range couplings effects (2 and 3 bonds away) are removed. This is done by irradiating the protons at their appropriate precessional frequency with radio frequency (rf). This effectively removes the proton coupling effects on the carbon-13 signals and the resulting carbon signals appear as single sharp lines. There is no mutual coupling from other carbon-13 nuclei since each line which is observed arises from an ensemble of molecules in solution which is rich in carbon-13 *at only one carbon site*; i.e., each resonance line in the spectrum arises from molecules in solution which have a single carbon-13 in the molecule all located at a specific carbon site (isolated carbon-13 nuclei). Protonated carbon-13 nuclei generally give more intense signals due to a phenomenon known as the nuclear Overhauser effect (Claridge, 1999). This phenomenon aids in the detection of carbon-13 spectra and serves to improve the detectability of carbon-13 NMR data.

The next logical step after obtaining the 1-dimensional carbon spectrum would be to assign the multiplicity of the carbon-13 resonances. This is done by using one of two experiments: attached proton test (APT) (Patt, 1982) or distortionless enhancement by polarization transfer (DEPT) (Doddrell, 1982; Richarz, 1982). Both experiments provide information about the types of carbons typically found in a molecule (quaternary, methine, methylene, and methyl). Both experiments have advantages and disadvantages.

The APT is variation of a class of experiments known as spin-echo J-modulation experiments. In the APT experiment, the spectrum which is obtained consists of signals which are either phased up (quaternary and methylene) or phased down (methine and methyl) and relate to the odd or even number of protons attached to carbon. Quaternary and methylene carbons have an even number (0 and 2) of protons attached to them whereas methine and methyl carbons have an odd number (1 and 3) of protons attached to them. The multiplicity of the carbon atom is thus implied from the position of the carbon resonance (chemical shift) and the phasing of the resonance line. Figure 5.31 shows the results of an APT experiment together with a plot of a carbon spectrum (bottom) and the corresponding APT spectrum (top).

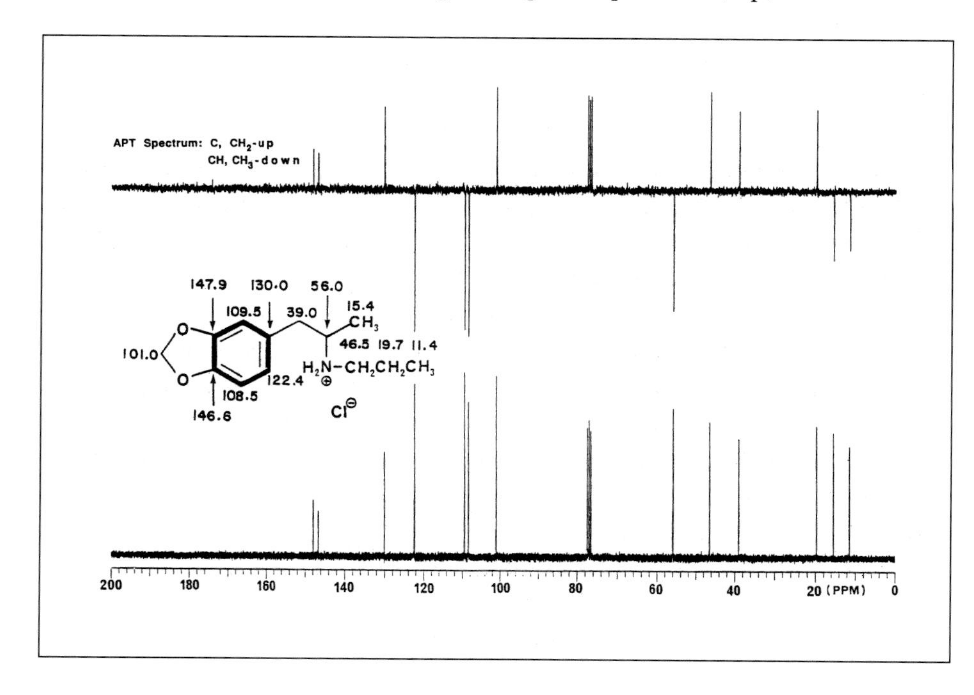

Figure 5.31

Carbon-13 NMR spectrum (75 MHz, CDCl3) of N-(n-propyl)MDA hydrochloride; (a) proton decoupled attached proton test (APT) spectrum, (b) proton decoupled carbon-13 spectrum.

A caveat to the APT experiment involves its sensitivity (slightly less than a normal proton decoupled carbon-13 spectrum) and the fact that since this is a spin echo experiment, the spin echo scheme discriminates with respect to broad lines. If you have a sample whose carbon resonances are broadened due to some exchange process (known as exchange broadening) the carbon resonances can be eliminated completely. By changing solvents, adding a proton source (H^+), or changing the temperature of the probe (sample), you can alter the dynamics of the experiment. The practitioner should be aware of this as it represents a potentially serious problem.

The other standard experiment for determining carbon multiplicity, the DEPT experiment (Doddrell, 1982), involves a polarization transfer scheme in which proton polarization (abundant nucleus) is transferred to the carbon-13 nucleus (rare spin nucleus) and modulated by the one bond coupling constant (^1JCH).

In this experiment, only protonated carbons will appear (quaternary carbons including residual solvent resonances are suppressed). The "raw" DEPT data results in three types of sub-spectra:

1. an all protonated carbon spectrum;
2. a methine carbon only spectrum; and
3. a spectrum in which the methine and methyl carbon resonances are phased up and the methylene carbon resonances are phased down.

These spectra may be subsequently combined to generate a set of what are referred to as "edited" DEPT spectra. For the purposes of identification, a normal proton decoupled carbon-13 spectrum needs to be collected and plotted along with, in this case, the edited DEPT data. Figure 5.32 illustrates a typical result including proton decoupled carbon-13 spectrum together with the corresponding edited DEPT data.

The advantage of the DEPT experiment is that data can be obtained very quickly with good sensitivity relative to the simple proton decoupled carbon-13 spectra and the solvent peak in this instance is suppressed. The solvent behaves like a quaternary carbon resonance. There is no proton attached to the solvent and therefore no polarization transfer takes place. This is an advantage since the solvent signal may very well obscure carbon resonances that are of interest which are present in the carbon spectrum of the sample. A disadvantage of the DEPT experiment generally occurs when the proton resonances of the sample are dynamically broadened due to some exchange process (exchange broadening). The spin-spin relaxation (T_2), which is related to the line width-at-half-height ($\upsilon_{1/2}$), is short and the dynamics of the polarization transfer in the DEPT experiment competes with T_2 relaxation. The result is that the carbon signals

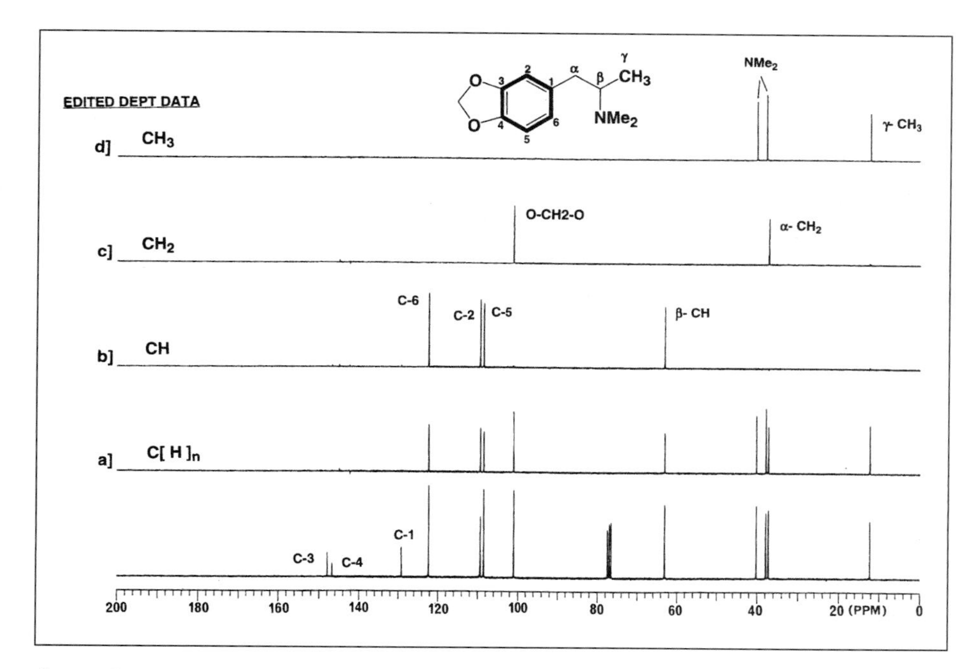

Figure 5.32
Carbon-13 NMR spectra (75 MHz, CDCl3) of N,N-dimethyl MDA. Bottom spectrum. proton decoupled carbon-13 spectrum showing all carbon resonances. Upper spectra (a)–(d). Proton decoupled edited DEPT data (a) all protonated carbon resonances (Note: the absence of the quaternary carbons C-1, C-3, and C-4 as well as the solvent 3-line pattern centered at 77 ppm), (b) CH – subspectrum, (c) CH_2 – subspectrum, (d) CH_3 – subspectrum

from these protonated carbons can disappear (no polarization transfer takes place). This is a potentially serious caveat to the DEPT experiment since, as we saw with the APT experiment, chemical information about the sample is being deleted. The practitioner needs to be aware of this.

Survey proton and carbon-13 data together with the carbon-13 multiplicity information (APT and/or DEPT) represent the essential minimum data necessary for the identification and structural characterization of a complete unknown. For a sample containing a substance of potentially known structure, a comparative analysis of the simple proton and/or carbon-13 spectrum with spectra obtained from an authentic sample may be all that is necessary.

The development of 2-dimensional (2-D) NMR (Claridge, 1999; Richarz, 1982; Ernst, 1992; van der Ven, 1995) in the late 1970s and to the present day has revolutionized the process of structure elucidation by NMR. The 2-D experiment allows the correlation information to be obtained via a variety of mechanisms and spread into two frequency dimensions which involves manipulation (Freeman, 1997) of the three basic characteristics of any NMR spectrum: chemical shift, spin-spin coupling, and spin lattice relaxation (T_1). Table 5.1 provides a brief summary of a few of the useful 2-D experiments, which are currently in use together with the chemical information that is associated with the NMR experiment.

Experiments with a "g" in front are a pulsed field gradient variant. They provide the same chemical information as their carbon detected counterpart but data collection is generally faster and lower levels (small) samples can be

Table 5.1

Summary of useful 2-D experiments

2-D Experiment	Correlation Observed	Information Content
COSY, gCOSY	H,H correlation	H,H spin coupling
DQCOSY, gDQCOSY		
TOSCY, gTOCSY		
ECOSY, PECOSY		
HOM2DJ	H, J correlation	H,H couplings
HET2DJ	C, J correlation	$^{1}J_{C\text{-}H}$ couplings
HETCOR, *HMQC*	H,C correlation	direct $^{1}J_{C\text{-}H}$ bonding
gHMQC		
FLOCK, COLOC	H,C correlation	long-range $^{n}J_{C\text{-}H}$
HMBC, gHMBC	H,C correlation	long-range $^{n}J_{C\text{-}H}$
NOESY, ROESY	H,H correlation	stereochemistry,
		H-H distance
INADEQUATE	^{13}C,^{13}C correlation	carbon framework

done. Experiments that are in italics are proton (as opposed to carbon) detected experiments.

All 2-D experiments share common elements. There is a preparation period, an incremented evolution period (t_1), a mixing period, and a detection period (t_2). The preparation period can be as simple as a fixed delay and a pulse, applied repetitively and in a reproducible manner for each increment of t_1 or it can be as complex as the double quantum filter elements found in the 2-D INADEQUATE (Buddrus, 1987) experiment. The evolution time (t_1, usually an incremented evolution time, but there are constant time evolution periods used in certain heteronuclear 2-D experiments, e.g. COLOC) is a time period during which evolution of an NMR measurable parameter occurs based on what information is being sought. The mixing time can be a simple fixed delay (e.g. as in NOESY) or a pulse (e.g. COSY). A pulse usually (but not always) precedes the detection period (t_2) and the final FID is collected. The resulting information is Fourier transformed to create a 3-dimensional plot that resembles a topographical map.

The 2-D experiments permit elucidation of structural details, sometimes subtle details that show themselves. For simple assignments of ^1H and ^{13}C resonances, one of the appropriate homonuclear proton correlations would be run in order to deduce proton coupling patterns and tell which protons are spin coupled to one another. It is this kind of information that would be characteristic of a particular structure. Figures 5.33 and 5.34 illustrate a simple COSY (Claridge, 1999) spectrum together with an expansion of the high field region of a PCP analogue (PCM). The proton resonances for the morpholine ring are clearly evident. There is no "cross coupling" between the cyclohexane ring protons and the protons on the morpholine ring. The complexity of the cyclohexane ring protons is also clearly evident from Figure 5.34.

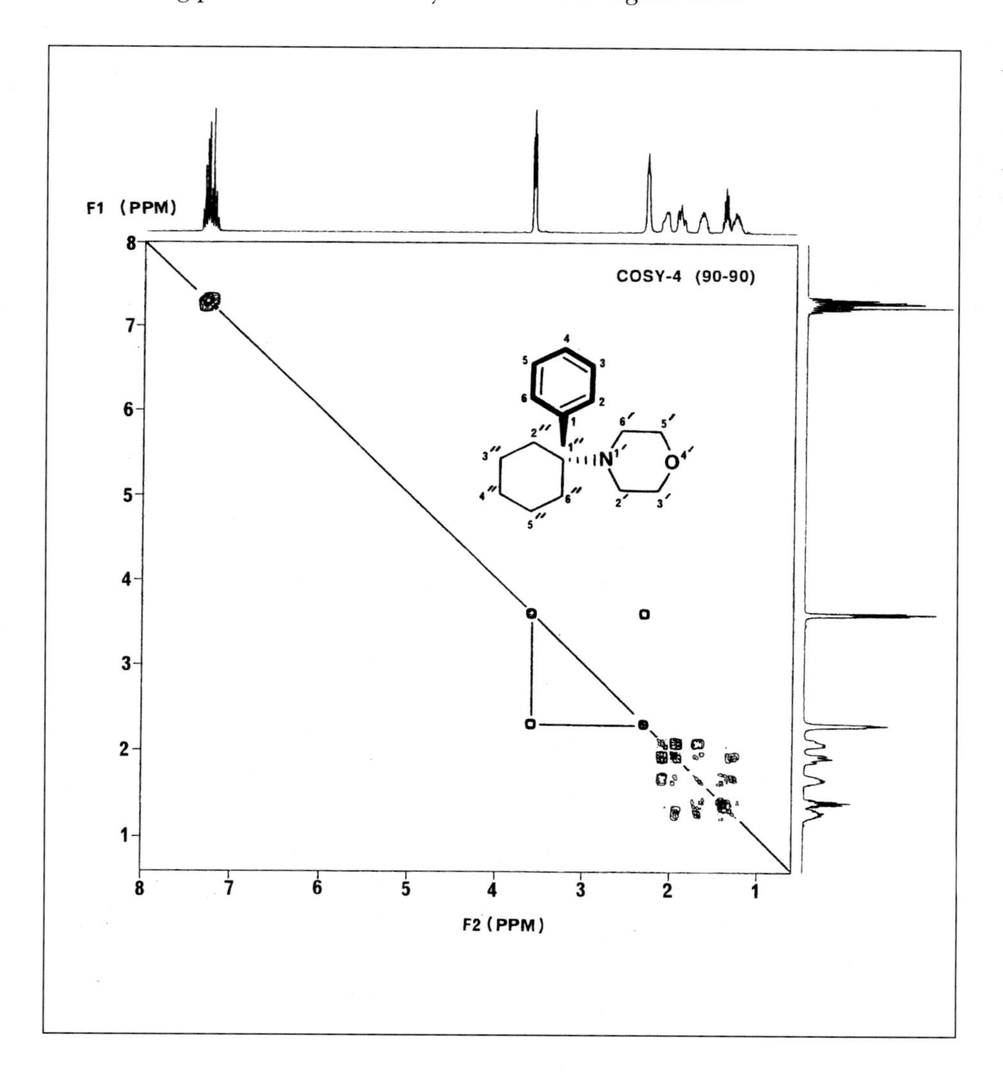

Figure 5.33

The 2-dimensional ^1H, ^1H - COSY spectrum (300 MHz, $CDCl_3$) of the phencyclidine analogue (PCM)

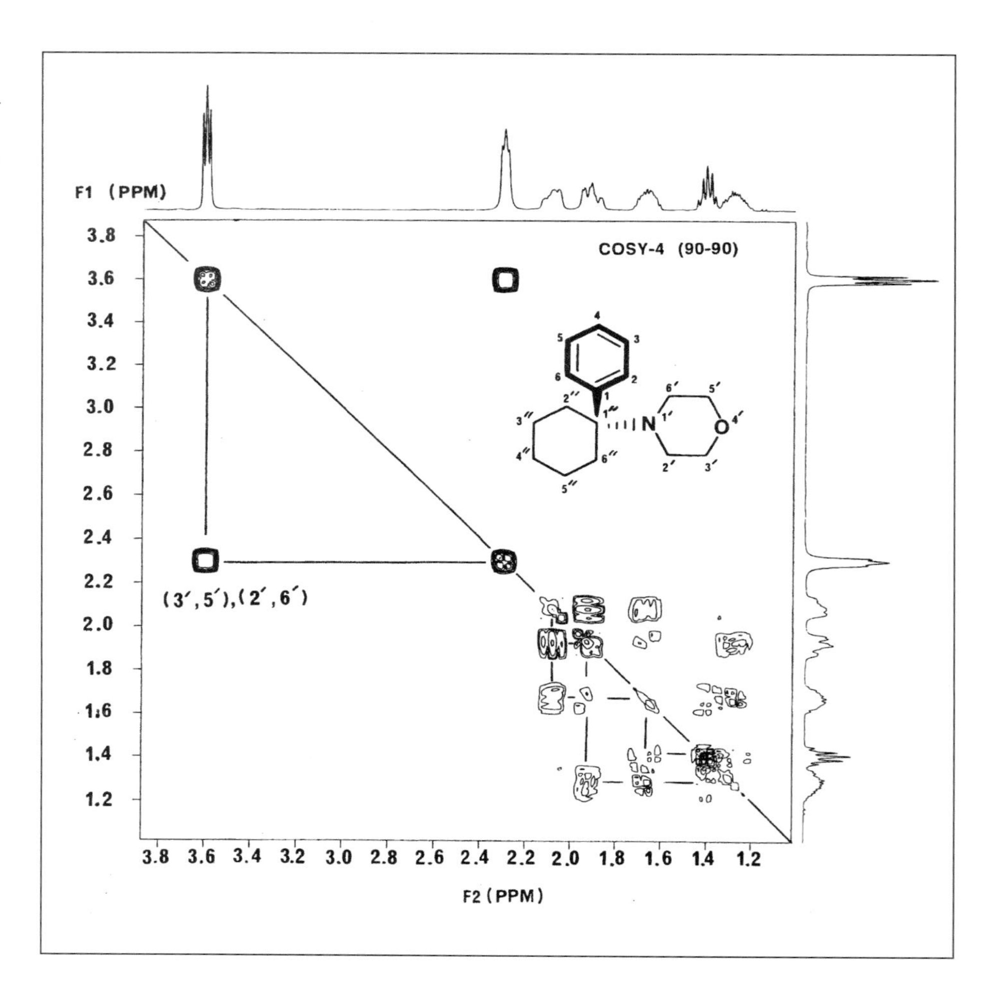

Figure 5.34

Expansion of the 2-dimensional COSY spectrum from Figure 5.33 showing the detailed correlation information from the respective non-aromatic six-membered rings

The coupling complexity of the cyclohexane ring protons can be partially resolved and in the process a self-consistent assignment of both the ^{1}H spectrum and the ^{13}C spectrum can be confirmed.

Figure 5.35 shows the results of the carbon-detected proton-carbon heterocorrelation experiment HETCOR (Bax, 1981; Bax, 1983; Rutar, 1984; Martin, 1988) (J = 140Hz) for the PCP analogue (PCM). The connectivity (one-bond) correlation of the protons and the carbons of the morpholine ring are quite obvious. Expansion of the high-field region of the HETCOR data (Figure 5.36) actually allows both the carbon-13 resonances to be assigned and in the process the proton assignments become apparent. The geminal pairs of protons attached to C-2 and C-6 can be identified from the two correlation cross peaks corresponding to two protons having different chemical shifts but which correlate to the *same* carbon resonance. Similarly, the shift position of the geminal pairs of protons attached to C-3 and C-5 are also evident and are

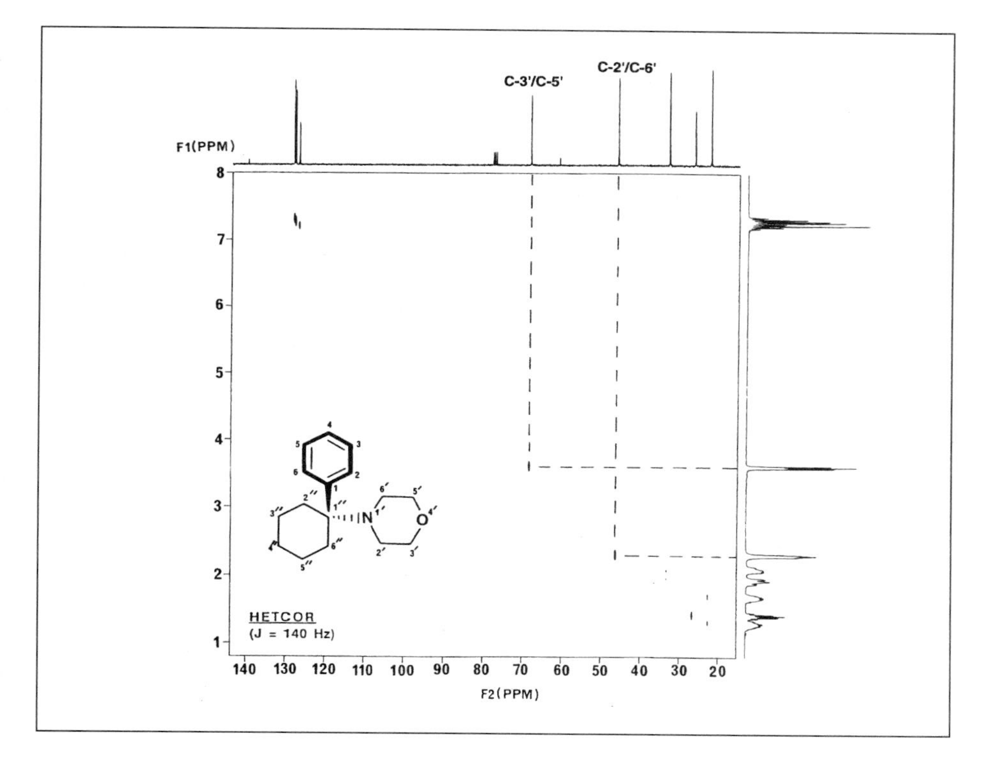

Figure 5.35

The carbon-detected ^1H, ^{13}C *correlation spectrum (HETCOR) (75 MHz,* $CDCl_3$, *J=140 Hz) for the phencyclidine analogue (PCM)*

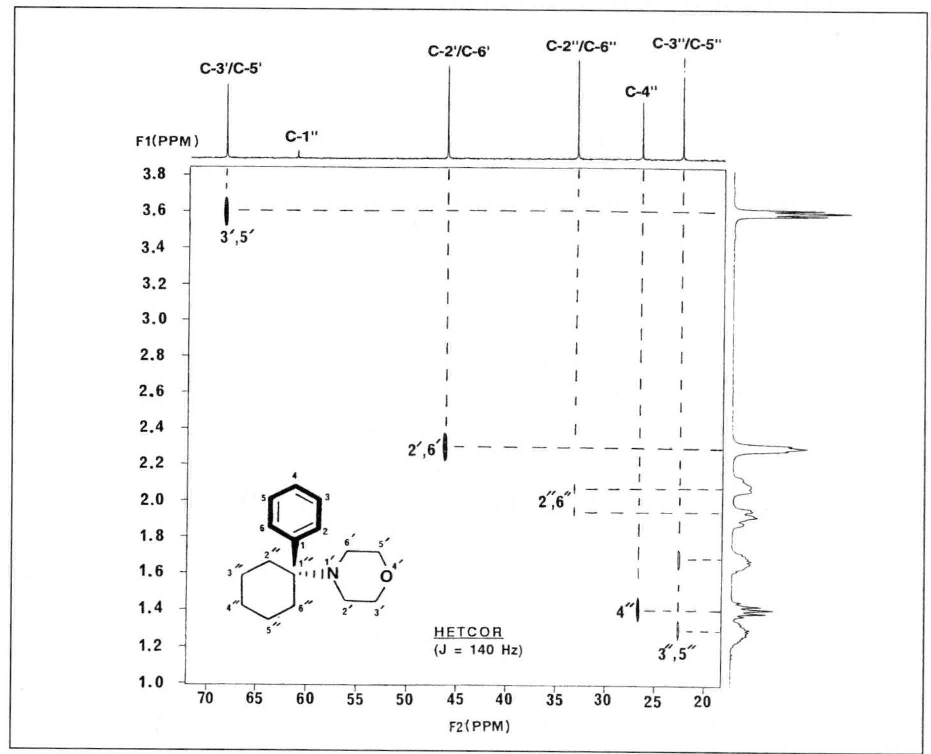

Figure 5.36

Expansion of the HETCOR spectrum from Figure 5.35 showing the detailed correlation information for the two non-aromatic six-membered rings characterized by two correlation cross peaks, both of which correlate with the same carbon resonance. The C-4 carbon resonance may be assigned because of its intensity relative to the intensities of C-3/C-5 and C-2/C-6.

Clearly, the ability of 2-dimensional NMR experiments to establish structure and stereochemistry and to confirm, in this instance, proton and carbon-13 spectral assignments and, in a self-consistent manner, serve as a reliable structural tool for the structural elucidation of complete unknowns, is of considerable importance to the forensic community. In the next few sections, you will see some of the specific applications that have been addressed in the area of forensic analysis of hallucinogenic drugs and the impact that 2-dimensional NMR is making in this important area.

Pharmaceutical laboratories and other researchers conducted much of the early NMR structural work reported on various drug molecules. Some of the compounds that exhibited pharmacological activity were ultimately investigated and results reported. A number of studies on these molecules have been summarized in comprehensive review articles that include literature citations to the work on the NMR of Drugs of Abuse (Groombridge, 1996).

Hallucinogens represent one of the challenges to forensic identification as most of the analogues seen have occurred in this area. This is a challenge because the primary identification tools, IR and GC/MS, used in forensic laboratories today have difficulty in providing definitive confirmation of identity of a substance when the compounds are structurally closely related. More than 200 psychotropic drugs have been reported (Shulgin, 1997, 2000) and their synthesis listed. Many of these are very similar in chemical structure. Many of the compounds do not have published spectral information. Without the use of NMR, these compounds could not be reliably and rapidly identified. The rapid growth of the Internet has made this information more readily available to a large portion of the population.

5.15 PHENCYCLIDINE AND RELATED SUBSTANCES

Parke Davis marketed phencyclidine (1- [1-phenylcyclohexyl]piperidine, PCP), Figure 5.37, as an anesthetic from 1958 to 1967. Because of its unpredictable adverse effects, PCP was subsequently withdrawn from the commercial market. Since 1967, PCP, because of the hallucinogenic and stimulant qualities, has been manufactured primarily in clandestine laboratories. There are, however, neurotoxic side-effects of PCP, which can lead to "a psychosis clinically indistinguishable from schizophrenia" (Reynolds, 1989).

Most of the early NMR work on phencyclidines was directed toward establishing the conformational equilibrium dynamics of structurally modified phencyclidines. These investigations attempted to control receptor binding

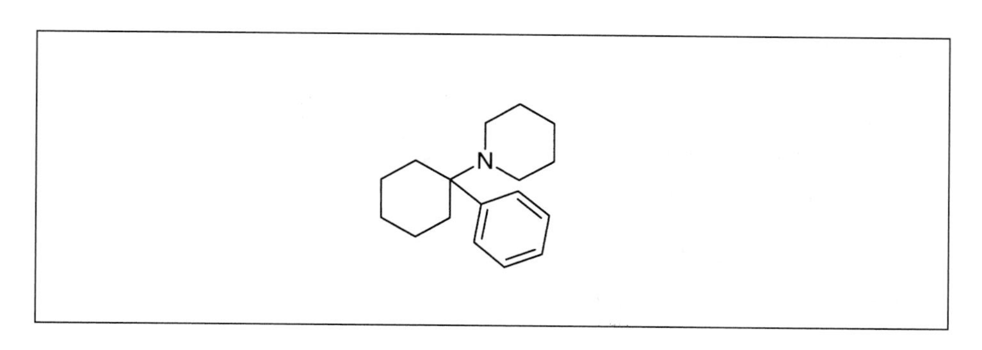

Figure 5.37
Chemical structure of PCP

through appropriate substitution on the ring components. This provided valuable information about the biological effects of various substituted phencyclidine analogues that has been of considerable forensic value. As new analogues of PCP appeared on the street, they were subsequently identified and the spectra (IR, MS, NMR) reported. Other analogues were studied because of their ease of manufacture and for the potential to be seen in forensic samples.

The ^{1}H spectrum of PCP as well as some of its analogues exhibits complex overlapping resonances because of the presence of the two saturated rings (Eaton, 1983). Only the advent of 2-D NMR techniques has the complexity of the patterns been sorted out. The free base of PCP shows line broadening due to certain molecular motions (ring flipping dynamics) at room temperature. The spectra of PCP and its analogues can easily be differentiated and can be compared against spectra of standards for identification. The ^{13}C spectrum is conceptually easier to comprehend (Eaton, 1983), although the chemical shifts of phenyl ring carbons were demonstrated to exhibit solvent and temperature dependence, presumably due to their influence on the conformational populations within the PCP molecule (Manoharan, 1983; Kamenka, 1987).

Often, samples submitted to forensic laboratories are for analysis and are not highly purified. Thus, the starting ingredients, intermediates and byproducts of the synthesis are frequently present. The immediate precursor, 1-piperidino-cyclohexylcarbonitrile (PCC), Figure 5.38, has been characterized

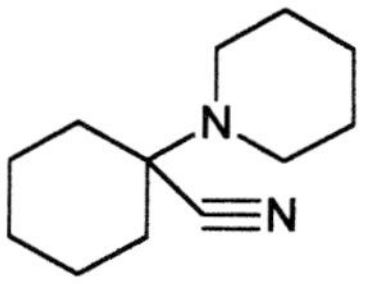

Figure 5.38
Chemical structure of PCC

by proton (Bailey, 1976; Gagné, 1977) and carbon-13 NMR (Bailey, 1981). NMR has also been used to characterize the starting ingredients and by-products. Often the presence of one by-product can help in the identification of the synthetic route used to prepare the PCP or its analogue.

More recently, the NMR spectra of PCP, PCC and analogues obtained at higher magnetic fields has been reported in the literature thus allowing better discrimination among the various compounds. Some of the analogues have appeared on the illicit drug market and some have been synthesized in an effort to facilitate the identification of additional new analogues. Figures 5.39, 5.40, and 5.41 show the ^{1}H spectra of PCP, 1-thienylcyclohexylpiperidine (TCP) and 1-phenylcyclohexylpyrrolidine (PCPy) respectively. Figures 5.42, 5.43 and 5.44 show the ^{13}C spectra of PCP, TCP and PCPy respectively.

Figure 5.39

^{1}H spectra of phencyclidine hydrochloride

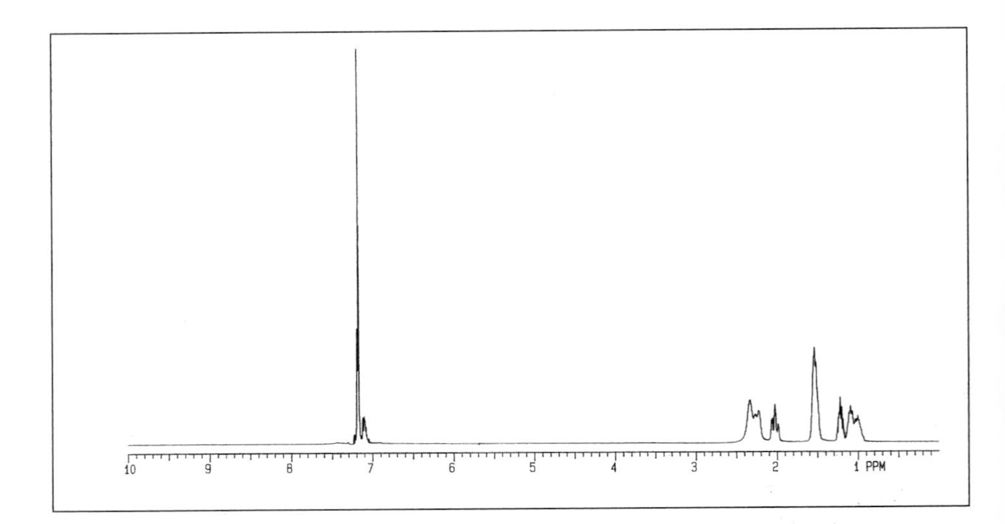

Figure 5.40

^{1}H spectra of thiophene analogue of phencyclidine hydrochloride

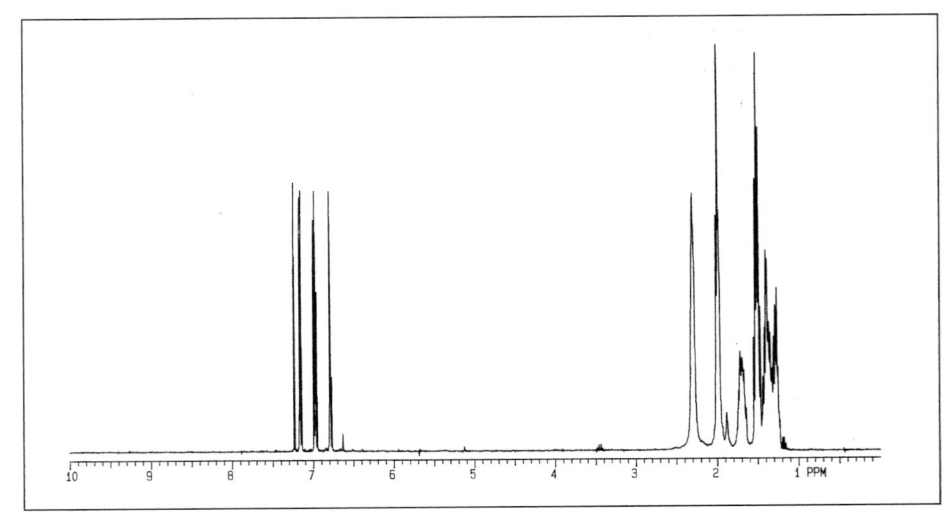

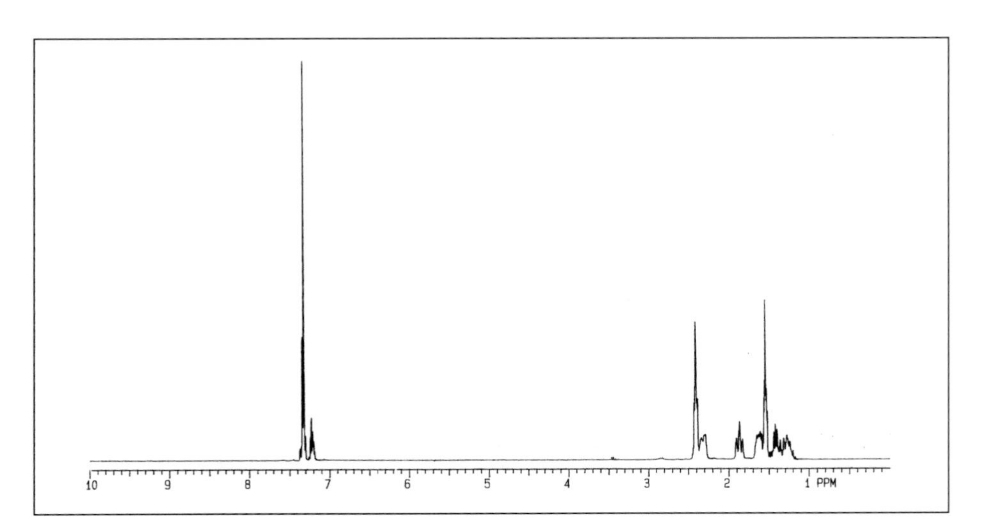

Figure 5.41

^{1}H spectra of morpholine analogue of phencyclidine hydrochloride

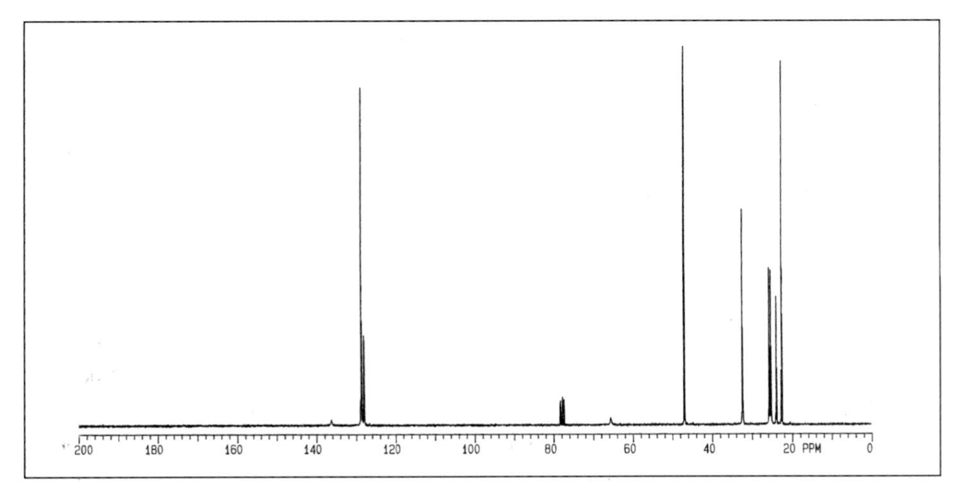

Figure 5.42

^{13}C spectra of phencyclidine hydrochloride

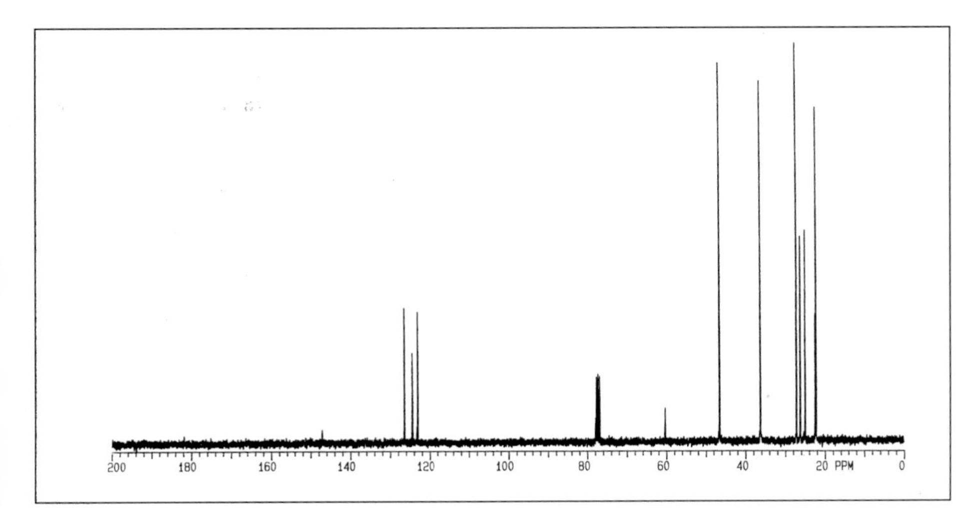

Figure 5.43

^{13}C spectra thiophene analogue of phencyclidine hydrochloride

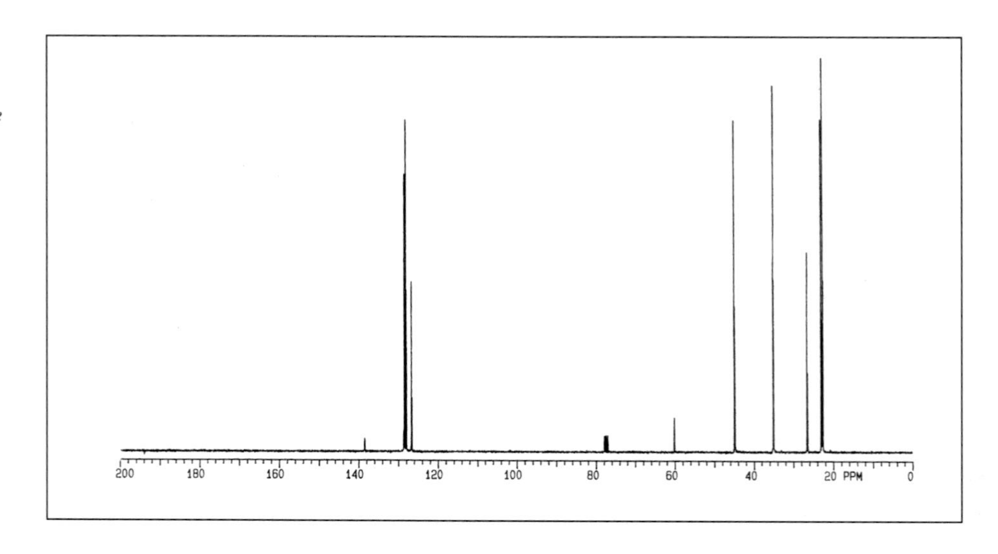

Figure 5.44
^{13}C spectra morpholine analogue of phencyclidine hydrochloride

5.16 MDA AND ANALOGUES

In the early 1970s Bailey *et al.* conducted a systematic investigation of methoxy- and methyl-substituted amphetamines. Initially, only the ^{1}H NMR (60MHz) (Bailey, 1971, 1974, 1975, 1976, 1977), data was reported followed by reports of ^{13}C NMR (Bailey, 1981, 1983) data in the 1980s. In many of these papers, NMR was used to confirm the structures suggested by IR and MS data. This is especially true in the identification of aromatic ring substitution patterns (Dawson, 1987, 1989; Delliou, 1983). This body of information is of particular importance when new analogues such as 4-bromo-2, 5-dimethoxyamphetamine (DOB) (Figure 5.45), or phenethylamines such as 4-bromo-2,5-dimethoxyphenethylamine (2C-B) (Figure 5.46), were introduced into the illicit drug market (Ragan, 1985). The use of NMR, especially employing 2-D techniques, can reduce the time for identification of new compounds without the need for the synthesis of all possible combinations. NMR can be used to direct what compounds need to be synthesized for authentic comparison (Dal Cason, 1997).

Figure 5.45
Chemical structure of 4-bromo-2,5-dimethoxyamphetamine (DOB)

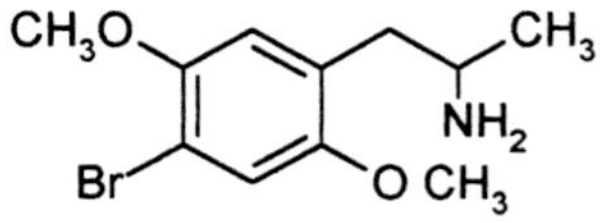

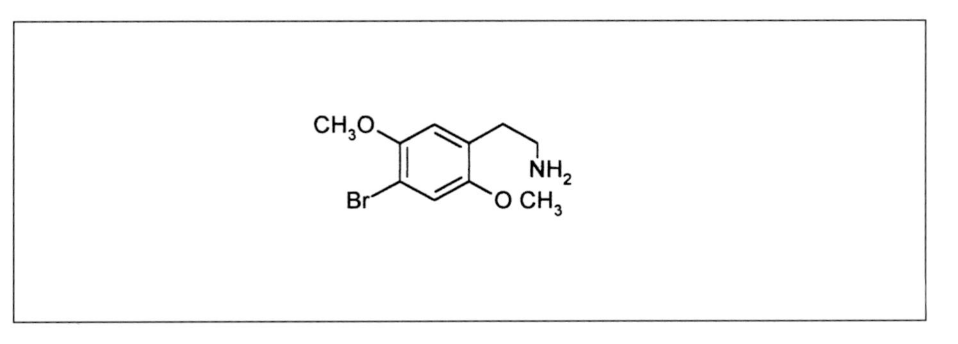

Figure 5.46

Chemical structure of 4-bromo-2,5-dimethoxyphenylethyl-amine (2C-B)

Dimethoxyamphetamines were reported as early as 1967. The earliest use of NMR to identify a new drug was to aid in the identification of 4-methyl-2,5-dimethoxyamphetamine (DOM or STP) (Martin, 1968). The various isomeric forms 2,3-, 2,4-, 2,5-, 3,4-, 3,5- were investigated to determine whether or not they could be distinguished (Bailey, 1971). 2-, 3-, 4-methoxy amphetamines were reported in 1971. All of the ^{1}H work was performed at 60MHz. (Bailey, 1974). Later, other substituted dimethoxyamphetamines namely: 4-methyl-2,5-dimethoxy amphetamine (Ono, 1970), 2,5-dimethoxy-4-bromoamphetamine, 2,5-dimethoxy-4-ethyloxyamphetamine, 2,5-dimethoxy-4-propylamphetamine, 2,5-dimethoxy-4-methylthioamphetamine were reported (Bailey, 1971, 1974, 1975,1976, 1977, 1981, 1983).

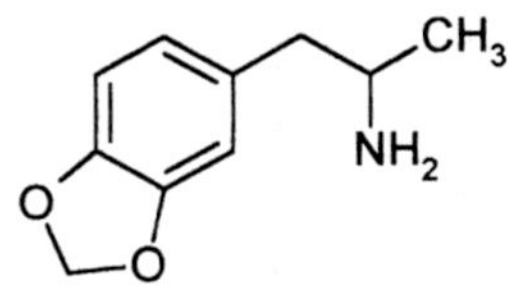

Figure 5.47

Chemical structure of MDA

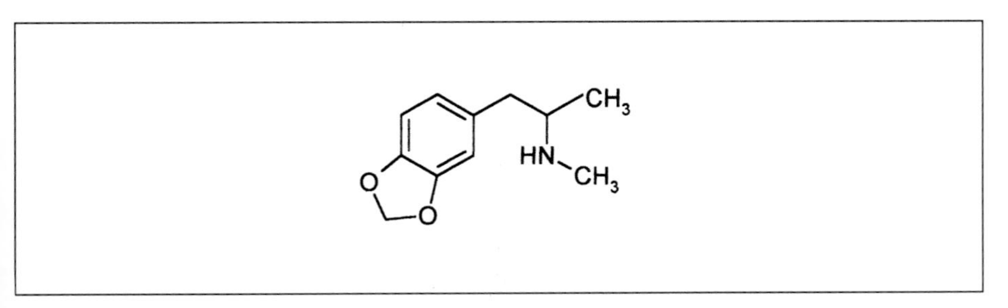

Figure 5.48

Chemical structure of MDMA

3,4-Methylenedioxyamphetamine (MDA) (Figure 5.47), first appeared on the illicit drug market in the 1960s with its NMR data being reported as early as 1970 (Bellman, 1970; Lukaszewski, 1978). 3,4-Methylenedioxy-methampheta-mine (MDMA) (Bailey, 1975) (Figure 5.48), and 4-methoxyamphetamine

(PMA) (Bailey, 1974) (Figure 5.49), among other MDA analogues have been seen more recently with a huge increase in the number of tablets seized in the late 1990s. Other analogues such as 3,4-methylenedioxyethylamphetamine (MDEA) (Figure 5.50), *N*-propyl-MDA, *N*-isopropyl-MDA, *N*-butyl-MDA, *N*-isobutyl-MDA, and *N*-neobutyl-MDA have been reported (Noggle, 1986). It was shown that the early NMRs (^{1}H 90 MHz) did not have the resolution sufficient to differentiate between MDA and 2,3-methylene dioxyamphetamine. As higher magnetic field NMR instrumentation became available this problem was resolved. Table 5.2 and Figure 5.51 provides a summary of the assigned 1H chemical shifts for some MDA analogues and ^{13}C chemical shifts for the free bases and the hydrochloride salts respectively. Table 5.3 provides a summary of the assigned ^{13}C chemical shifts for some MDA analogues.

As can be expected, the proton spectra of MDA and *N*-hydroxy-MDA are similar (Dal Cason, 1989; Shimamine, 1990, 1993) but can be clearly differentiated by NMR spectroscopy. The hydroxy proton resonance was too broad to observe which contrasted the resonance measurements observed earlier for *N*-hydroxy-amphetamines, dimethoxyamphetamines, and others (Beckett, 1975; Mourad, 1985).

One of the areas where NMR is more readily suited is the identification of isomers. This has been demonstrated by the occurrence in Europe of *N*-methyl-1-(1,3-benzodioxol-5-yl)-2-butanamine (MBDB) (Figure 5.52). When compared to the isomeric forms 3,4-methylenedioxy-*N*-ethylamphetamine (MDEA) or *N,N*-dimethyl-MDA (Figure 5.53), the three compounds can be differentiated by careful comparison of MS but the ^{1}H NMR spectra are very different (Nichols, 1986; Azafonov, 1990).

Figure 5.49

Chemical structure of PMA

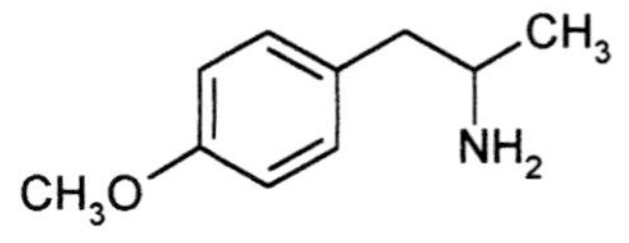

Figure 5.50

Chemical structure of MDEA

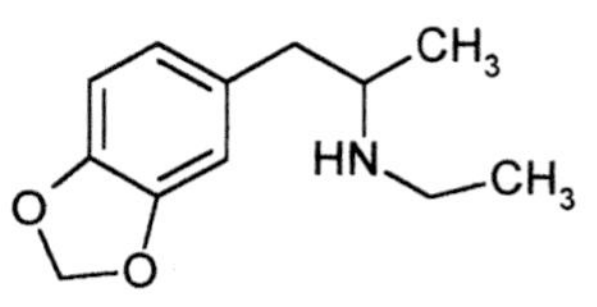

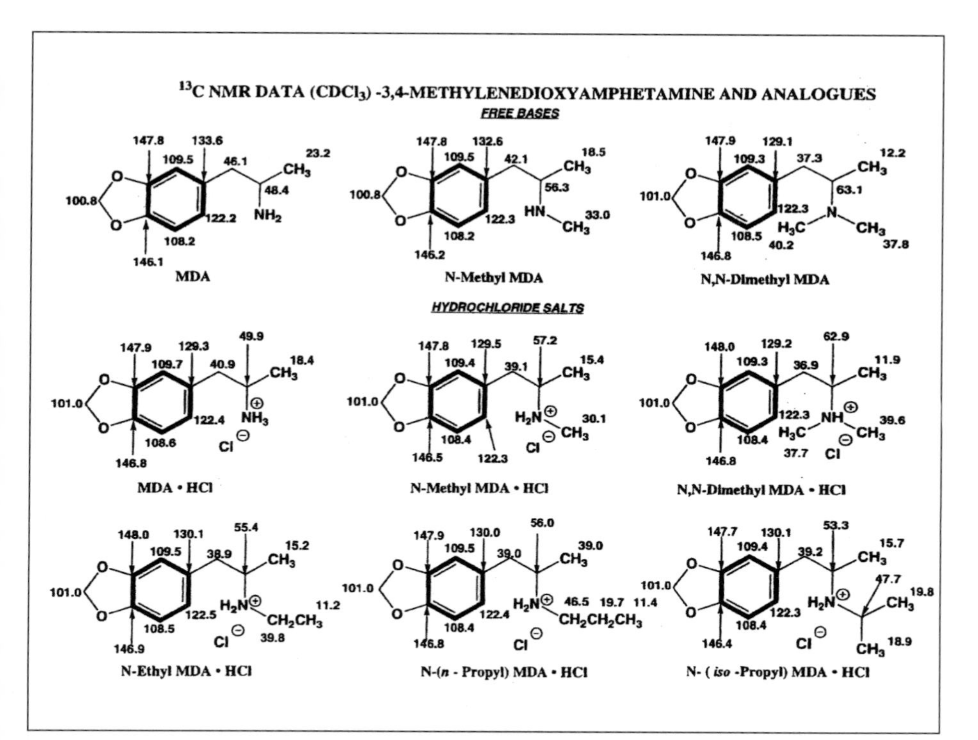

Figure 5.51

Provides a summary of the assigned ^{13}C chemical shifts for some MDA analogues

Table 5.2

^{1}H table of chemical shifts for several MDA analogues as the free bases

3,4-Methylenedioxyamphetamines — ^{1}H-NMR data (CDCl$_3$, 25°C) free bases

Proton	Proton chemical shifts, ppm (coupling constants, Hz) (N=2)	(N=1)	(N=0)
H-2	6.61 d (1.2)	6.65 d (1.2)	6.66 d (1.6)
H-5	6.67 d (7.8)	6.67 d (7.8)	6.66 d (7.9)
H-6	6.56 dd (7.8, 1.2)	6.56 dd (7.8, 1.2)	6.60 dd (7.9, 1.6)
H-α	2.54 dd (5.2, −13.4)		3.41 dd (3.1, −12.8)
H-α	2.36 dd (8.2, −13.4)	2.4–2.7 m	2.46 dd (11.0, −12.8)
H-β	3.03 br.m. (6.2, 5.2, 8.2)		3.39 br.m.
H-γ	1.03 d (6.2)	1.00 d (6.2)	1.19 d (7.0)
–O–CH$_2$–O–	5.85 s	5.88 s	5.86 s
N–CH$_3$	—	2.34 s	2.74 s 2.72 s 2.71 s 2.69 s
N–H	1.4 br.s.	1.7 br.s.	—

Table 5.3

^{13}C table of chemical shifts for several MDA analogues as the free bases

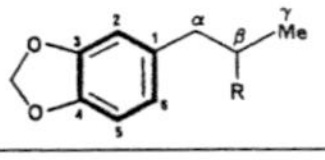

3,4-Methylenedioxyamphetamines — ^{1}H-NMR data ($CDCl_3$, 25°C) hydrochloride salts

R	Chemical shifts, ppm (coupling constants, Hz)										
	H-2	H-5	H-6	H-α	H-α	H-β	H-γ	$-OCH_2O-$	–NH	$-N-CH_3$	Other
$-N^{+}H_3\ Cl^{-}$ [a]	6.66 d (1.4)	6.72 d (7.8)	6.62 dd (7.8, 1.4)	2.89 dd (7.5)	2.81 dd (7.0)	3.58 br.m.	1.37 d (6.6)	5.90 (s)	7.1 br.s.	—	—
$-N^{+}H_2-CH_3\ Cl^{-}$	6.65 d (1.6)	6.69 d (7.8)	6.61 dd (7.8, 1.6)	3.30 dd (−13.0) (4.0)	2.70 dd (−13.0) (4.0)	3.22 br.m.	1.28 d (6.5)	5.88 (s)	9.6 br.s.	2.65 (s)	—
$-N^{+}H(CH_3)_2\ Cl^{-}$	6.66 d (1.5)	6.66 d (7.8)	6.60 dd (7.8, 21.5)	3.40 dd (−12.4) (2.9)	2.45 dd (−12.4) (10.9)	3.35 br.m.	1.18 d (6.8)	5.86 (s)	12.3 br.s.	2.72 s 2.73 s 2.71 s 2.70 s	—
$-N^{+}H_2-CH_2CH_3\ Cl^{-}$	6.66 d (1.5)	6.68 d (7.9)	6.62 dd (7.9) (1.5)	3.42 dd (−13.0) (3.6)	2.72 dd (−13.0) (11.0)	3.22 br.m	1.29 d (6.5)	5.89 (s)	9.65 br.s	—	CH_3, 1.48 d (7.3) CH_2, 3.35 br.m
$-N^{+}H_2-CH_2CH_2CH_3\ Cl^{-}$	6.70 dd (1.8, 0.5)	6.15 dd (7.8, 0.5)	6.66 dd (7.8, 1.8)	3.48 dd (−12.8) (3.5)	2.76 dd (−12.8) (11.0)	3.30 br.m.	1.34 d (6.5)	5.90 (s)	9.64 br.s.	—	CH_3, 0.99 d (7.4) CH_2, 1.99 dq (7.4, 7.8) CH_2, 2.98 br.m.
$-N^{+}H_2-CH(CH_3)_2\ Cl^{-}$	6.36 dd (1.8, 0.8)	6.68 dd (7.8, 0.8)	6.62 dd (7.8, 1.8)	3.43 dd (−13.0) (3.5)	2.80 dd (−13.0) (11.0)	3.30 br.m.	1.29 d (6.4)	5.87 (s)	9.39 br.s.	—	CH, 3.35 br.m. CH_3, 1.51 d (6.5) CH_3, 1.43 d (6.5)

[a] CF_3CO_2H added to enhance solubility in $CDCl_3$ solution.

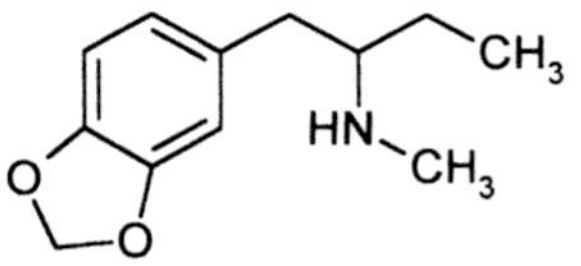

Figure 5.52

Chemical structure of MBDB

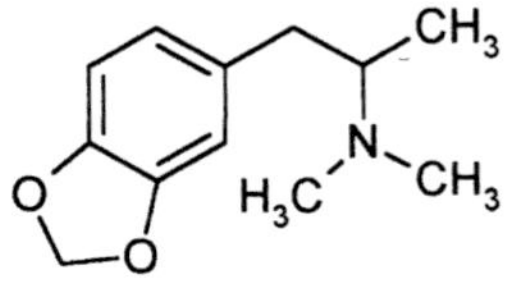

Figure 5.53

Chemical structure of N,N-dimethyl MDA

Other related compounds: mescaline (3,4,5-trimethoxyphenethylamine) (Ono, 1970) (Figure 5.54), 2,4,5-trimethoxyamphetamine (Foster, 1992) (Figure 5.55), 4-bromo-2,5-dimethoxyphenethylamine (2C-B) (Ragan, 1985) (Figure 5.46), as well as other ring substituted compounds have been studied and data reported. NMR provides a definitive solution to the identification of these compounds provided the non-proton compound or group is identified by other means (e.g. halogens). 2,4,5-Trisubstituted permutations (16 possible combinations) have been resolved using lanthanide shift behavior (Dawson,

1987, 1989) or nuclear Overhauser effect (NOE) difference methods (Dawson, 1989). NMR is the only technique that can provide a variety of different means to attack and resolve an identification problem. Figure 5.56 and Figure 5.57 show the ^{1}H and ^{13}C spectra of Mescaline respectively.

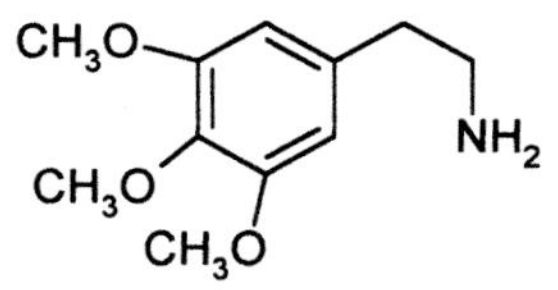

Figure 5.54

Chemical structure of mescaline

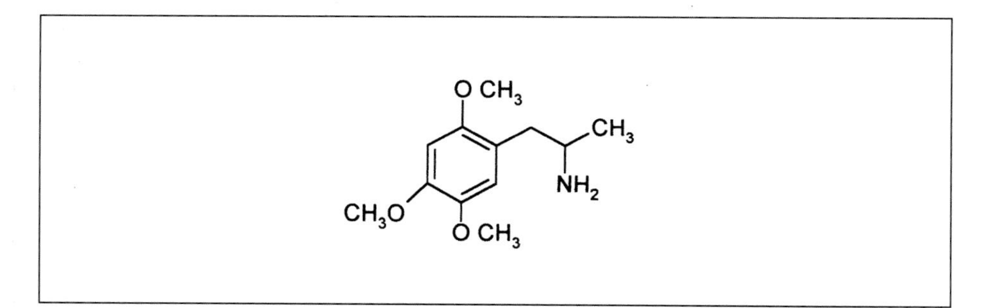

Figure 5.55

Chemical structure of 2,4,5-TMA

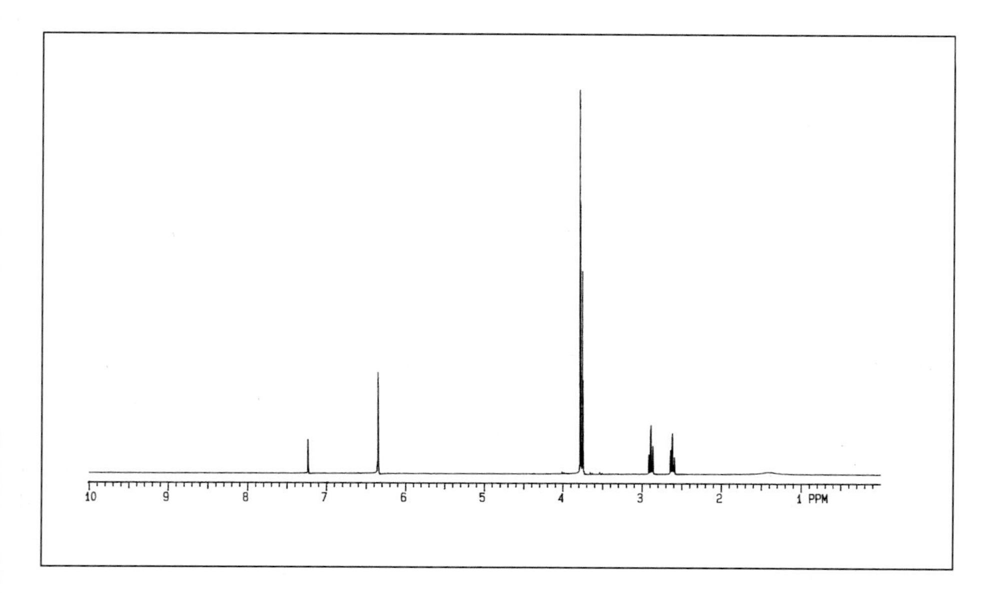

Figure 5.56

^{1}H spectra of mescaline

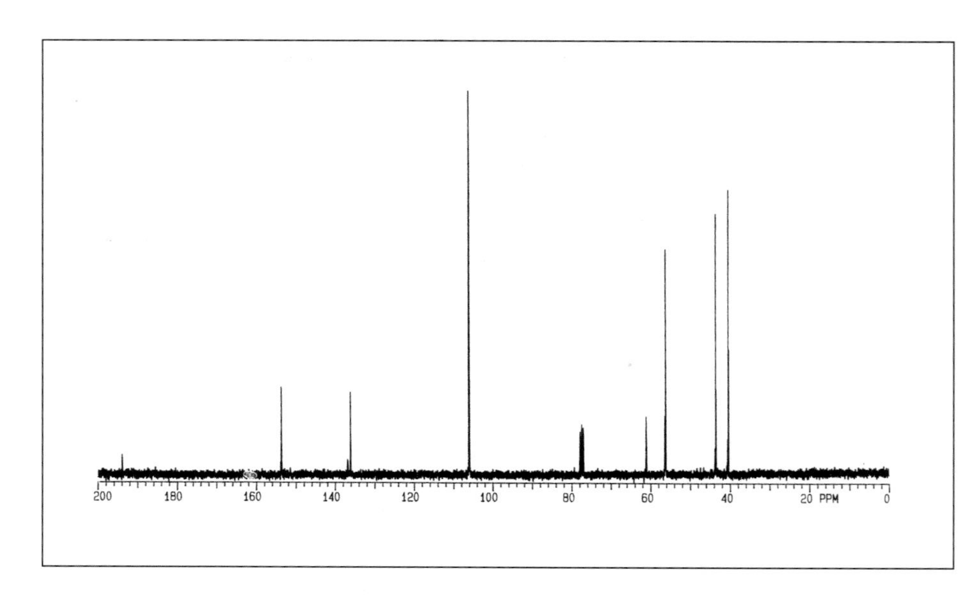

Figure 5.57
^{13}C spectra of mescaline

Figure 5.58
Chemical structure of LSD

5.17 ERGOT AND OTHER INDOLE ALKALOIDS

Ergot is the dried material of the parasitic fungus *Claviceps purpurea*, which grows on rye and other grain. It yields four main alkaloid classes: clavines, lysergic acids, lysergic amides, and ergot peptides. Ergot alkaloids have also been found in many plant species, with the *Convolvulacea* (morning glories) also having mixed lysergic acid substances. Lysergic acid diethylamide (LSD) (Figure 5.58), is derived from ergot. Its discovery by Hofmann in 1943 was the result of investigation of lysergic acid derivatives for the treatment of migraine headaches.

The abuse of LSD for its hallucinogenic quality was extensive in the 1960s, encompassing worldwide use. There was a subsequent decline in the 1970s

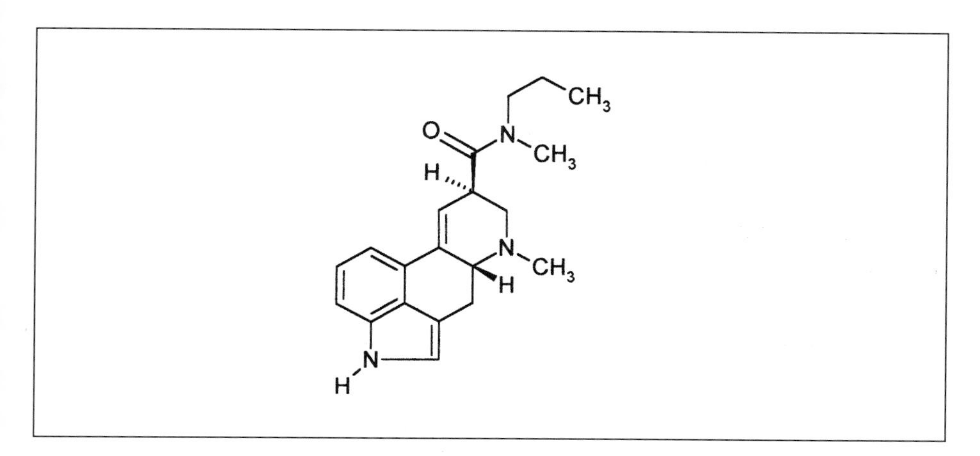

Figure 5.59

Chemical structure of LAMPA

when it was controlled in virtually every country around the world. However, it re-emerged in the 1980s more often on impregnated paper ("blotter acid") or gelatin squares. The analysis of LSD poses a unique set of problems owing to, first, the relative small amounts required for a typical dosage (10–100μg) as well as its sensitivity to light and/or high temperatures and to the presence of moisture. Detection of amounts of LSD below 10μg is feasible with ^{1}H NMR at 400 MHz using 1–2 hour accumulation on a 5mm probe. Application of micro-probe technology in conjunction with cryoprobe technology can significantly reduce the analysis time at this level and permit a whole host of proton-detected 2-D homonuclear- and heteronuclear experiments to be run on LSD to further confirm structure. Applications of this technology would also push the limits of detection of LSD even lower (e.g. Varian and Nalorac website).

The ^{1}H spectrum of LSD has been discussed in only a few publications, with the ^{1}H NMR assignments suggested by Hoffman and Nichols (Hoffman, 1985) based on the earlier detailed work of Bailey and Grey (Bailey, 1972). Rings C and D both adopt half-chair conformations, with the D ring in a "flap-up" mode. Coupling constants between H-8 and the two protons H-7 provide evidence for the conformational disposition. The spectra are made more complex due to the ergoline framework which exhibits significant long-range coupling pathways.

Proton spectra of LSD in aqueous solution are more poorly resolved than for the free base, and it is apparent that there is some variation from sample to sample. This variation reflects small changes in pH evident in the *N*6 protonation at near neutral pH.

Proton NMR data has been used to reveal a problem with supposed pure reference material of LSD tartrate. Neville *et al.* (1992) showed that a reference standard supplied from a commercial firm contained a stoichiometric excess of tartrate (65%). It is customary for forensic laboratories to use certified commercial samples for qualitative and quantitative analyses. Problems such as this can lead to the overestimation of LSD quantity. NMR can be used to verify the validity of the standard and provide an alternate method to check the accuracy of the quantitative determination.

The ^{13}C spectra of LSD (free base and tartrate) have been reported by Neville *et al.* (1992) and there were earlier partial data and assignments given by Kidric and Kocjan (1982). Shift differences between LSD and iso-LSD were up to 5.5 ppm.

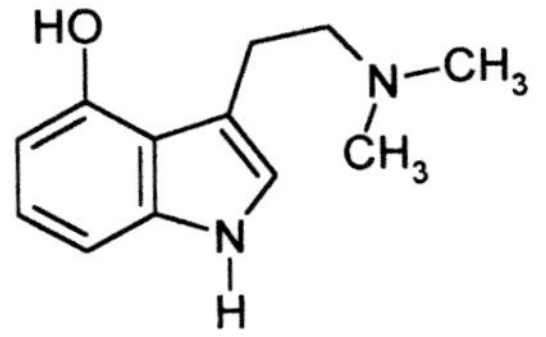

HO
N–CH$_3$
CH$_3$
N
H

Figure 5.60
Chemical structure of psilocin

Because forensic evidence can be challenged in court by an isomer defense, work was done to establish the analysis for the differentiation of LSD and lysergic acid methylpropylamide (LAMPA). The chromatographic analytical techniques had problems distinguishing between the two; MS fragmentation patterns had only significant small differences (Clark, 1989); the ^{1}H NMR showed a simple distinction between LSD and LAMPA (Figure 5.59), even at lower magnetic fields (Bailey, 1972, 1973; Neville, 1992). The same is true for the epimers LSD and iso-LSD. Although no data has been presented for the methyl-isopropylamide analogue of LSD, new prediction software has demonstrated the possibility for ease of identification.

5.18 TRYPTAMINES

Psilocin (Figure 5.60), psilocybin (Figure 5.61) (from *Psilocybe* mushroom) and bufotenine (Figure 5.62), are the best-known naturally occurring indoles. Psilocin and bufotenine are isomers, which give similar MS fragmentation patterns and have small differences in GC retention times. They should give readily discernable ^{1}H spectra, but the only reported spectrum of bufotenine (Bailey, 1975) is unclear.

Other research into natural product isolation has reported ^{1}H and ^{13}C spectra of other indoles. Among others, the ^{13}C shift data for *N*-methyl- and *N,N*-dimethyltryptamine (Mills, 1993) and the ^{1}H spectra of *N,N*-dimethyltryptamine (Poupat, 1976). Morales-Rios (1987) has published an extensive review

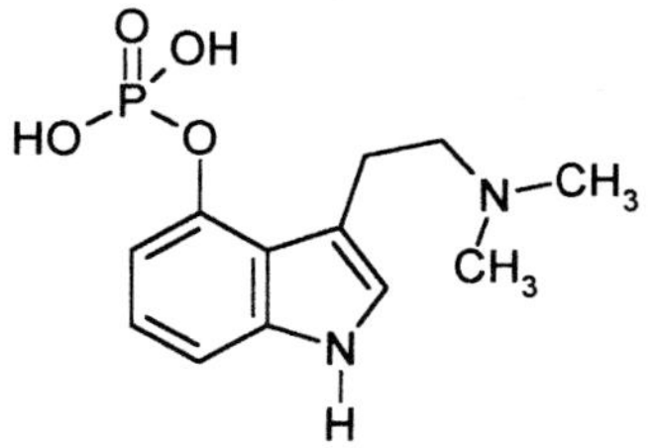

Figure 5.61

Chemical structure of psilocybin

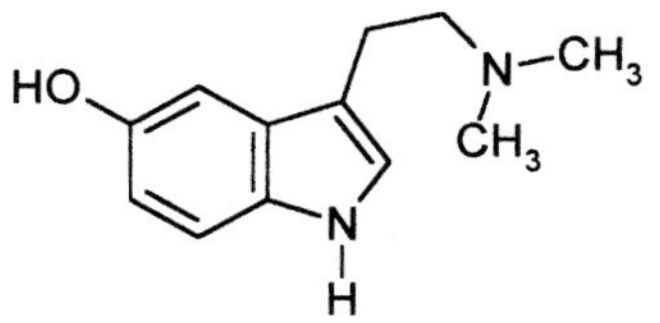

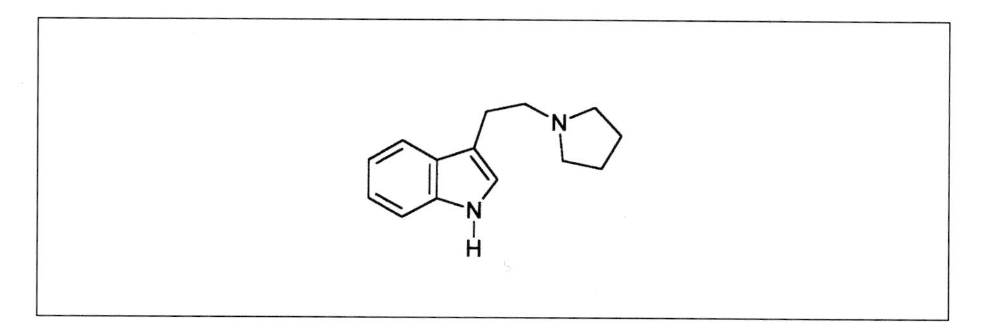

Figure 5.62

Chemical structure of bufotenine

Figure 5.63

Chemical structure of tetramethylene tryptamine

of indole ^{13}C NMR data. Ranc and Jurs (1993) developed models for the prediction of ^{13}C shifts for this class of compounds.

Tetramethylene tryptamine (Figure 5.63), a highly unusual tryptamine substance was identified by Cowie *et al.* (1982) using MS and ^{1}H NMR. An intermediate from incomplete reduction was also identified, *N*-[1-hydroxy-2(3-indolyl)ethyl]pyrrolidine.

REFERENCES

Azafonov, N. E., Sedishev, I. P. and Zhulin, V. M. (1990) *Bull. Acad. Sci. USSR Div. Chem. Sci.*, 738. English translation of *Izv. Akad. Nauk SSSR, Ser. Khim.* (1990), p. 829.

Bailey, K. (1971). *J. Pharm. Sci.*, 60, p. 1232.

Bailey, K. and Grey, A. A. (1972) *Can. J. Chem.*, 50, p. 3876.

Bailey, K. and Legault, D. (1981) *Anal. Chim. Acta.*, 123, p. 75.

Bailey, K. and Legault, D. (1981) *J. Forens. Sci.*, 26, p. 27.

Bailey, K. and Legault, D. (1981) *J. Forens. Sci.*, 26, p. 368.

Bailey, K. and Legault, D. (1981) *Org. Magn. Reson.*, 15, p. 68.

Bailey, K. and Legault, D. (1983) *Org. Magn. Reson.*, 21, p. 391.

Bailey, K., Beckstead, H. D., Legault, D. and Verner, D. (1974) *J. Assoc. Off. Anal. Chem.*, 57, p. 1134.

Bailey, K., By, A. W., Graham, K. C. and Verner, D. (1971) *J. Assoc. Off. Anal. Chem.*, 49, p. 3143.

Bailey, K., By, A. W., Legault, D. and Verner, D. (1975) *J. Assoc. Off. Anal. Chem.*, 58, p. 62.

Bailey, K., Chow, A. Y. K., Downie, R. H. and Pike, R. K. (1976) *J. Pharm. Pharmacol.*, 28, p. 713.

Bailey, K., Gagné, D. R., and Pike, R. K. (1976) *J. Assoc. Off. Anal. Chem.*, 59, p. 1162.

Bailey, K., Gagné, D. R., Legault, D. and Pike, R. K. (1977) *J. Assoc. Off. Anal. Chem.*, 60, p. 642.

Bailey, K., Legault, D. and Verner, D. (1974) *J. Assoc. Off. Anal. Chem.*, 57, p. 70.

Bailey, K., Verner, D. and Legault, D. (1973) *J. Assoc. Off. Anal. Chem.*, 53, p. 88.

Bax, A. and Morris, G. A. (1981) *J. Magn. Res.*, 42, p. 501.

Bax, A. (1983) *J. Magn. Res.*, 53, p. 517.

Beckett, A. H., Haya, K., Jones, G. R. and Morgan, P. H. (1975) *Tetrahedron*, 31, p. 1531.

Bellman, S. W., Turczan, J. W. and Kram, T. C. (1970) *J. Forens. Sci.*, 15, p. 261.

Buddrus, J., Bauer, H. (1987) *Angew. Chem. Int. Ed. Engl.*, 26, p. 625.

Claridge, T. D. W. (1999) *High Resolution NMR Techniques in Organic Chemistry*, Oxford: Pergamon (Elsevier).

Claridge, T. D. W. (1999) in *High Resolution NMR Techniques in Organic Chemistry*, Chapter 8, pp. 277–339. Pergamon (Elsevier), Oxford.

Claridge, T. D. W. (1999) in *High Resolution NMR Techniques in Organic Chemistry*, Chapter 5, pp. 148–220. Pergamon (Elsevier), Oxford.

Clark, C. C. (1989) *J. Forens. Sci.*, 34, p. 532.

Cowie, J. S., Holtman, A. L. and Jones, L. V. (1982) *J. Forens. Sci.*, 27, p. 527.

Dal Cason, T. A. (1989) *J. Forens. Sci.*, 34, p. 928.

Dal Cason, T. A., Meyers, J. A. and Lankin, D. C. (1997) *Forens. Sci. Int.*, 86, pp. 15–24.

Dawson, B. A. and Avdovich, H. W. (1987) *Can. Soc. Forens. Sci. J.*, 20, p. 29.

Dawson, B. A. and Neville, G. A. (1989) *Can. Soc. Forens. Sci. J.*, 22, p. 195.

Delliou, D. (1983) *Forens. Sci. Int.*, 21, p. 259.

Doddrell, D. M, Pegg, D. T. and Bendall, M. R. (1982) *J.Magn.Res.*, 48, p. 323.

Eaton, T. A., Houk, K. N., Watkins, S. F. and Fronczek, F. R. (1983) *J. Med. Chem.*, 26, p. 479.

Ernst, R. R, Bodenhausen, G. and Wokaun, A. (1992) *Principles of Nuclear Magnetic Resonance in One and Two Dimensions*, Oxford: Clarendon Press.

Foster, B. C., McLeish, J., Wilson, D. L., Whitehouse, L. W., Zamecnik, J. and Lodge, B. A. (1992. *Xenobiotica*, 22, p. 1383.

Freeman, R. (1997) *Spin Choreography. Basic Steps in High Resolution NMR*, Oxford: Spektrum Academic Publishers and University Press.

Fyfe, C. (1983) *Solid State NMR for Chemists*, Guelph, Ontario, Canada: C.F.C. Press.

Gagné, D. R. and Pike, R. K. (1977) *J. Assoc. Off. Anal. Chem.*, 60, p. 32.

Groombridge, C. J. (1996) *NMR Spectroscopy in Forensic Sciences*, Annual Reports on NMR Spectroscopy, Volume 12.

Hoffman, A. J. and Nichols, D. E. (1985) *J. Med. Chem.*, 28, p. 1252.

Kamenka, J. M. and Chicheportiche, R. (1987) *Eur. J. Med. Chem.*, 22, p. 193.

Kidrič, J. and Kocjan, D. (1982) *Stud. Phys. Theor. Chem.*, 18, p. 35.

Lukaszewski, T. (1978) *J. Assoc. Off. Anal. Chem.*, 61, p. 1978.

Manoharan, M., Eliel, E. L. and Carroll, F. I. (1983) *Tetrahedron Lett.*, 24, p. 1855.

Martin, G. E. and Zektzer, A. S. (1988) *Two Dimensional NMR Methods for Establishing Molecular Connectivity. A Chemist's Guide to Experiment Selection, Performance, and Interpretation*, NY: VCH Publishers.

Martin, G. E. and Zektzer, A. S. (1988) *Magn. Res. Chem.*, 26, p. 631

Martin, R. J. and Alexander, T. G. (1968) *J. Assoc. Off. Anal. Chem.*, 51, p. 159.

Mehring, Michael (1983) *High Resolution NMR in Solids*, Berlin/Heidelberg/New York: Springer-Verlag.

Mills III, T. and Roberson, J. C. (1993) *Instrumental Data for Drug Analysis*, Vols 1–5, 2nd edn, NY: Elsevier.

Morales, M. S., Espinera, J. and Joseph-Nathan, P. (1987) *Magn. Reson. Chem.*, 25, p. 377.

Mourad, M. S., Varma, R. S. and Kabalka, G. W. (1985) *J. Org. Chem.*, 50, p. 133.

Neville, G. A., Beckstead, H. D., Black, D. B., Dawson, B. A. and Ethier, J.-C. (1992) *Can. J. Appl. Spectrosc.*, 37, p. 149.

Nichols, D. E., Hoffman, A. J., Oberlender, R. A., Jacob, P. and Shulgin, A. T. (1986) *J. Med. Chem.*, 29, p. 2009.

Noggle, F. T., DeRuiter, J. and Long, M. J. (1986) *J. Assoc. Off. Anal. Chem.*, 69, p. 681.

Ono, M. (1970) *Nippon Hoigaku Zasshi*, 33, p. 339.

Patt, S. L. and Shoolery, J. N. (1982) *J. Magn. Res.*, 46, p. 535.

Poupat, C., Ahond, A. and Sévenet T., (1976) *Phytochem.*, 15, p. 2019.

Ragan, F. A., Hite, S. A., Samuels, M. S. and Garey, R. E. (1985) *J. Anal. Toxicol.*, 9, p. 91.

Ranc, M. L. and Jurs, P. C. (1993) *Anal. Chim. Acta*, 280, p. 145.

Reynolds, J. E. F. (ed.) (1989) *Martindale. The Extra Pharmacopoeia*, London: The Pharmaceutical Press.

Richarz, R., Ammann, W. and Wirthlin, T. (1982) No. Z-15 in *Varian Instruments at Work*, pp. 1–19 and references cited therein.

Rutar, V. (1984) *J. Magn. Res.*, 58, p. 306.

Shimamine, M., Takahashi, K. and Nakahara, Y. (1990) *Eisei Shikensho Hokoku*, 108, p. 118.

Shimamine, M., Takahashi, K. and Nakahara, Y. (1993) *Eisei Shikensho Hokoku*, 111, p. 66.

Shulgin, A. T. and Shulgin, A. (1991) *PIHKAL, A Chemical Love Story*, Berkeley, CA: Transform Press.

Shulgin, A. T. and Shulgin, A. (1997) *TIHKAL, The Continuation*, Berkeley, CA: Transform Press.

van de Ven, F. J. M. (1995) in *Multidimensional NMR in Liquids. Basic Principles and Experimental Methods*, New York: Wiley-VCH.

Wehrli, F. W., and Marchand, A.P. (1988) in *Interpretation of Carbon-13 NMR Spectra*, 2nd edn, New York: John Wiley & Sons.

INDEX